Leaf margins 葉緣

crenate 鈍齒緣   dentate 有齒緣   serrate 鋸齒緣   fimbriate 縫緣   entire 全緣   lobed 淺裂緣

Leaf apices 葉尖

acute 急尖   subacute 稍尖   obtuse 鈍尖   rounded 圓形   cuspidate 硬尖

acuminate 漸尖   mucronate 銳尖   aristate 芒狀尖   retuse 微缺的   emarginate 倒心形

Leaf bases 葉基

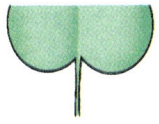 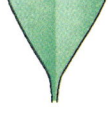 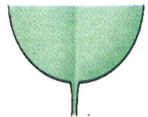 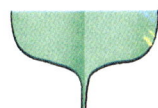

cordate 心形   cuneate 楔形   rounded 圓形   truncate 平截狀

前言

　　朗文英漢科學系列圖解詞典包括生物、化學、科學、植物和地質五冊。這是一套內容既有聯繫而又各自獨立成冊的系列詞典，《朗文英漢植物圖解詞典》為其中的一冊。

　　本書收詞1200多個，包括植物化學、細胞學、器官與組織、生長與發育、植物生理學、植物分類學、植物解剖學、植物形態學、繁殖、遺傳學、進化及生態學等全部植物學的基本原理詞彙。

　　這些詞按詞義分科目編排，英漢雙解對照，釋義簡明，概念準確。每一科目的上、下各個詞條，內容互有關聯。釋義深入淺出，易於理解；其中又標出頁碼和箭嘴號，引導讀者查找相關詞條，運用更多資料作比較，以加深理解，掌握更多詞彙。

　　詞典中收入300多幅印刷精美的彩色插圖和圖表，直觀式顯示所闡釋的題目和原理，有助於讀者更好理解釋義，但釋義並不依賴這些插圖。

　　詞典後部的索引，按英文字母順序排列，標注K.K.音標，自成一個英漢植物詞彙表，方便讀者檢索。

　　本詞典適合具高中至大學一、二年級程度的學生，以及需要深入了解植物學術語的非植物學專業的讀者查閱使用。

朗文出版（遠東）有限公司
一九九一年十一月

朗文英漢植物圖解詞典
LONGMAN ENGLISH-CHINESE ILLUSTRATED DICTIONARY OF BOTANY

　YORK PRESS

English edition © Librairie du Liban 1984
This bilingual edition © Librairie du Liban & Longman Group (Far East) Ltd 1992

Longman Group (Far East) Ltd
18/F., Cornwall House
Tong Chong Street
Quarry Bay
Hong Kong
Tel: 811 8168
Fax: 565 7440
Telex: 73051 LGHK HX

First published 1992

ISBN 962 359 054 7 (Hong Kong Edition)
ISBN 962 359 723 1 (Taiwan Edition)

All rights reserved. No part of this publication may be reproduced, stored in a retrieval system, or transmitted in any form or by any means, electronic, mechanical, photocopying, recording, or otherwise, without the prior written permission of the Publishers.

Produced by Longman Group (Far East) Ltd.
Printed in Hong Kong

朗文出版（遠東）有限公司
香港鰂魚涌糖廠街
康和大廈十八樓
電話：811 8168
圖文傳真：565 7440
電傳：73051 LGHK HX

一九九二年初版

國際書號 962 359 054 7（香港版）
國際書號 962 359 723 1（台灣版）

本書任何部分之文字及圖片，如未獲得本社之書面同意，不得用任何方式抄襲、節錄或翻印。
本書如有缺頁、破損或裝訂錯誤，請寄回本公司更換。

出版：朗文出版（遠東）有限公司
印刷：香港

Contents 目錄

page 頁次

How to use the dictionary — 本詞典的用法 — 5

Phytochemistry — 植物化學 — 8
The atom; molecules, ions; compounds; reactions; solutions; metabolism; enzymes
原子；分子、離子；化合物；反應；溶液；新陳代謝；酶

Cells — 細胞 — 16
General; cell walls; membranes, organelles
概述；細胞壁；膜、細胞器

Respiration — 呼吸作用 — 22
General; glycolysis; fermentation, Krebs cycle; photorespiration, phosphorylation; ADP, ATP
概述；糖酵解；醱酵、三羧酸循環；光呼吸、磷酸化；二磷酸腺苷、三磷酸腺苷

Carbohydrates — 碳水化合物 — 28
Sugars; starch
糖類；澱粉

Lipids — 脂類 — 31

Photosynthesis — 光合作用 — 32
General; dark reaction; Calvin cycle; CO_2 fixation pathways; light reaction, pigments, cytochromes; chlorophyll and light; phosphorylation; electron transfer
概述；暗反應；卡爾文循環；二氧化碳固定途徑；光反應、色素；細胞色素；葉綠素與光；磷酸化；電子傳遞

Genetics — 遺傳學 — 41
General; Mendel's laws; loci, dominance, inheritance
概述；孟德爾定律；基因位、顯性、遺傳

Cell division — 細胞分裂 — 45
Mitosis; chromosomes; meiosis; haploid, diploid, polyploid
有絲分裂；染色體；減數分裂；單倍體、二倍體、多倍體

Nucleic acids — 核酸 — 51
DNA, RNA; nucleotides, codons; genetic code, mutation
脫氧核糖核酸、核糖核酸；核苷酸、密碼子；遺傳密碼、突變

Proteins — 蛋白質 — 56
General; protein synthesis; structure
概述；蛋白質合成；結構

Reproduction — 生殖 — 59
Sexual, asexual; vegetative reproduction; gametes, zygotes; fertilization; breeding; alternation of generations; gametes and gametangia; spores and sporangia; propagation
有性生殖、無性生殖；營養生殖；配子、合子；受精；培育；世代交替；配子及配子囊；孢子及孢子囊；繁殖

Flower biology — 花的生物學 — 70
Flower parts; flower types; male parts; pollen, pollination; female parts; ovaries; ovules; flower sexes; inflorescences
花的各部分；花的形式；雄性部分；花粉、授粉作用；雌花部分；子房；胚珠；花的性別；花序

Fruits and seeds — 果實及種子 — 83
Fruits; seeds; germination
果實；種子；萌發

Anatomy and morphology — 解剖學與形態學 — 88
General; roots; tissues, shoots; trees; growth, wood; leaf tissues; leaves; spines, hairs
概述；根；組織、枝條；喬木；生長、木材；葉的組織；葉片；刺、毛

Vascular systems — 維管系統 — 101
Translocation, osmosis; osmotic processes; tissues; xylem; phloem
轉移作用、滲透作用；滲透過程；組織；木質部；韌皮部

Growth and physiology — 生長與生理學 — 109
Meristems; growth; physiology; glands, hormones; tropisms; photoperiodism; growth period
分生組織；生長；生理學；腺體、激素；向性；光週期性；生長週期

	page 頁次
Plant kingdom 植物界	118
General, viruses; bacteria, algae; bryophytes; pteridophytes; spermatophytes; gymnosperms; angiosperms	
概述、病毒;細菌、藻類;苔蘚植物;蕨類植物;種子植物;裸子植物;被子植物	
Classification 分類	132
General; taxonomy; taxa; variation	
概述;分類學;分類單元;變異	
Habits 習性	136
Evolution 進化	139
General; natural selection; adaptation; speciation, palaeobotany; geological time	
概述;自然選擇;適應;物種形成、古植物學;地質時代	
Interactions 相互作用	144
General; mycorrhizae; nitrogen-fixing bacteria; lichens; defense and attack	
概述;菌根;固氮細菌;地衣;防衛與進攻	
Ecology 生態學	149
General; colonization; succession; food webs; nitrogen cycle; soils; forests; scrub; grasslands; aquatic habitats; climates	
概述;羣落形成;演替;食物網;氮循環;土壤;森林;密灌叢、草地;水生生境;氣候	
Fungi 真菌類	163
General; Phycomycetes, Ascomycetes; Basidiomycetes; Zygomycetes; Chytridiomycetes, Myxomycetes	
概述;藻菌綱、子囊菌綱;擔子菌類;接合菌類;壺菌綱、黏菌綱	
General and technical words in botany 植物學一般詞彙及技術詞彙	170
Appendixes: 附錄	
One: Understanding botanical words	
一、理解意義的植物學用詞彙	174
Two: International System of Units (SI)	
二、國際單位制	179
Index 索引	181

How to use the dictionary 本詞典的用法

This dictionary contains over 1200 words used in the botanical sciences. These are arranged in groups under the main headings listed on pp. 3-4. The entries are grouped according to the meaning of the words to help the reader to obtain a broad understanding of the subject.

At the top of each page the subject is shown in bold type and the part of the subject in lighter type. For example, on pp. 18 and 19:

> **18 · CELLS**/MEMBRANES, ORGANELLES
> **CELLS**/MEMBRANES, ORGANELLES **· 19**

In the definitions the words used have been limited so far as possible to about 1500 words in common use. These words are those listed in the 'defining vocabulary' in the *New Method English Dictionary* (fifth edition) by M. West and J. G. Endicott (Longman 1976). Words closely related to these words are also used: for example, *characteristic*, defined under *character* in West's *Dictionary*.

本詞典共收 1200 多個植物學用詞，這些詞按照第 3-4 頁所列主要標題分類。所有詞條均按詞義歸類，旨在幫助讀者對所查找科目獲得概括的瞭解。

每一頁的上方用黑體字印出有關科目的名稱，並以秀麗體印出該科目下的分段。例如在第 18 頁和第 19 頁上：

> 18 · 細胞/膜、細胞器
> 細胞/膜、細胞器 · 19

釋意部分所使用的詞盡可能限於常用的 1500 個詞左右。這些詞列於 M · 韋斯特及 J.G. 恩廸科特合編的《新法英語詞典》（第五版）（朗文公司 1976 年）中的釋義詞彙表內。本詞典也使用和這些詞密切相關的一些詞，例如：使用韋斯特詞典中在 character（特徵）條下解釋的 characteristic（特徵的）這個詞。

1. To find the meaning of a word

Look for the word in the alphabetical index at the end of the book, then turn to the page number listed.

The description of the word may contain some words with arrows in brackets (parentheses) after them. This shows that the words with arrows are defined near by.

(↑) means that the related word appears above or on the facing page;

(↓) means that the related word appears below or on the facing page.

A word with a page number in brackets (parentheses) after it is defined elsewhere in the dictionary on the page indicated. Looking up the words referred to in either of these two ways may help in understanding the meaning of the word that is being defined.

The explanation of each word usually depends on knowing the meaning of a word or words above it. For example, on p. 80 the meaning of *peduncle*, *pedicel*, and the words that follow depends on the meaning of the word *inflorescence*, which appears above them. Once the earlier words are understood those that follow become easier to understand. The illustrations have been designed to help the reader understand the definitions but the definitions are not dependent on the illustrations.

1. 查明詞的意義

在本詞典末尾的字母順序索引中找出欲查的詞，然後翻到該詞旁註明的頁碼。

詞的釋義中遇到一些詞後面帶有箭號括在括弧（圓括號）內，表示這個詞的解釋就在附近。

（↑）表示這有關詞出現在本詞條之前或在對面頁上。

（↓）表示這有關詞出現在本詞條之後或在對面頁上。

詞後面有頁碼括在括弧（圓括弧）內表示該詞的解釋在所註明的頁碼上。再查閱所標註的詞可幫助理解原先所解釋詞的詞義。

對每一個詞的闡釋通常都依賴於理解該詞前面一個詞或幾個詞的詞義。例如在第 **80** 頁，"總花梗"、"花梗"以及接着出現的那些詞，其詞義都依賴於理解在這些詞前面出現的"花序"這個詞的詞義。理解了前面出現的詞的詞義，就較容易理解接着出現的詞的詞義。本詞典中的插圖可幫助讀者理解釋義，但釋義並不依賴這些插圖。

2. To find related words

Look in the index for the word you are starting from and turn to the page number shown. Because this dictionary is arranged by ideas, related words will be found in a set on that page or one near by. The illustrations will also help to show how words relate to one another.

For example, words relating to cell division are on pp. 45-50. On p. 45 *cell division* is followed by words used to describe mitosis; pp. 46 and 47 give words used in describing chromosomes; pp. 48 and 49 illustrate and explain meiosis; and p. 50 gives words relating to chromosome numbers.

3. As an aid to studying or revising

The dictionary can be used for studying or revising a topic. For example, to revise your knowledge of photosynthesis, you would look up *photosynthesis* in the alphabetical index. Turning to the page indicated, p. 32, you would find *photosynthesis, autotrophic, heterotrophic, chloroplast,* and so on; on p. 33 you would find *grana, lamellae,* and so on. Turning over to p. 34 you would find *Calvin cycle* etc.

In this way, by starting with one word in a topic you can revise all the words that are important to this topic.

4. To find a word to fit a required meaning

It is almost impossible to find a word to fit a meaning in most dictionaries, but it is easy with this book. For example, if you had forgotten the word for the outer whorl of the perianth of a flower, all you would have to do would be to look up *perianth* in the alphabetical index and turn to the page indicated, p. 70. There you would find the word *calyx* with a diagram to illustrate its meaning.

2. 查找相關的詞

在索引中找出你作為起頭的詞，然後翻到所標示的頁碼。由於本詞典是按概念編排詞條，因此可以在同一頁或前後的一頁上查到一組相關的詞。插圖用於幫助說明各詞之間的關係。

例如，與細胞分裂有關的各個詞，均列在第45頁至第50頁上。在第45頁，"細胞分裂"詞條之下列出一些用於闡述有絲分裂的詞；第46頁及第47頁列出闡述染色體方面用的詞；第48頁及第49頁用插圖說明並解釋減數分裂；第50頁列出與染色體數目有關的各個詞。

3. 作為學習或複習的輔助工具

本詞典可用於學習或複習某一項課題。例如，你在複習有關光合作用的知識時，可以在字母順序索引中查出"光合作用"這個詞條，翻到所示第32頁，就可以找到"光合作用"、"自養的"、"異養的"、"葉綠體"等詞；在第33頁，你可以找到"基粒"、"片層"等詞；翻到第34頁，你可以找到"卡爾文循環"等詞。

因此，你由某一項課題的一個詞著手，即可複習與此課題有關的全部重要詞彙。

4. 查找適當的詞表達確切的意義

在大多數詞典中，想查找一個能確切表達意義的詞幾乎是不可能做到的，但用這本詞典就容易達此目的。例如，你忘了表達花的花被外輪所用的詞，你只需從字母順序索引中查出"花被"(Perianth)這個詞，再翻到所示的頁碼第70頁，你會找到"花萼"(calyx)這個詞及一個說明其意義的插圖。

5. Abbreviations used in the definitions

5. 詞義解釋中所用的縮寫

abbr.	abbreviated as 縮寫為	i.e.	id est (that is to say) 即是	p.p.	pages 頁數（複數）
adj.	adjective 形容詞（形）	n.	noun 名詞（名）	sing.	singular 單數
e.g.	exempli gratia (for example) 例如	p.	page 頁數	v.	verb 動詞（動）
etc.	et cetera (and so on) 等等	pl.	plural 複數	=	the same as 等同

THE
DICTIONARY
詞典正本

phytochemistry (n) the chemistry of plants.
phytochemical (adj).
atom (n) the smallest unit of a chemical element (↓). Atoms contain electrons (↓), protons (↓) and neutrons (↓). The numbers of electrons and protons in an atom are equal. **atomic** (adj).

植物化學 (名) 研究植物的化學的一門學科。(形容詞為 phytochemical)
原子 (名) 化學元素 (↓) 的最小單元。原子包含電子 (↓)、質子 (↓)、及中子 (↓)。在一個原子中，電子數目和質子數目相等。(形容詞為 atomic)

the four commonest atoms in biological compounds
生物學中最常見的 4 種原子

hydrogen (1 proton, 1 electron) 氫 (含 1 個質子 1、1 個電子)

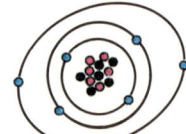

carbon (6 protons, 6 neutrons, 6 electrons) 碳 (含 6 個質子、6 個中子、6 個電子)

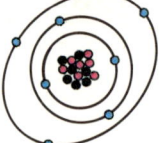

nitrogen (7 protons, 7 neutrons, 7 electrons) 氮 (含 7 個質子、7 個中子、7 個電子)

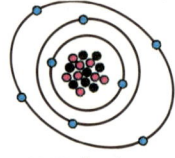

oxygen (8 protons, 8 neutrons, 8 electrons) 氧 (含 8 個質子、8 個中子、8 個電子)

element (n) a substance consisting of atoms (↑) all of the same kind. An element cannot be changed into another element except by splitting atoms. Each element, e.g. oxygen, carbon, or nitrogen, has its own characteristic number of protons (↓) in its atoms.
proton (n) a particle with a positive electric charge, found in all atoms (↑). The electric charge of a proton is exactly equal and opposite to that of an electron (↓), so an atom has no charge. A proton is a hydrogen ion (↓), since hydrogen atoms consist of only one proton and one electron.
electron (n) a particle with a negative electric charge, found in all atoms (↑). The electric charge on an electron is exactly equal and opposite to that on a proton (↑). The addition or removal of electrons from atoms creates ions (↓). Electrons are 1840 times lighter than protons.
neutron (n) a particle with no electric charge, found in all atoms (↑) except hydrogen atoms. Neutrons have the same weight as protons (↑).

元素 (名) 由若干個同類原子 (↑) 組成的物質。一種元素除非使其原子分裂，否則不能轉變為另一種元素。每一種元素，例如氧、碳或氮，其原子中都有本身特有的質子 (↓) 數目。

質子 (名) 一切原子 (↑) 中帶正電荷的粒子。在一個原子中，質子與電子 (↓) 的電荷完全相等，電性則相反，因而原子不帶電荷。質子是一個氫離子 (↓)，因為氫原子僅有一個質子和一個電子。

電子 (名) 一切原子 (↑) 中帶負電荷的粒子。在一個原子中，電子與質子 (↑) 的電荷完全相等，電性則相反。給原子加上或移去電子即產生離子 (↓)。質子比電子重 1840 倍。

中子 (名) 一切原子中 (↑)(氫原子除外) 不帶電荷的粒子。中子與質子 (↑) 的重量相等。

PHYTOCHEMISTRY/MOLECULES, IONS 植物化學／分子、離子

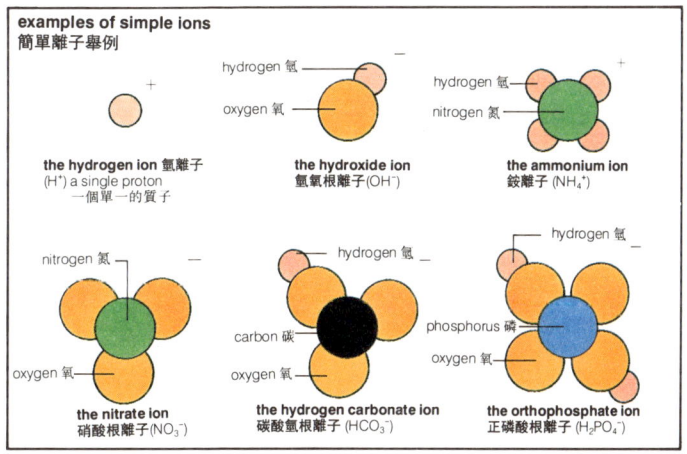

ion (*n*) an atom (↑) or molecule (↓) with an electric charge, due to having an unequal number of protons (↑) and electrons (↑). An ion with a positive charge has more protons than electrons, and an ion with a negative charge has less protons than electrons. **ionization** (*n*).

molecule (*n*) the smallest unit of an element (↑) or compound which occurs naturally. Molecules consist of more than one atom (↑). A molecule of hydrogen consists of two hydrogen atoms (H_2), and a molecule of carbon dioxide has one atom of carbon and two of oxygen (CO_2).
molecular (*adj*).

離子（名） 由於質子（↑）及電子（↑）數目不相等而帶電荷的一個原子（↑）或分子（↓）。帶正電荷的離子所含的質子多於電子，而帶負電荷的離子所含的質子少於電子。（名詞 ionization 意為電離）

分子（名） 元素（↑）或化合物自然存在的最小單元。分子由一個以上的原子（↑）組成。一個氫分子由兩個氫原子（H_2）組成，而一個二氧化碳（CO_2）的分子含有一個碳原子和兩個氧原子。（形容詞為 molecular）

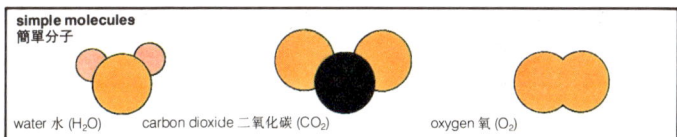

macromolecule (*n*) a large molecule (↑) consisting of many atoms, e.g. proteins (p. 56), nucleic acids (p. 51), polysaccharides (p. 30).
crystal (*n*) a solid symmetrical (p. 71) structure made of molecules (↑) all of the same kind and size. **crystalline** (*adj*).

大分子（名） 由許多原子組成的一個巨大的分子（↑），如蛋白質（第 56 頁）、核酸（第 51 頁）、多醣類（第 30 頁）。

結晶體（名） 由同類及同大小的許多分子（↑）所組成的一種結構對稱（第 71 頁）的固體。（形容詞為 crystalline）

compound[1] (*n*) a molecule (p.9) which consists of more than one kind of atom (p.8).

化合物(名) 由一種以上的元素原子(第8頁)所構成的分子(第9頁)。

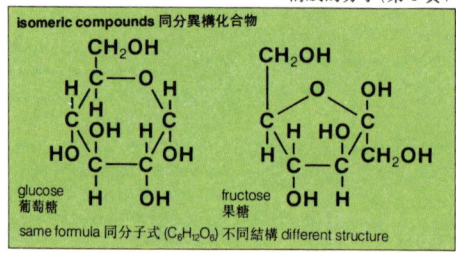

isomers (*n*) two or more molecules (p.9) with the same numbers and kinds of atoms (p.8), but with different arrangements of atoms and sometimes different chemical properties. **isomeric** (*adj*).

同分異構體(名) 含相同數目及同類的原子(第8頁),但原子排列不同,往往呈現不同化學性質的兩個或多個分子(第9頁)。(形容詞為 isomeric)

polymer (*n*) a chemical substance formed by the joining together of many molecules (p.9) of the same kind, e.g. polysaccharides (p.30), polypeptides (p.56) and nucleic acids (p.51).

聚合體(名) 由許多同類的分子(第9頁)連接在一起而形成的一種化學物質。例如多醣(第30頁)、多肽(第56頁)及核酸(第51頁)。

monomer (*n*) one of the units in a polymer (↑).

單體(名) 聚合體(↑)分子中的一個單元。

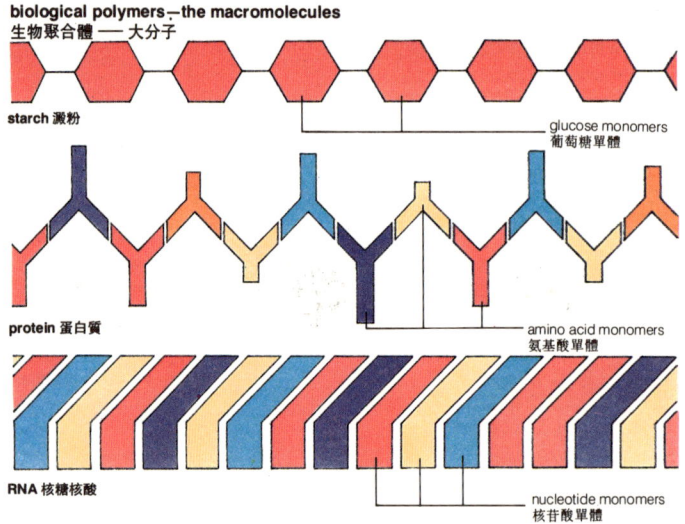

reduction (n) the process in which a substance (1) gains electrons (p. 8); (2) has oxygen removed from it; or (3) has hydrogen added to it. **reduce** (v), **reductive** (adj).

oxidation (n) the process in which a substance (1) loses electrons (p. 8); (2) has oxygen added to it; or (3) has hydrogen removed from it. **oxidize** (v), **oxidative** (adj).

redox (adj) of chemical reactions that involve oxidation (↑) and reduction (↑).

還原作用(名) 指物質的以下過程之一：(1)獲得電子(第8頁)；(2)除去物質所含的氧；(3)物質中加入氫。(動詞為 reduce，形容詞為 reductive)

氧化作用(名) 指物質的以下過程之一：(1)失去電子(第8頁)；(2)物質中加入氧；(3)除去物質中所含的氫。(動詞為 oxidize，形容詞為 oxidative)

氧化還原作用(形) 包括氧化作用(↑)和還原作用(↑)的化學反應。

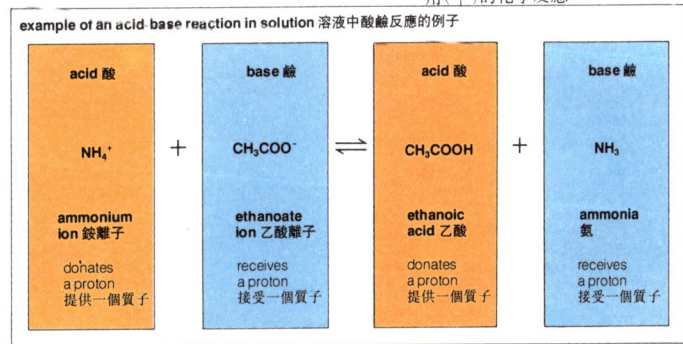

reaction (n) chemical process in which two or more compounds act on each other, by exchanging atoms (p. 8) or electrons (p. 8), to produce different compounds.

potential energy energy that is stored in a molecule (p. 9) and can be released to drive chemical reactions. Potential energy is usually measured in terms of electric charge.

organic (adj) of compounds which contain atoms (p. 8) of carbon. The compounds synthesized (p. 13) by living organisms (p. 118) are organic.

inorganic (adj) of compounds which do not contain carbon.

acid (n) any chemical compound which can donate protons (p. 8) to water molecules (p. 9). The acidity of a solution (p. 12) is measured on the pH scale (-log H⁺ concentration). **acidic** (adj).

base¹ (n) any substance which can accept protons (p. 8) from water molecules (p. 9). **basic** (adj).

反應(名) 兩個或多個化合物相互作用，交換原子(第8頁)或電子(第8頁)產生各種化合物的過程。

勢能 分子(第9頁)中所蓄的能量，此能量可釋放出來以推動化學反應的進行。勢能通常是以電荷來量度。

有機的(形) 指含有碳原子(第8頁)的化合物。由有生命的有機體(第118頁)合成(第13頁)的化合物都是有機的。

無機的(形) 指不含碳的化合物。

酸(名) 指能提供質子(第8頁)給水分子(第9頁)的任何化合物。溶液的酸度(第12頁)是以 pH 標度(-logH⁺ 的濃度)來量度。(形容詞為 acidic)

鹼(名) 指能從水分子(第9頁)接受質子(第8頁)的任何物質。(形容詞為 basic)

12 · PHYTOCHEMISTRY/SOLUTIONS 植物化學／溶液

solution (n) a liquid with substances dissolved (↓) in it.
dissolve (v) of solids to break down into molecules (p. 9) or ions (p. 9) when placed in a liquid.
solute (n) a substance which is dissolved (↑) in a liquid.
solvent (n) a liquid in which substances are dissolved (↑).
soluble (adj) of substances which can be dissolved (↑), e.g. sugar in water. **solubility** (n).
insoluble (adj) not soluble (↑).
aqueous (adj) of solutions (↑) in which the solvent (↑) is water.
concentration (n) the amount of a substance dissolved (↑) in a given volume of liquid.
evaporation (n) the process by which molecules (p. 9) of a liquid become a gas. **evaporate** (v).

溶液（名） 有物質溶解(↑)於其中的一種液體。
溶解（動） 使固體放入液體中裂解為分子（第9頁）或離子（第9頁）。
溶質（名） 溶解(↑)於液體中的物質。
溶劑（名） 能使物質溶解(↑)的液體。
可溶解的（形） 指可被溶解(↑)的物質。例如糖能溶解在水中。（名詞為 solubility）
不溶解的（形） 指不可溶解的(↑)。
水的（形） 指以水為溶劑(↑)的溶液(↑)。
濃度（名） 在一定體積的液體中所溶解(↑)的物質的量。
蒸發作用（名） 液體分子（第9頁）成為氣體的過程。（動詞為 evaporate）

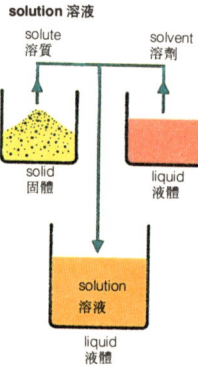

hydrolysis (n) a chemical reaction involving the breakdown of a molecule (p. 9) into two molecules, with the addition of the parts of a molecule of water. **hydrolyze** (v), **hydrolytic** (adj).

水解作用(名) 加入一克分子份量的水使一克分子(第9頁)物質分裂為兩克分子物質的反應。(動詞為 hydrolyze，形容詞為 hydrolytic)

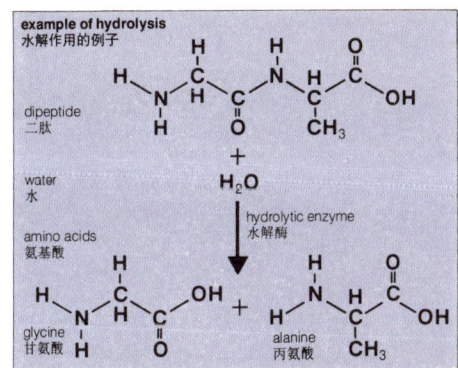

synthesis (n) the process of building chemical compounds from small molecules (p. 9), e.g. carbohydrates (p. 28) from carbon dioxide and water in photosynthesis (p. 32), or proteins (p. 56) from amino acids (p. 56) in protein synthesis (p. 57). **synthesize** (v), **synthetic** (adj).

phosphate (n) an inorganic (p. 11) ion (p. 9), present in the soil, which is an important nutrient (p. 111) for plants. Phosphate (PO_4^{3-}), is used in the synthesis (↑) of ATP (p. 26) during photosynthesis (p. 32) and respiration (p. 22). It is also used in the nucleotide (p. 52) molecules (p. 9) of nucleic acids (p. 51).

nitrate (n) an inorganic (p. 11) ion (p. 9), present in the soil, which is an important nutrient (p. 111) for plants. Nitrate (NO_3^-) provides nitrogen for the synthesis (↑) of amino acids (p. 56) and other nitrogen-containing compounds, e.g. nucleotides (p. 52).

orthophosphate (n) another term for inorganic (p. 11) phosphate (↑) ion (p. 9). Pi (abbr.).

ammonia (n) an inorganic (p. 11) molecule (p. 9) with one atom (p. 8) of nitrogen and three atoms of hydrogen (NH_3).

合成作用(名) 從小分子(第9頁)構成化合物的過程。例如，在光合作用(第32頁)中由二氧化碳和水合成碳水化合物(第28頁)；或者在蛋白質合成(第57頁)中由氨基酸(第56頁)合成蛋白質(第56頁)。(動詞為 synthesize，形容詞為 synthetic)

磷酸根(名) 土壤中存在的一種無機(第11頁)鹽離子(第9頁)，是植物的重要營養物(第111頁)。磷酸根(PO_4^{3-})在光合作用(第32頁)及呼吸作用(第22頁)時供合成(↑)三磷酸腺苷(ATP)(第26頁)。亦用於合成核酸(第51頁)的核苷酸(第52頁)分子(第9頁)。

硝酸根(名) 土壤中存在的一種無機(第11頁)鹽離子(第9頁)，是植物的重要營養物(第111頁)。硝酸根(NO_3^-)提供氮素用於合成(↑)氨基酸(第56頁)和其他含氮化合物。例如核苷酸(第52頁)。

正磷酸根(名) 無機(第11頁)磷酸根(↑)離子(第9頁)的另一名稱，縮寫為 Pi。

氨(名) 含有一個氮原子(第8頁)及三個氫原子(NH_3)的無機(第11頁)物分子(第9頁)。

metabolism (n) the sum of the chemical reactions which occur in an organism or a cell. Metabolism involves the breakdown of organic (p. 11) compounds, releasing energy that is used in the synthesis (p. 13) of other compounds. **metabolize** (v).

metabolite (n) a substance produced by metabolism (↑).

metabolic pathway a set of chemical reactions which follow each other in a sequence. Each reaction uses the product of the reaction before it. *See also* **metabolism** (↑).

inhibitor (n) a substance which prevents or slows down a chemical reaction or process. Some inhibitors can slow down enzyme (↓) reactions by blocking the active site (↓) of the enzyme. **inhibit** (v).

inhibition (n) the prevention or slowing down of a metabolic (↑) reaction, e.g. by an inhibitor (↑) or by temperatures that are too high or too low.

feedback (n) the process by which a product at or near the end of a metabolic pathway (↑) affects the reactions (p. 11) at the beginning of the same pathway. Feedback can be either positive or negative.

新陳代謝；代謝作用（名） 在生物體或細胞內所發生化學反應的總和。新陳代謝包括有機化合物（第 11 頁）分解，釋放能量用於合成（第 13 頁）別的化合物。（動詞為 metabolize）

代謝產物（名） 指新陳代謝（↑）所產生的物質。

代謝途徑 在一個反應系列中連續出現的一組化學反應。每一個反應都利用其前面一個反應所產生的產物。參見**新陳代謝**（↑）。

抑制劑（名） 阻止或延緩化學反應或過程的一種物質。某些抑制劑能封阻酶的活性部位（↓）而延緩酶（↓）的反應。（動詞為 inhibit）

抑制作用（名） 阻礙或延緩新陳代謝（↑）的反應，例如由於有抑制劑（↑）或者溫度過高或過低而起抑制作用。

反饋（名） 在代謝途徑（↑）結束或接近結束時，一種產物在同一代謝途徑開始時影響反應（第 11 頁）的過程。反饋可能是正作用也可能是負作用。

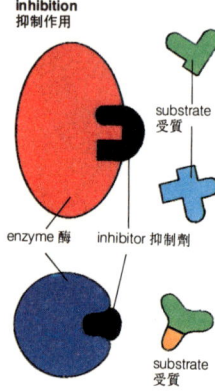

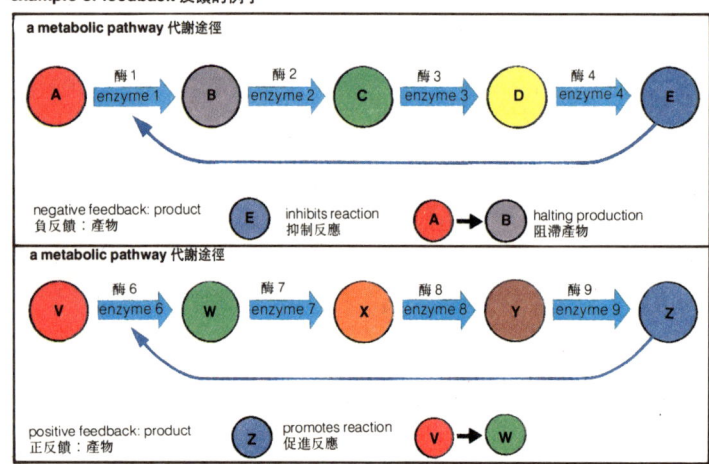

example of feedback 反饋的例子

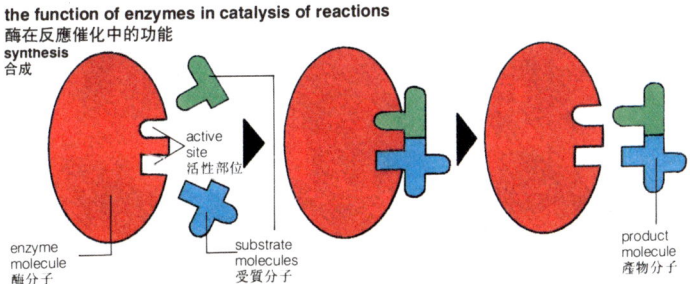

the function of enzymes in catalysis of reactions
酶在反應催化中的功能
synthesis
合成

enzyme molecule 酶分子
substrate molecules 受質分子
active site 活性部位
product molecule 產物分子

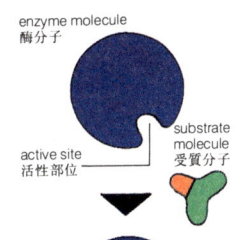

breakdown 裂解
enzyme molecule 酶分子
active site 活性部位
substrate molecule 受質分子

product molecules 產物分子

enzyme (*n*) a protein (p. 56) which, in very small quantities, catalyses (↓) and controls the natural chemical reactions of metabolism (↑). Enzymes are usually large complex molecules (p. 9), and most are responsible for one or two particular reactions in the cell. Cells contain many thousands of different enzymes.

substrate[1] (*n*) the general name for the substance on which an enzyme (↑) acts.

catalysis (*n*) the process in which natural chemical reactions are made more rapid, e.g. by enzymes (↑). **catalyze** (*v*), **catalytic** (*adj*).

catalyst (*n*) a substance that increases the rate of a chemical reaction without itself being changed in the process, e.g. enzymes (↑).

active site the part of an enzyme (↑) molecule (p. 9) to which substrate (↑) molecules become attached and where catalysis (↑) takes place.

coenzyme (*n*) a non-protein (p. 56) substance which some enzymes (↑) require to make them active. Different enzymes have different coenzymes, e.g. vitamins (↓).

vitamin (*n*) an organic (p. 11) substance required as a coenzyme (↑) in many of the chemical reactions of metabolism (↑). There are many different kinds of vitamin, and they are required by organisms in very small amounts.

multi-enzyme complex a set of enzymes (↑), often grouped into a particular arrangement in an organelle (p. 16) or a membrane (p. 18), which catalyze (↑) different reactions in the same metabolic pathway (↑).

酶(名) 一種以很少的量,就能催化(↓)和控制新陳代謝(↑)自然化學反應的蛋白質(第56頁)。酶的分子(第9頁)通常巨大而複雜,大多數酶是在細胞中對一兩種特定反應起作用。細胞中含有幾千種不同的酶。

受質;基質(名) 酶(↑)對之起作用的物質的通稱。

催化作用(名) 使自然化學反應加速的過程。例如被酶(↑)所作用。(動詞為 catalyze,形容詞為 catalytic)

催化劑(名) 能提高反應速率而在反應過程中本身不起變化的一種物質。例如酶(↑)。

活性部位 酶(↑)分子(第9頁)中能為受質(↑)分子接上並在此處發生催化作用(↑)的部位。

輔酶(名) 某些酶(↑)要求使其變為活性的一種非蛋白質(第56頁)物質。不同的酶有不同的輔酶。例如維生素(↓)。

維生素(名) 在許多新陳代謝(↑)的化學反應中要求作為一種輔酶(↑)的有機(第11頁)物質。維生素有許多不同的種類,生物體對維生素的需求量非常微小。

多酶複合物 指一組酶(↑),常在細胞器(第16頁)或細胞膜(第18頁)裏組成一個特別的排列,在同一個代謝途徑(↑)中催化(↑)不同的反應。

cell (n) a unit of protoplasm (↓) surrounded by a membrane (p. 18). Nearly all living organisms are made of one or more cells. Cells are either prokaryotic (↓) or eukaryotic (↓). Plant cells differ from animal cells by having cell walls (↓), and plastids (p. 18) in the case of eukaryotic cells. **cellular** (adj).

細胞(名) 被膜(第18頁)包圍著的一個原生質(↓)單元。幾乎一切有生命的有機體都是由一個或多個細胞所組成。這些細胞或者是原核的(↓)或者是真核的(↓)。植物細胞因有細胞壁(↓)而別於動物細胞。質體(第18頁)亦屬真核細胞。(形容詞為 cellular)

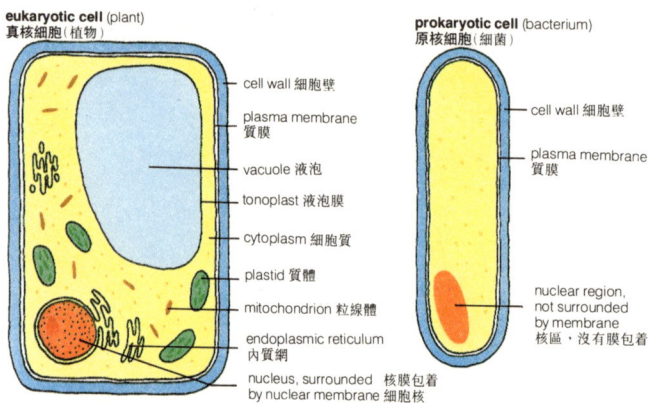

prokaryotic (adj) of cells having no organelles (↓) and no membrane (p. 18) surrounding the nuclear (p. 19) material of the cell, e.g. bacteria (p. 119). **prokaryote** (n).

eukaryotic (adj) of cells with a nucleus (p. 19) surrounded by a nuclear membrane (p. 19), and distinct organelles (↓). **eukaryote** (n).

organelle (n) a body inside a eukaryotic (↑) cell, usually with a membrane (p. 18) around it. There are usually several kinds of organelle in any cell, and each kind has a special function. For example, the function of chloroplasts (p. 32) is photosynthesis (p. 32), the function of mitochondria (p. 21) is respiration (p. 22).

cytology (n) the study of cells by microscopy.
intracellular (adj) inside a cell.
extracellular (adj) outside a cell.
protoplasm (n) the general name for all the substances and bodies inside a cell. All living organisms are made of protoplasm.

原核的(形) 指不具細胞器(↓),也沒有膜(第18頁)包圍著細胞核(第19頁)質的細胞。例如細菌(第119頁)。(名詞 prokaryote 意為原核生物)

真核的(形) 指有核膜(第19頁)包圍著細胞核(第19頁)並有明顯細胞器(↓)的細胞。(名詞 eukaryote 意為真核生物)

細胞器(名) 真核(↑)細胞內的一個小體,常有一層膜(第18頁)包著。任何細胞內常有好幾種細胞器,而每種細胞器都有其特定的功能。例如葉綠體(第32頁)的功能是行光合作用(第32頁),粒線體(第21頁)的功能是呼吸作用(第22頁)。

細胞學(名) 用顯微鏡研究細胞的學科。
細胞內的(形) 指在一個細胞的內部。
細胞外的(形) 指在一個細胞的外部。
原生質(名) 細胞內全部的物質及物體的通稱。一切有生命的有機體都是由原生質組成的。

CELLS/CELL WALLS 細胞／細胞壁 · 17

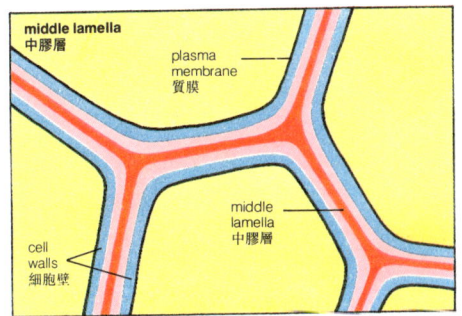

cell wall the rigid wall which surrounds a plant cell, lying outside the cell membrane (p. 18). Cell walls are made mainly of carbohydrate (p. 28) polymers (p. 10) such as cellulose (↓). All plants, fungi (p. 163) and bacteria (p. 119) have cell walls, but animals do not.

microfibril (n) one of the threads of carbohydrate (p. 28) polymer (p. 10) of which cell walls (↑) are made.

middle lamella the thin young cell wall (↑) formed between two new eukaryotic (↑) cells after cell division (p. 45). The middle lamella is made of pectin (↓), and the thicker layers of cellulose (↓) are laid down on either side of it.

cellulose (n) a carbohydrate (p. 28) polymer (p. 10) made of glucose (p. 28) molecules, which is the most important substance in plant cell walls (↑).

細胞壁　包圍著一個植物細胞的堅固的壁，位於細胞膜(第 18 頁)的外側。細胞壁主要是由碳水化合物(第 28 頁)的聚合體(第 10 頁)，如纖維素(↓)組成。所有植物、真菌類(第 163 頁)及細菌類(第 119 頁)的細胞都有細胞壁，動物的細胞則無細胞壁。

微纖絲(名)　構成細胞壁(↑)的線狀碳水化合物(第 28 頁)聚合體(第 10 頁)的一種。

中膠層　細胞分裂(第 45 頁)之後兩個新的真核(↑)細胞之間所形成的初生薄細胞壁(↑)。中膠層是由果膠(↓)組成而在細胞兩側的任何一側都有較厚的纖維素(↓)層。

纖維素(名)　由葡萄糖(第 28 頁)分子構成的碳水化合物(第 28 頁)的聚合體(第 10 頁)所組成，它是植物細胞壁(↑)的最重要的物質。

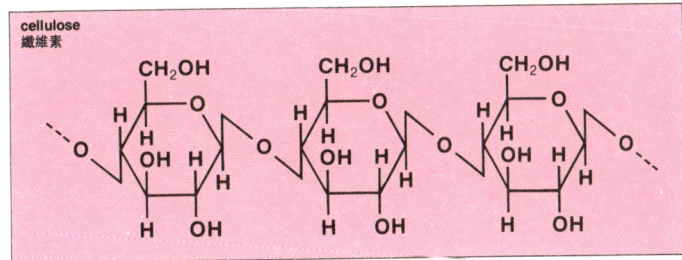

pectin (n) an acidic polysaccharide (p. 30) found in young cell walls (↑). **pectic** (adj).

果膠(名)　一種酸性的多醣類(第 30 頁)，存在於初生的細胞壁(↑)。(形容詞為 pectic)

18 · CELLS/MEMBRANES, ORGANELLES 細胞／膜、細胞器

membrane (*n*) a thin sheet of soft material which protects and encloses cells and organelles (p. 16). Membranes control the movement of substances in and out of cells and organelles. Biological membranes are made of protein (p. 56) and phospholipid (p. 31).
cell membrane the membrane (↑) which encloses a cell.
plasmalemma (*n*) = the cell membrane (↑).
plasma membrane = the cell membrane (↑).
protoplast (*n*) a plant cell or a bacterial (p. 119) cell, not including the cell wall (p. 17).
cytoplasm (*n*) all parts of a cell outside the nucleus (↓) and inside the cell membrane (↑).

膜(名) 保護並包裹著細胞及細胞器(第 16 頁)的柔軟物質的薄片。膜控制著細胞及細胞器內部和外部的物質運動。生物膜是由蛋白質(第 56 頁)和磷脂(第 31 頁)所組成。
細胞膜 包裹著一個細胞的膜(↑)。
原生質膜(名) 同細胞膜(↑)。
質膜 同細胞膜(↑)。
原生質體(名) 一個植物細胞或一個細菌(第 119 頁)細胞，不包括細胞壁(第 17 頁)。
細胞質(名) 細胞核(↓)外側及細胞膜(↑)內側的一切部分。

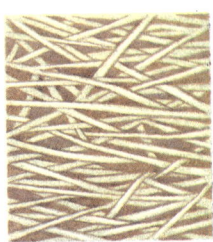

cellulose microfibrils in surface view of plant cell wall (x24,000)
植物細胞壁的纖維素微纖絲的表面觀(x24,000)

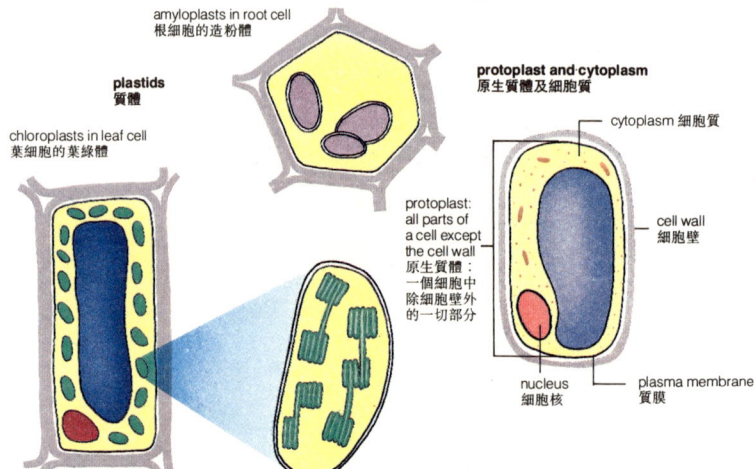

plastid (*n*) the general name for the type of organelle (p. 16) in plant cells which is surrounded by a double membrane (↑) and contains plastoglobuli (↓) and a network of internal membranes and vesicles (p. 20). There are several different kinds of plastids, each with a special function, e.g. chloroplasts (p. 32), chromoplasts (↓), amyloplasts (↓).
plastoglobuli (*n.pl.*) small, round droplets of lipid (p. 31), found in plastids (↑).

質體(名) 植物細胞內的細胞器(第 16 頁)的通稱，它為一層雙層膜(↑)所包裹，含有胞質球體(↓)、內膜網絡和泡囊(第 20 頁)。質體有幾種不同類型。例如葉綠體(第 32 頁)、有色體(↓)、造粉體(↓)。它們各具有特殊的功能。
胞質球體(名、複) 細小的脂類(第 31 頁)液態小圓滴，存在於質體(↑)中。

CELLS/MEMBRANES, ORGANELLES 細胞／膜、細胞器 · 19

chromoplast (n) a plastid (↑) containing pigment (p. 36), e.g. the coloured plastids in the cells of the tissues (p. 88) of petals (p. 70) and fruits.

amyloplast (n) a plastid (↑) in the cortical (p. 89) cells of roots in many plants. The function of amyloplasts is to store starch (p. 30).

leucoplast (n) a plastid (↑) containing no pigment (p. 36). Leucoplasts may form pigments under certain conditions, e.g. leucoplasts in root cells form chlorophyll (p. 36) if exposed to light.

pore (n) any small hole in a surface or membrane (↑), which allows substances to pass through it, e.g. pores in nuclear membranes (↓).

nucleoplasm (n) the protoplasm (p. 16) inside the nucleus (↓) of a cell. The nucleoplasm contains the chromosomes (p. 46) and the nucleoli (↓).

nucleolus (n) small, dark body inside the nucleus (↓), which can only be seen during interphase (p. 45). It consists mostly of RNA (p. 51).

有色體（名） 含有色素（第36頁）的質體（↑）。例如花瓣（第70頁）及果實組織（第88頁）的細胞內所含的有色質體。

造粉體；澱粉體（名） 許多植物根部皮層（第89頁）細胞內存在的一種質體（↑）。造粉體的功能是貯存澱粉（第30頁）。

白色體（名） 不含色素（第36頁）的質體（↑）。在一定條件下白色體可以形成色素。例如根細胞內的白色體在光照下可以形成葉綠素（第36頁）。

孔（名） 指膜（↑）的表面的任何小孔，孔容許物質通過。例如核膜（↓）上的孔。

核質（名） 細胞核（↓）內部的原生質（第16頁），核質含有染色體（第46頁）及核仁（↓）。

核仁（名） 細胞核（↓）內部的細小的黑色體，只在分裂間期（第45頁）能見到。核仁主要含有RNA（第51頁）。

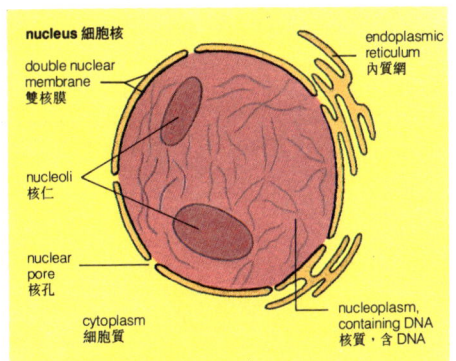

nucleus (n) an organelle (p. 16) of a eukaryotic (p. 16) cell, containing the nucleoplasm (↑), nucleoli (↑) and chromosomes (p. 46). Cells usually have only one nucleus, which controls most of the activities of the cell. **nuclear** (adj).

nuclear membrane the membrane (↑) around the nucleus (↑) of a cell. Nuclear membranes have two layers, and many pores (↑) connecting the nucleoplasm (↑) with the cytoplasm (↑).

細胞核（名） 真核（第16頁）細胞的一個細胞器（第16頁），含有核質（↑）、核仁（↑）及染色體（第46頁）。細胞通常只有一個細胞核，它操縱細胞的大部分活動。（形容詞為 nuclear）

核膜 圍繞細胞核（↑）的膜（↑）。核膜有兩層和許多核孔（↑），以聯繫細胞質（↑）和核質（↑）。

CELLS/MEMBRANES, ORGANELLES 細胞／膜、細胞器

vacuole (n) a liquid-filled space in a cell, surrounded by a membrane (p. 18). Many plant cells, especially in leaves, have a single large vacuole and a thin layer of cytoplasm (p. 18) between it and the cell membrane (p. 18).
vacuolar (adj).
tonoplast (n) the membrane (p. 18) enclosing the vacuole (↑) of a plant cell.

液泡(名) 細胞內充滿液體的空間，有一層膜(第18頁)包著。許多植物細胞，特別是葉內的細胞都有單個的大液泡，在液泡和細胞膜(第18頁)之間有一層薄的細胞質(第18頁)。(形容詞為 vacuolar)

液泡膜(名) 植物細胞內包圍著液泡(↑)的膜(第18頁)。

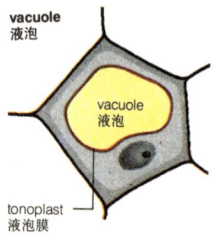

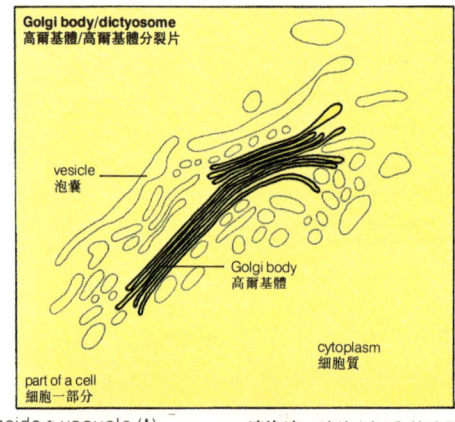

vacuolar sap the liquid inside a vacuole (↑).
Golgi body an organelle (p. 16) consisting of a group of membranes (p. 18) and vesicles (↓). The Golgi body is often important in the synthesis (p. 13) of carbohydrates (p. 28) and the secretion (p. 112) of substances, especially glycoproteins (p. 58), from the cell. In plants, it is usually called a dictyosome (↓).
dictyosome (n) the Golgi body (↑) of a plant cell.
vesicle (n) any small body in a cell or organelle (p. 16) which is surrounded by a membrane (p. 18) and contains products of metabolism (p. 14). Vesicles in the cytoplasm (p. 18) are produced mainly by the Golgi body (↑).
lysosome (n) an organelle (p. 16) surrounded by a membrane (p. 18), in which hydrolytic (p. 13) enzymes (p. 15) are stored. Lysosomes are common in animal cells, but probably not in plant cells.

液泡液 液泡(↑)內的液體。

高爾基體 由一組膜(第18頁)和泡囊(↓)所組成的細胞器(第16頁)。高爾基體對碳水化合物(第28頁)的合成(第13頁)及物質的分泌作用(第112頁)，特別是細胞內糖蛋白(第58頁)的合成很重要。植物學上常稱之為高爾基體分裂片(↓)。

高爾基體分裂片(名) 植物細胞的高爾基體(↑)。

泡囊(名) 一個細胞或細胞器(第16頁)內的任何小體，它被一層膜(第18頁)包圍著，並含有新陳代謝(第14頁)的產物，細胞質(第18頁)內的泡囊主要是由高爾基體(↑)產生的。

溶酶體(名) 為一層膜(第18頁)包著的細胞器(第16頁)，其內貯存有水解(第13頁)酶(第15頁)。溶酶體通常存在於動物細胞內，植物細胞內則可能不存在。

CELLS/MEMBRANES, ORGANELLES 細胞／膜、細胞器 · 21

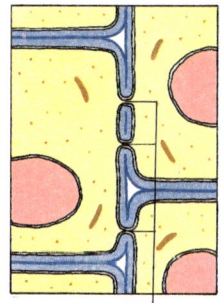
plasmodesmata
胞間連絲

peroxisome (n) a small organelle (p. 16) surrounded by a membrane (p. 18), containing the enzyme (p. 15) catalase, which catalyses (p. 15) the breakdown of hydrogen peroxide (H_2O_2) to water and oxygen. Catalase prevents the build-up of H_2O_2 in cells; H_2O_2 is toxic (p. 148), and is known to be a product of some metabolic (p. 14) reactions. Peroxisomes also contain enzymes involved in the oxidation (p. 11) of glycolic acid ($COOHCH_2OH$).

plasmodesmata (n.pl.) threads of protoplasm (p. 16) which run through the cell wall (p. 17) between cells, along which substances can be passed. **plasmodesma** (sing.).

過氧化酶體（名） 一種由膜（第18頁）包著的微小細胞器（第16頁），含有酶（第15頁）過氧化氫酶，它能催化（第15頁）過氧化氫（H_2O_2）使之分解成水和氧。過氧化氫酶能阻止細胞內生成 H_2O_2。H_2O_2 有毒（第148頁），是某些新陳代謝（第14頁）反應的產物。過氧化氫酶體也含有介入羥基乙酸（$COOHCH_2OH$）氧化作用（第11頁）的酶。

胞間連絲（名、複） 在細胞之間穿過細胞壁（第17頁）的原生質（第16頁）絲，使物質得以沿著胞間連絲通過。（單數為 plasmodesma）

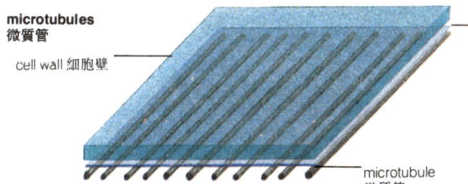

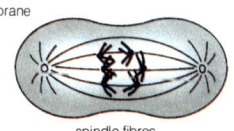
spindle fibres in dividing cells
細胞分裂時的紡縋絲

microtubule (n) a very thin, hollow thread of protein (p. 56) in a cell. Microtubules have several different functions. They form the spindle (p. 46) in mitosis (p. 45), control the formation of microfibrils (p. 17) in cell walls (p. 17), and form a structural part of the flagella (p. 121).

mitochondrion (n) a round or rod-shaped organelle (p. 16), in which the reactions of the Krebs cycle (p. 24) and the electron transfer chain (p. 40) take place. Mitochondria (pl.) have a smooth outer membrane (p. 18) and an inner membrane which is folded into cristae (↓).

cristae (n.pl.) the folds of the inner membrane (p. 18) of a mitochondrion (↑). **crista** (sing.).

matrix (n) the liquid inside a mitochondrion (↑).

微質管（名） 細胞內的一種極細的中空蛋白質（第56頁）微管。微管有幾種不同的功能。它們在有絲分裂（第45頁）時形成紡縋體（第46頁），控制細胞壁（第17頁）內微纖絲（第17頁）的形成，並形成纖毛（第121頁）的構造部分。

粒線體；線粒體（名） 圓形或棒狀的細胞器（第16頁），在粒線體出現克雷伯氏循環（第24頁）和電子傳遞鏈（第40頁）。粒線體具有平滑的外膜（第18頁），內膜則摺疊成崤（↓）。

崤（名、複） 由粒線體（↑）的內膜（第18頁）摺疊而成。（單數為 crista）

基質（名） 粒線體（↑）內部的液體。

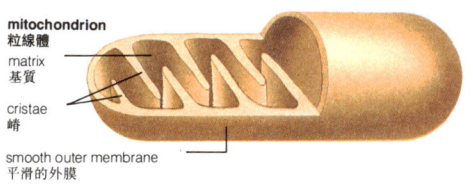

respiration (n) the process in which cells release energy stored in carbohydrates (p. 28) in order to drive the chemical reactions of metabolism (p. 14). Aerobic (↓) respiration involves glycolysis (↓), the Krebs cycle (p. 24), and the electron transfer chain (p. 40); this process uses oxygen, and produces carbon dioxide and ATP (p. 26). The Krebs cycle and electron transfer chain take place in the mitochondria (p. 21). Anaerobic (↓) respiration involves glycolysis (↓) and fermentation (p. 24), in which oxygen is not used. **respire** (v), **respiratory** (adj).

呼吸作用（名） 細胞釋放出貯存在碳水化合物（第 28 頁）中的能量，從而推動新陳代謝（第 14 頁）化學反應的過程。需氧（↓）呼吸包括醣酵解（↓），克雷伯氏循環（第 24 頁）及電子傳遞鏈（第 40 頁），這過程需要氧，並產生二氧化碳和 ATP（第 26 頁），克雷伯氏循環及電子傳遞鏈發生在粒線體（第 21 頁）內。厭氧（↓）呼吸包括糖酵解（↓）及醱酵作用（第 24 頁），這些過程不需要氧。（動詞為 respire，形容詞為 respiratory）

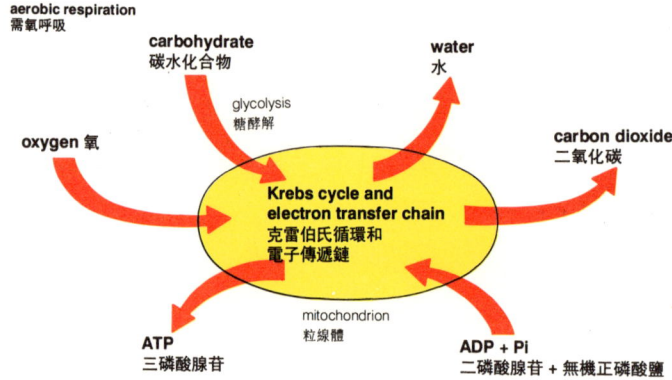

aerobic (adj) of respiration using molecular oxygen and involving oxidative (p. 11) processes. Also, of organisms which respire (↑) aerobically. **aerobe** (n).

anaerobic (adj) of respiration (↑) in which molecular oxygen is not used, that is, in glycolysis (↓) and fermentation (p. 24). Also, of organisms which can live without molecular oxygen, e.g. the bacteria (p. 119) that live in mud or in the gut of animals; these organisms are sometimes called anaerobes.

glycolysis (n) the anaerobic (↑) sequence of reactions in the breakdown of glucose (p. 28) during respiration (↑), ending with pyruvic acid (p. 24).

需氧的；好氧的（形） 指利用分子氧，並涉及各種氧化作用（第 11 頁）過程的呼吸作用。也指需氧呼吸（↑）的生物。（名詞 aerobe 意為需氧生物、好氧生物）

厭氧的；缺氧的（形） 指不需要分子氧的呼吸作用（↑）。例如糖酵解（↓）及醱酵作用（第 24 頁）。也指在無分子氧下能生活的生物。例如細菌（第 119 頁）生活在淤泥中或動物腸道中，這些生物體往往稱之為厭氧生物。

糖酵解（名） 呼吸作用（↑）時，葡萄糖（第 28 頁）分解的厭氧（↑）反應順序，反應結束產生丙酮酸（第 24 頁）。

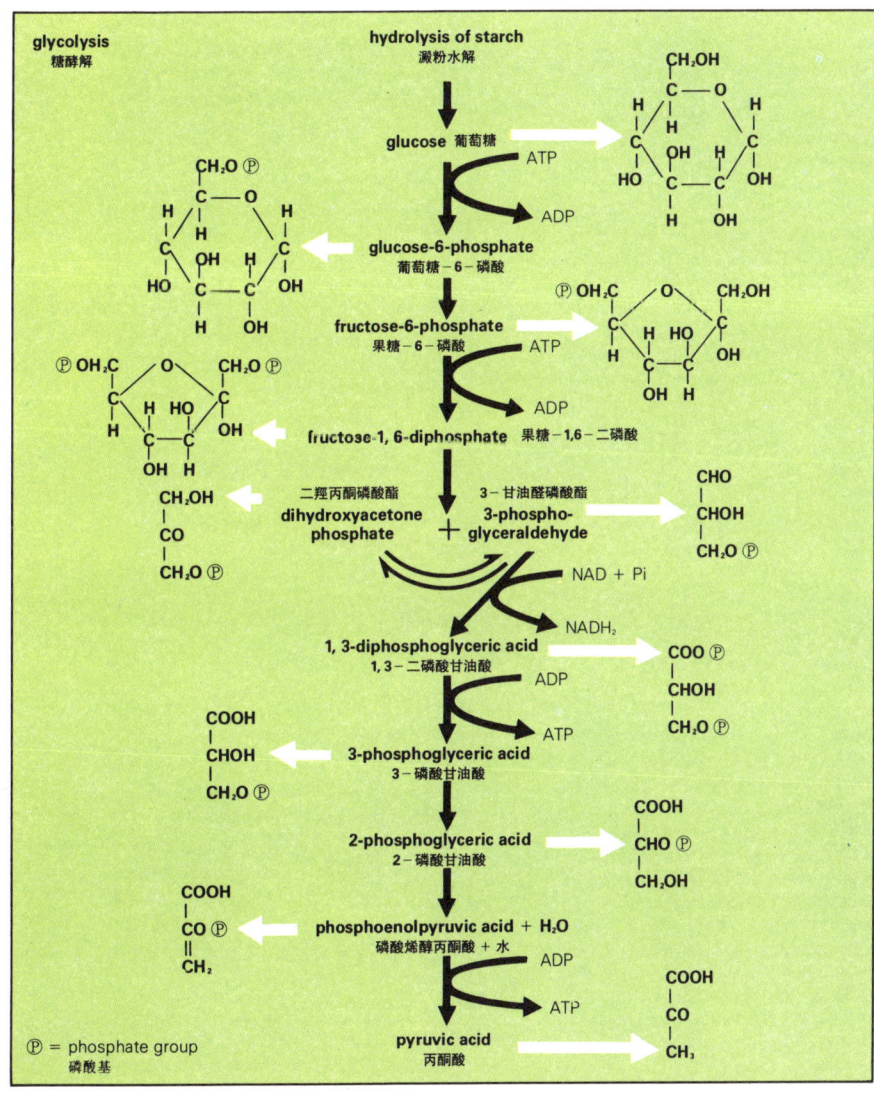

fermentation (*n*) the breaking down of organic (p.11) molecules, especially by yeast (p.164) and bacteria (p.119) under anaerobic (p.22) conditions, to produce carbon dioxide and alcohol (↓) or lactic acid (↓). **ferment** (*v*).

alcohol (*n*) any one of a class of organic (p.11) compounds with one or more hydroxyl groups (—OH), e.g. ethanol (CH_3CH_2OH).

lactic acid $CH_3CHOHCOOH$. One of the end products of fermentation (↑).

pyruvic acid $CH_3COCOOH$. The end product of glycolysis (p.22). Pyruvic acid is the fuel for the Krebs cycle (↓) in aerobic (p.22) organisms.

Krebs cycle a series of metabolic (p.14) reactions in aerobic (p.22) respiration (p.22), in which pyruvic acid (↑) is broken down to carbon dioxide and water. The energy released in this process is used to produce ATP (p.26) from ADP (p.26) and orthophosphate (p.13). The Krebs cycle takes place in the mitochondria (p.21).

tricarboxylic acid cycle TCA cycle = Krebs cycle (↑).

citric acid cycle = Krebs cycle (↑).

GTP guanosine triphosphate. A nucleotide (p.52) similar to ATP (p.26), involved in the reactions of the Krebs cycle (↑).

NAD nicotinamide adenine dinucleotide. A hydrogen carrier in the Krebs cycle (↑).

FAD flavin adenine dinucleotide. A hydrogen carrier in the Krebs cycle (↑).

chemiosmosis (*n*) a process in which energy from the hydrolysis (p.13) of ATP (p.26) or the oxidation (p.11) of organic (p.11) molecules (p.9) can be used to make an electrical and chemical gradient (↓) of protons (p.8) across a membrane (p.18). This gradient can be used to drive energy-requiring reactions such as the uptake of ions or the synthesis (p.13) of ATP.

gradient (*n*) an increase or decrease of a measurable quantity over a given distance, e.g. a chemical gradient, the increase in concentration (p.12) of a solute (p.12) from one part of a plant to another; environmental gradient, the decrease in temperature with increasing height on a mountain.

醱酵作用（名）有機（第11頁）分子的分解，特別是酵母（第164頁）及細菌（第119頁）在缺氧（第22頁）條件下分解生成二氧化碳及醇（↓）或乳酸（↓）。（動詞為 ferment）

醇（名）任何一類含有一個或幾個羥基(–OH)的有機（第11頁）化合物例如乙醇（CH_3CH_2OH）。

乳酸（$CH_3CHOHCOOH$）醱酵作用（↑）的終產物之一。

丙酮酸（$CH_3COCOOH$）糖酵解（第22頁）的終產物，丙酮酸是需氧（第22頁）生物體克雷伯氏循環（↓）的燃料。

克雷伯氏循環 為有氧（第22頁）呼吸（第22頁）中的一序列新陳代謝（第14頁）反應循環，在反應過程中丙酮酸（↑）分解為二氧化碳和水。在此過程所釋放的能量用於將 ADP（第26頁）轉變成 ATP（第26頁）及正磷酸鹽（第13頁）。克雷伯氏循環在粒線體（第21頁）中進行。

三羧酸(**TCA**)循環 同克雷伯氏循環（↑）。

檸檬酸循環 同克雷伯氏循環（↑）。

鳥嘌呤核苷三磷酸(GTP) 一種類似 ATP（第26頁）的核苷酸（第52頁），它參與克雷伯氏循環（↑）的各個反應。

菸鹼醯胺腺嘌呤二核苷酸(NAD) 為克雷伯氏循環（↑）的載氫體。

黃素腺嘌呤二核苷酸(FAD) 為克雷伯氏循環（↑）的載氫體。

化學滲透（名）ATP（第26頁）水解（第13頁）的能量，或有機（第11頁）分子（第9頁）氧化作用（第11頁）的能量，可用於造成質子（第8頁）越過膜（第18頁）的電梯度或化學梯度（↓）的過程。此梯度可用於推動要求能量的那些反應。例如離子的吸收或 ATP 的合成（第13頁）。

梯度（名）指越過一定距離的可計量量的增加或減少。例如，從植物的某一個部位到另一部位的溶質（第12頁）濃度（第12頁）增加是化學梯度；溫度隨着山的海拔升高而降低是環境梯度。

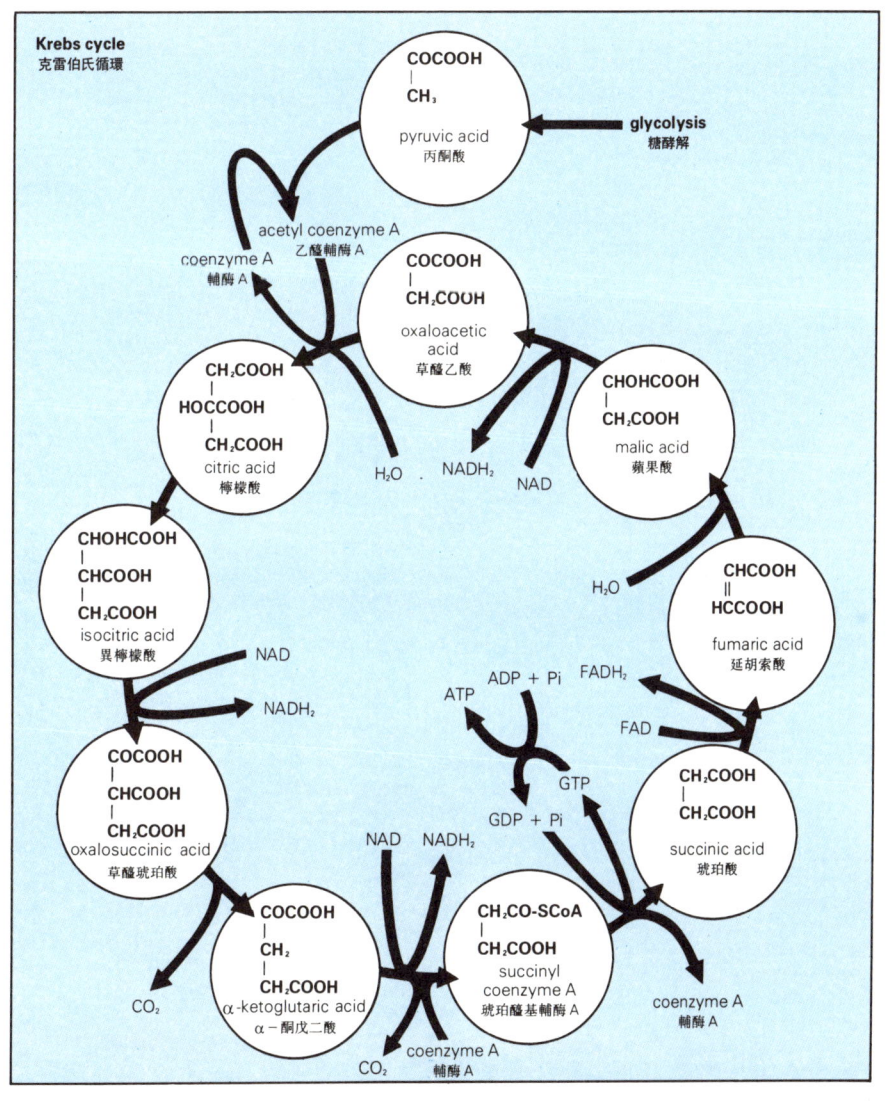

photorespiration (n) the process in which plants, in the presence of light, high oxygen concentration (p. 12) and low carbon dioxide concentration, will take up oxygen and give off carbon dioxide, due to the oxidation of organic (p. 11) compounds produced by CO_2 fixation (p. 33). Photorespiration involves the chloroplasts (p. 32), mitochondria (p. 21) and peroxisomes (p. 21). Its function is unclear.

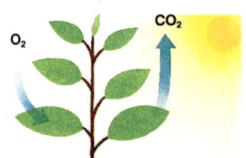

photorespiration
光呼吸作用

high O_2 concentration in plant tissue
植物組織內氧的濃度高

low CO_2 concentration in plant tissue
植物組織內二氧化碳的濃度低

ADP adenosine diphosphate. The nucleotide (p. 52) which results from the hydrolysis (p. 13) of ATP (↓), and which, with the addition of an orthophosphate (p. 13) group, is used to synthesize (p. 13) ATP.

ATP adenosine triphosphate. The nucleotide (p. 52) which stores energy in the bonds between its three phosphate (p. 13) groups. This energy is released by hydrolysis (p. 13) to drive synthetic (p. 13) reactions in the cell.

metabolic poison any substance, e.g. cyanide, that prevents the production of ATP (↑) in cells. Without ATP as a source of energy for metabolism (p. 14), cells and organisms quickly die.

phosphorylation (n) the reaction in which a phosphate (p. 13) group is added to a molecule (p. 19), e.g. the phosphorylation of ADP (↑) to give ATP (↑).

oxidative phosphorylation the production of ATP (↑) from ADP (↑) and orthophosphate (p. 13), using energy released in the oxidation (p. 11) of organic (p. 11) compounds in the electron transfer chain (p. 40). Oxidative phosphorylation takes place in the mitochondria (p. 21), and is the main source of ATP in heterotrophic (p. 32) organisms. In the final reaction of the process, molecular oxygen (O_2) is reduced (p. 11) to water.

光呼吸作用(名) 植物在光照及氧氣濃度(第12頁)高和二氧化碳濃度低時，吸入氧氣並放出二氧化碳的過程，這是由於二氧化碳固定(第33頁)所產生的有機(第11頁)化合物發生氧化作用。光呼吸作用和葉綠體(第32頁)、粒線體(第21頁)及過氧化物酶體(第21頁)有關。它的功能還不清楚。

二磷酸腺苷(ADP) ATP(↓)水解(第13頁)所產生的核苷酸(第52頁)，ADP加入正磷酸(第13頁)基團而用於合成(第13頁)ATP。

三磷酸腺苷(ATP) 在其三個磷酸(第13頁)基團之間的鍵中貯存着能量的核苷酸(第52頁)。這種能量在水解作用(第13頁)時釋放出來以在細胞內推動合成(第13頁)反應。

代謝毒 能阻止細胞內產生ATP(↑)的任何物質。例如氰化物。細胞和生物體由於缺乏ATP作為新陳代謝(第14頁)的能源而迅速死亡。

磷酸化(名) 分子(第19頁)中加入磷酸(第13頁)基團的反應。例如ADP(↑)磷酸化生成ATP(↑)。

氧化磷酸化 利用有機(第11頁)化合物在電子傳遞鏈(第40頁)中氧化作用(第11頁)釋放的能量從ADP(↑)和正磷酸鹽(第13頁)合成ATP(↑)。氧化磷酸化在粒線體(第21頁)內進行，它是異養生物(第32頁)體內ATP的主要來源。在這過程的最終反應中，分子態氧(O_2)被還原(第11頁)成水。

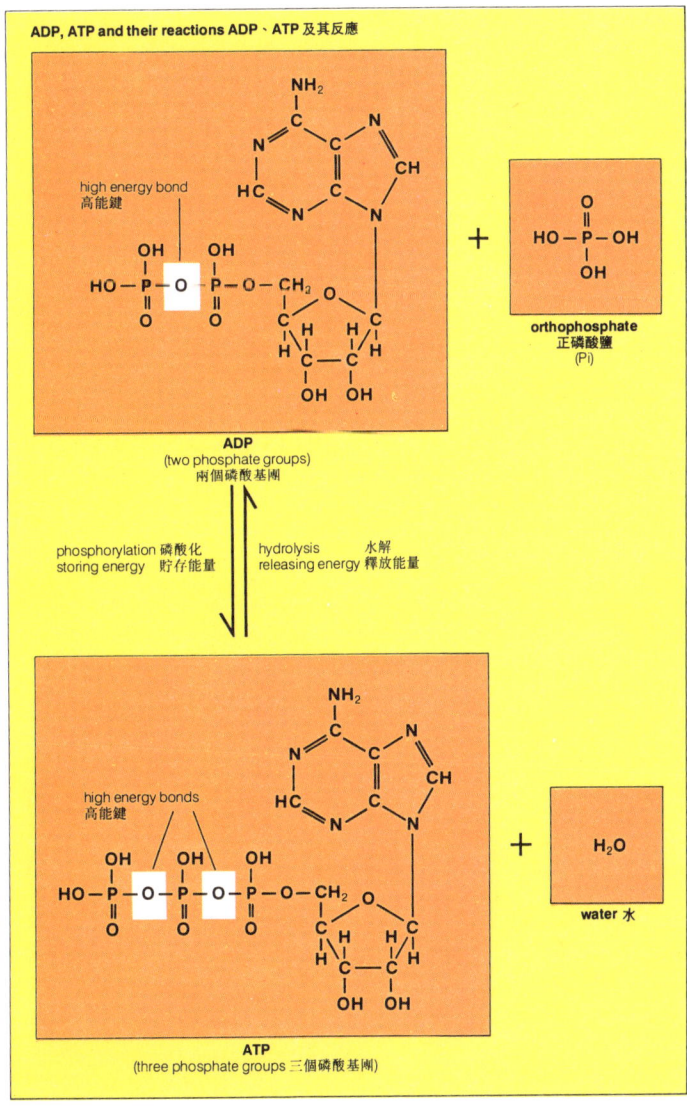

carbohydrate (n) an organic (p.11) compound containing carbon, hydrogen and oxygen in a ratio of 1:2:1. Starch (p.30) and all sugars, in which plants store the energy obtained from light during photosynthesis (p.32), are carbohydrates.

sugar (n) a carbohydrate (↑) which is soluble (p.12) in water and sweet to the taste, e.g. sucrose (↓), glucose (↓). Sugars are produced in photosynthesis (p.32), and the energy obtained from them during respiration (p.22) is used to drive the reactions of metabolism (p.14).

glycoside (n) an organic (p.11) compound consisting of a sugar molecule (p.9) bonded to another organic molecule by a glycosidic bond (↓).

monosaccharide (n) a simple sugar, with between three and seven carbon atoms.

hexose (n) any monosaccharide (↑) containing six carbon atoms (p.8), e.g. glucose (↓), fructose (↓).

pentose (n) any monosaccharide (↑) with five carbon atoms (p.8). Important pentoses are ribose and deoxyribose, which are found in RNA (p.51) and DNA (p.51), and ribulose, which, as ribulose-diphosphate (p.33), is used for CO_2 fixation (p.33) in photosynthesis (p.32).

triose (n) any monosaccharide (↑) with three carbon atoms (p.8).

glucose (n) a hexose (↑), aldose (↓) monosaccharide (↑), with formula $C_6H_{12}O_6$. Glucose is the unit in polysaccharides (p.30) such as starch (p.30) and cellulose (p.17). It is a product of photosynthesis (p.32). With fructose (↓), it forms the disaccharide (↓) sucrose (↓).

碳水化合物；醣(名) 含碳、氫及氧比例為1：2：1的一種有機(第11頁)化合物。醣包括澱粉(第30頁)和一切糖類，植物在光合作用(第32頁)時從陽光獲得的能量都貯存在醣中。

糖(名) 可溶解(第12頁)於水並有甜味的碳水化合物(↑)。例如蔗糖(↓)和葡萄糖(↓)。糖是光合作用(第32頁)產生的。呼吸作用(第22頁)時從糖獲得能量，用以推動新陳代謝(第14頁)反應。

葡萄糖苷(名) 糖分子(第9頁)藉一個糖苷鍵(↓)連接到另一個有機分子而成的有機(第11頁)化合物。

單醣；單糖(名) 介於三至七個碳原子之間的一種簡單醣類。

己醣；己糖(名) 含六個碳原子(第8頁)的任何單醣(↑)。例如葡萄糖(↓)、果糖(↓)。

戊醣；戊糖(名) 含五個碳原子(第8頁)的任何單醣(↑)，核糖及脫氧核糖是重要的戊醣，它們可見於RNA(第51頁)及DNA(第51頁)中，而核酮糖(例如核酮糖二磷酸(第33頁))在光合作用(第32頁)時被用於固定(第33頁)二氧化碳(CO_2)。

丙醣；丙糖(名) 含有三個碳原子(第8頁)的任何單醣(↑)。

葡萄糖(名) 化學式為 $C_6H_{12}O_6$，屬己醣(↑)，是醛糖(↓)單醣(↑)。葡萄糖是多醣類(第30頁)，如澱粉(第30頁)、纖維素(第17頁)等的結構單元，它是光合作用(第32頁)的產物，可與果糖(↓)組成雙醣(↓)如蔗糖(↓)。

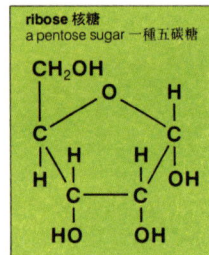

ribose 核糖
a pentose sugar 一種五碳糖

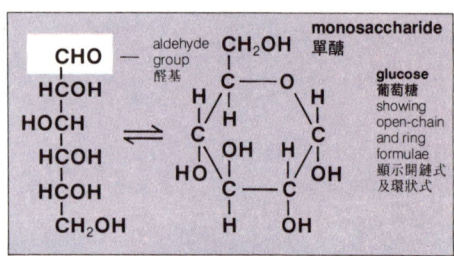

monosaccharide 單醣
glucose 葡萄糖
showing open-chain and ring formulae 顯示開鏈式及環狀式

aldehyde group 醛基

CARBOHYDRATES/SUGARS 碳水化合物／糖類 · 29

glyceraldehyde 甘油醛
a triose, aldose sugar 丙醣，醛醣
(3 carbon atoms) (三個碳原子)

aldehyde group 醛基

aldose (*n*) any monosaccharide (↑) in which one carbon atom is in an aldehyde (—CHO) group, e.g. glucose (↑).

ketose (*n*) any monosaccharide (↑) in which one carbon atom (p. 8) is in a ketone (—CO—) group, e.g. fructose (↓).

fructose (*n*) a hexose (↑), ketose (↑), monosaccharide (↑), with formula $C_6H_{12}O_6$. With glucose (↑), it forms the disaccharide (↓) sucrose (↓).

醛醣；醛糖（名） 指醛基(-CHO)上有一個碳原子的任何單醣(↑)。例如葡萄糖(↑)。

酮醣；酮糖（名） 指酮基(-CO-)上有一個碳原子（第8頁）的任何單醣(↑)。例如果糖(↓)。

果糖（名） 是一種己醣(↑)，屬酮醣(↑)，是單醣，化學式為 $C_6H_{12}O_6$。果糖和葡萄糖(↑)組成雙醣(↓)，如蔗糖(↓)。

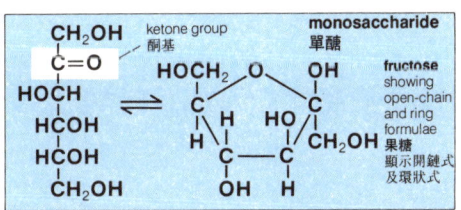

monosaccharide 單醣
fructose showing open-chain and ring formulae 果糖 顯示開鏈式及環狀式

glycosidic bond the chemical bond between the sugar monomers (p. 10) in a disaccharide (↓) or polysaccharide (p. 30). The formation of the glycosidic bond is due to the reaction of the —OH group on the first carbon atom (p. 8) of one sugar molecule (p. 9) with any —OH group of another sugar molecule; H_2O is produced, and the sugars become linked by an oxygen atom.

disaccharide (*n*) a sugar made of two monosaccharide (↑) units, e.g. sucrose (↓).

糖苷鍵 雙醣(↓)或多醣（第30頁）分子中各個糖單體（第10頁）之間的化學鍵。糖苷鍵是由於糖分子（第9頁）第1位碳原子（第8頁）上的-OH基和另一個糖分子任何位上的-OH基反應而形成的，反應生成水，糖則由一個氧原子連結。

雙醣；雙糖（名） 由兩個單醣(↑)單元組成的糖。例如蔗糖。

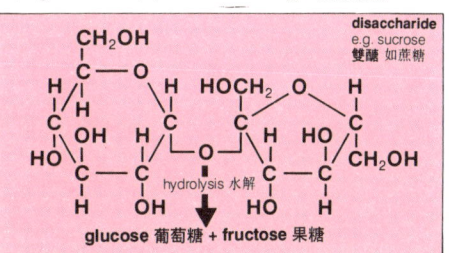

disaccharide e.g. sucrose 雙醣 如蔗糖
hydrolysis 水解
glucose 葡萄糖 + fructose 果糖

sucrose (*n*) $C_{12}H_{22}O_{11}$. A disaccharide (↑), formed from a molecule (p. 9) of glucose (↑) and a molecule of fructose (↑), found only in plants. It is the sugar which is obtained from sugar cane and sugar beet.

蔗糖 ($C_{12}H_{22}O_{11}$)（名） 由一個葡萄糖(↑)分子（第9頁）和一個果糖(↑)分子組成的雙醣(↑)，僅見於植物，蔗糖是從甘蔗及甜菜製得的糖。

oligosaccharide (*n*) any sugar made of anything between two and ten monosaccharide (p. 28) units.

polysaccharide (*n*) any polymer (p. 10) consisting of many monosaccharide (p. 28) units, e.g. starch (↓), cellulose (p. 17).

寡糖(名) 由二至十個單醣(第 28 頁)單元通過種種方式形成的任何一種糖。

多醣；多糖(名) 由許多單醣(第 28 頁)單元組成的任何聚合體(第 10 頁)。例如澱粉(↓)和纖維素(第 17 頁)。

polysaccharide 多醣
e.g. starch (amylopectin)
如澱粉（支鏈澱粉）

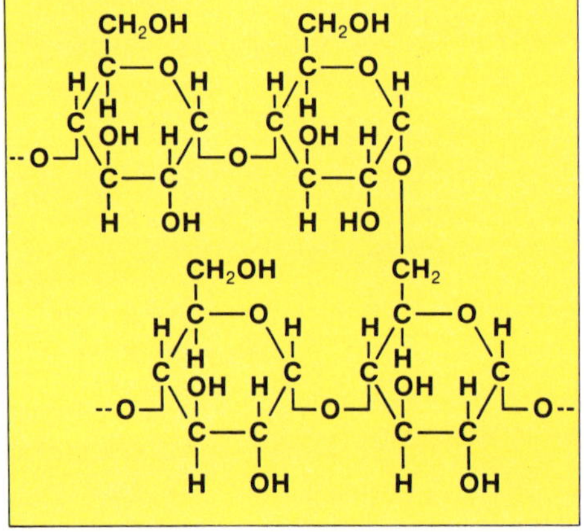

starch (*n*) a polysaccharide (↑) in which the carbohydrate (p. 28) produced during photosynthesis (p. 32) is stored in plants. Starch is a polymer (p. 10) of glucose (p. 28) units. It is deposited as small grains in chloroplasts (p. 32), and sometimes in amyloplasts (p. 19).

amylose (*n*) a form of starch (↑) made of straight chains of glucose (p. 28) monomers (p. 10).

amylopectin (*n*) a form of starch (↑) in which the glucose (p. 28) molecules (p. 9) are in branched chains.

amylase (*n*) an enzyme (p. 15) which catalyzes (p. 15) the breakdown of starch (↑) into monosaccharide (p. 28) units.

diastase (*n*) = amylase (↑).

inulin (*n*) a polysaccharide (↑) made of fructose (p. 29) monomers (p. 10). Inulin is a storage product in the roots of many plants.

澱粉(名) 為一種多醣(↑)。是植物行光合作用(第 32 頁)時產生的碳水化合物(第 28 頁)貯存在體內。澱粉是葡萄糖(第 28 頁)單元的聚合體(第 10 頁)。它以細小顆粒沉積在葉綠體(第 32 頁)內，有時沉積在造粉體(第 19 頁)內。

直鏈澱粉(名) 一種形式的澱粉(↑)，由直鏈葡萄糖(第 28 頁)單體(第 10 頁)構成的。

支鏈澱粉(名) 一種形式的澱粉(↑)，其葡萄糖(第 28 頁)分子(第 9 頁)在支鏈上。

澱粉酶(名) 一種能催化(第 15 頁)並使澱粉(↑)分解為單醣(第 28 頁)單元的酶(第 15 頁)。

澱粉酶製劑(名) 同澱粉酶(↑)。

菊芋多醣(名) 由果糖(第 29 頁)單體(第 10 頁)構成的一種多醣(↑)。菊芋多醣是許多種植物根內貯存的產物。

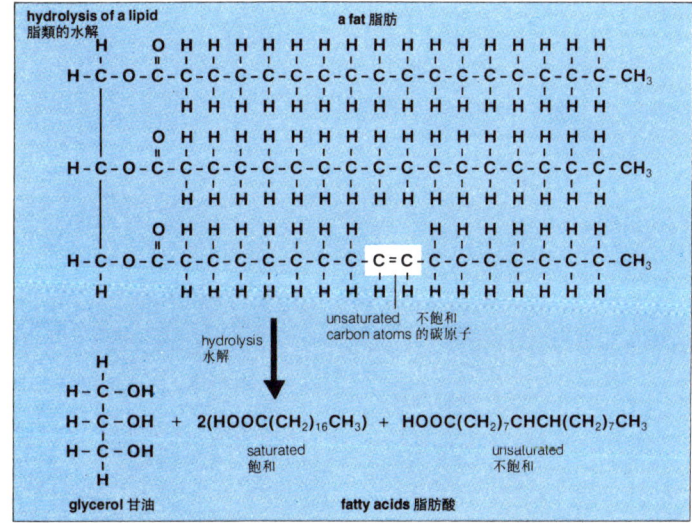

lipid (*n*) one of a group of chemical compounds, which contain glycerol (↓) and fatty acids (↓). Lipids are insoluble (p. 12) in water.

fatty acid an organic (p. 11) acid with the general formula $C_nH_{2n}O_2$. A fatty acid molecule (p. 9) is a straight chain and the number of carbon atoms (p. 8) is usually even.

saturated (*adj*) of any organic (p. 11) molecule (p. 9) with no double bonds between its carbon atoms (p. 8), e.g. the fatty acid (↑) palmitic acid, $(CH_2)_{15}COOH$.

unsaturated (*adj*) of any organic (p. 11) molecule (p. 9) with at least one double bond between carbon atoms (p. 8), e.g. the fatty acid (↑) oleic acid, $(CH_2)_8CHCH(CH_2)_7COOH$.

glycerol (*n*) $CH_2OHCHOHCH_2OH$. A compound which combines with fatty acids (↑) to form lipids (↑).

phospholipid (*n*) a lipid (↑) containing one or more phosphate (p. 13) groups.

aromatic (*adj*) of organic (p. 11) compounds in which carbon atoms (p. 8) are arranged in rings of six.

脂類(名) 含有甘油(↓)及脂肪酸(↓)的一類化合物。脂類不溶解(第12頁)於水。

脂肪酸(名) 通式為 $C_nH_{2n}O_2$ 的一種有機(第11頁)酸。脂肪酸分子(第9頁)為直鏈型,其碳原子(第8頁)數目通常為偶數。

飽和的(形) 指任何有機(第11頁)物分子(第9頁),在其碳原子之間不具雙鍵。例如屬脂肪酸(↑)的棕櫚酸($(CH_2)_{15}COOH$)。

不飽和的(形) 指任何一種有機(第11頁)物分子(第9頁),在其碳原子(第8頁)之間至少有一個雙鍵。例如屬脂肪酸(↑)的油酸($(CH_2)_8CHCH(CH_2)_7COOH$)。

甘油($CH_2OHCHOHCH_2OH$)(名) 能與多個脂肪酸(↑)分子結合成脂類(↑)的一種化合物。

磷脂(名) 含有一個或多個磷酸(第13頁)基團的脂類(↑)。

芳香的(形) 指有機(第11頁)化合物的碳原子排列在由六個碳原子(第8頁)構成的環中。

photosynthesis (n) the process by which plants use energy from sunlight to produce carbohydrates (p. 28) from carbon dioxide (CO_2) and water (H_2O), i.e. the conversion of simple inorganic (p. 11) compounds to complex organic (p. 11) compounds. Sunlight energy is captured by molecules (p. 9) of chlorophyll (p. 36) in the chloroplasts (↓) of cells in green leaves. The general equation for photosynthesis is $CO_2 + 4H_2O \rightarrow (CH_2O) + 3H_2O + O_2$. Some bacteria (p. 119), also use this process.
photosynthetic (adj), **photosynthesize** (v).

autotrophic (adj) able to synthesize (p. 13) food from simple chemical compounds, using energy from light or chemical reactions. Most plants are autotrophic. **autotroph** (n).

heterotrophic (adj) of organisms which need a supply of organic (p. 11) matter for growth. Such organisms cannot synthesize (p. 13) organic matter using energy from light. Fungi (p. 163) and animals are heterotrophic, as are many bacteria (p. 119). **heterotroph** (n).

chloroplast (n) a green plastid (p. 18) containing chlorophyll (p. 36). Chloroplasts are the site of photosynthesis (↑). They contain their own DNA (p. 51) and reproduce (p. 59) themselves. Chloroplasts are found in the cells of leaf tissues (p. 88), and those of green stems.

光合作用(名) 植物利用陽光的能量,將二氧化碳(CO_2)和水(H_2O)製造成碳水化合物(第28頁)的過程,即把簡單的無機(第11頁)化合物轉化為複雜的有機(第11頁)化合物的過程。綠葉細胞中葉綠體(↓)的葉綠素(第36頁)分子(第9頁)能從陽光獲取能量。光合作用的一般反應式是 $CO_2 + 4H_2O \rightarrow (CH_2O) + 3H_2O + O_2$。某些細菌(第119頁)也行光合作用。(形容詞為 photosynthetic,動詞為 photosynthesize)

自養的;自營的(形) 指能利用光能或化學反應的能,從簡單的化合物合成(第13頁)食物。大多數植物都是自養的。(名詞 autotroph 意為自養生物)

異養的;異營的(形) 指需要有機(第11頁)物質供生長的生物。此種生物不能利用光能合成(第13頁)有機物質。真菌(第163頁)及動物和許多種細菌(第119頁)一樣都屬異養的。(名詞 heterotroph 意為異養生物)

葉綠體(名) 含有葉綠素(第36頁)的綠色質體(第18頁)。葉綠體是光合作用(↑)的場所。它們有自己的DNA(第51頁),且本身能增殖(第59頁)。葉綠體存在於葉組織(第88頁)的細胞內,亦見於綠色的莖中。

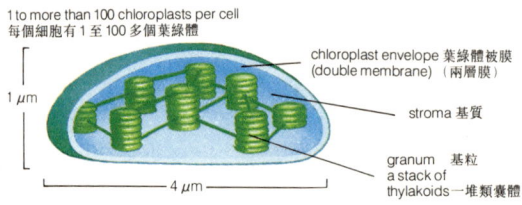

chloroplast 葉綠體

chloroplast envelope the double membrane (p. 18) surrounding the chloroplast (↑).

stroma (n) the parts of the chloroplast (↑) in between the stacks of grana (↓). The dark reaction (↓) of photosynthesis (↑) takes place in the stroma.

葉綠體被膜 包被著葉綠體(↑)的兩層膜(第18頁)。

基質(名) 是基粒(↓)堆之間的葉綠體(↑)部分。光合作用(↑)的暗反應(↓)在基質內進行。

PHOTOSYNTHESIS/DARK REACTION 光合作用／暗反應 · 33

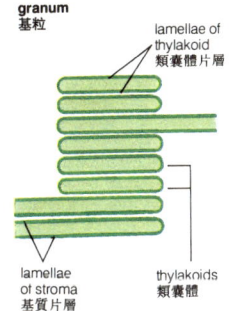

grana (*n.pl.*) the stacks of flat vesicles (p. 20) - or thylakoids (↓) - in the chloroplast (↑), where the pigments (p. 36) and enzymes (p. 15) of the light reaction (p. 36) of photosynthesis (↑) are found. **granum** (*sing.*).

lamellae (*n.pl.*) the membranes (p. 18) of the grana (↑) of chloroplasts (↑). **lamella** (*sing.*)

thylakoid (*n*) a flat vesicle (p. 20) in the grana (↑) of a chloroplast (↑).

CO_2 fixation the process in which CO_2 dissolved in the intercellular spaces (p. 95) is fixed into an organic (p. 11) molecule (p. 9) in the chloroplasts (↑) of plant cells. This is an important part of the dark reaction (↓). Usually, CO_2 reacts with ribulose-diphosphate (↓) to produce two molecules of PGA (↓).

reductive pentose pathway the set of reactions in photosynthesis (↑) in which CO_2 is fixed by ribulose diphosphate (↓), a pentose (p. 28) sugar, to give PGA (↓), which is used to produce a hexose (p. 28) sugar and more ribulose-diphosphate, which is used for further CO_2 fixation (↑). This pathway is driven by energy from ATP (p. 26) produced in the light reaction (p. 36). It also requires $NADPH_2$ (p. 38) from the light reaction for the reduction (p. 11) of PGA.

dark reaction the part of photosynthesis (↑) controlled by enzymes (p. 15) rather than light, i.e. CO_2 fixation (↑) and the reductive pentose pathway (↑).

ribulose-diphosphate a compound consisting of a molecule (p. 9) of the pentose (p. 28) sugar ribulose and two phosphate (p. 13) groups. It is the main compound involved in CO_2 fixation (↑) during photosynthesis (↑). It is also called ribulose-bis-phosphate and RuDP.

ribulose-diphosphate carboxylase an enzyme that catalyzes (p. 15) CO_2 fixation (↑) by ribulose-diphosphate (↑).

phosphoglyceric acid PGA. A compound with three atoms (p. 8) of carbon, which is the first product of the reaction between CO_2 and ribulose-diphosphate (↑) in the reductive pentose pathway (↑) of photosynthesis (↑).

PGA = phosphoglyceric acid.

葉綠餅；基粒(名、複) 在葉綠體(↑)內扁平的泡囊(第20頁)，或類囊體(↓)堆，其內含有光合作用(↑)光反應(第36頁)的色素(第36頁)及酶(第15頁)。(單數為granum)

片層(名、複) 葉綠體(↑)的葉綠餅(↑)內的膜(第18頁)。(單數為lamella)

類囊體(名) 葉綠體(↑)的葉綠餅(↑)內的泡囊(第20頁)。

CO_2 固定作用 CO_2 溶解在胞間空隙(第95頁)固定於植物細胞葉綠體(↑)內的有機(第11頁)物分子(第9頁)的過程。這是暗反應(↓)的一個重要部分。CO_2 通常與核酮糖二磷酸(↓)反應生成兩個分子的PGA(↓)。

還原性戊醣途徑 指光合作用(↑)的一系列反應，反應中CO_2被戊醣(第28頁)即核酮糖二磷酸(↓)固定，產生PGA(↓)，光合作用產生己醣(第28頁)和更多的核酮糖二磷酸，使更多的CO_2被固定(↑)。這途徑在光反應(第36頁)中由ATP(第26頁)產生的能量所推動，同時要求從光反應得到$NADPH_2$(第38頁)以還原(第11頁)PGA。

暗反應 這是受酶(第15頁)而不是受光控制的光合作用(↑)部分，即是 CO_2 的固定作用(↑)，同時是還原性戊醣的途徑(↑)。

核酮糖二磷酸 是由一個戊醣(第28頁)核酮糖分子(第9頁)及兩個磷酸(第13頁)基團組成的化合物。它是光合作用(↑)時參與CO_2固定作用(↑)的主要化合物，亦稱核酮糖-雙-磷酸或RuDP。

核酮糖二磷酸-羧化酶 是由核酮糖二磷酸(↑)催化(第15頁)CO_2固定作用(↑)的一種酶。

磷酸甘油酸(**PGA**) 在光合作用(↑)的還原性戊醣途徑中(↑)，CO_2 與核酮糖二磷酸(↑)之間反應的初始產物，是具有三個碳原子(第8頁)的一種化合物。

磷酸甘油酸的英文縮寫為**PGA**。

glyceric acid-3-phosphate = phosphoglyceric acid (p. 33).

Calvin cycle the reductive pentose pathway (p. 33) of photosynthesis (p. 32), named after one of its discoverers. The cycle was worked out in the 1940s and 1950s. It is also known as the C_3 pathway (↓).

3-磷酸甘油酸　同磷酸甘油酸(第33頁)。

卡爾文循環　光合作用(第32頁)的還原性戊醣途徑(第33頁)，是以發現者的名字之一命名的，這個循環是本世紀四、五十年代期間發現的，也稱為三碳(C_3)途徑(↓)。

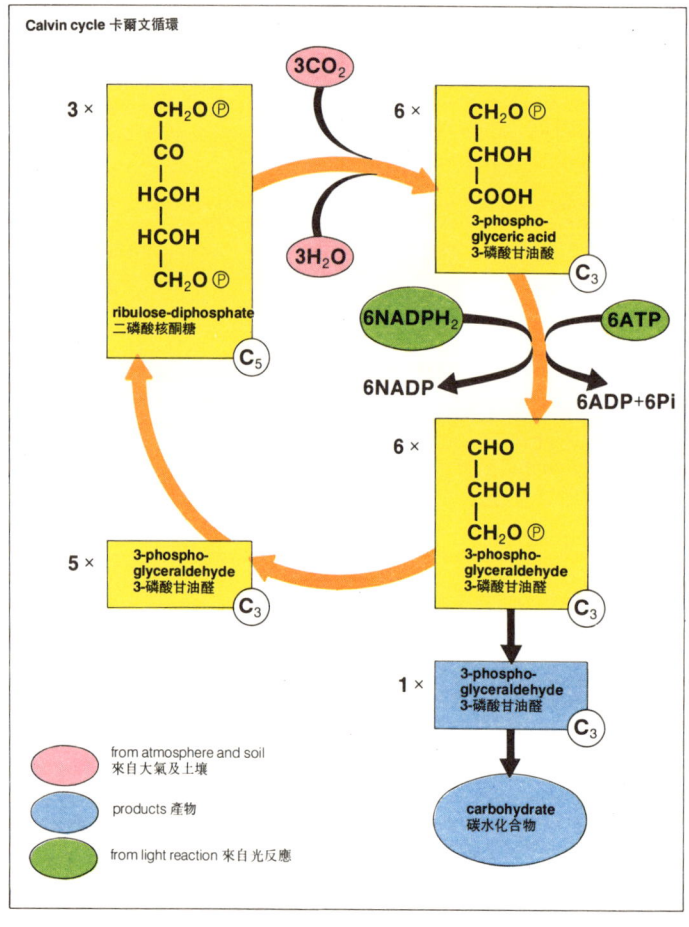

PHOTOSYNTHESIS/CO_2 FIXATION PATHWAYS 光合作用／CO_2固定途徑・35

sedoheptulose (*n*) a monosaccharide (p. 28) with seven carbon atoms (p. 8), produced in the Calvin cycle (↑).

C_3 pathway the fixation of CO_2 (p. 33) by ribulose-diphosphate (p. 33) to produce two molecules (p. 9) of a compound with three carbon atoms (p. 8) (PGA (p. 33)). Most plants use this pathway, and are called C_3 plants. It is also known as the reductive pentose pathway (p. 33) and the Calvin cycle (↑).

景天庚酮糖（名） 卡爾文循環（↑）中產生的具有七個碳原子（第 8 頁）的一種單醣（第 28 頁）。

三碳（C_3）途徑 核酮糖二磷酸（第 33 頁）使 CO_2 固定（第 33 頁）時產生兩個含有三個碳原子（第 8 頁）的兩個化合物分子（第 9 頁）（PGA（第 33 頁））。多數植物行這種途徑，此種植物稱為 C_3 植物。也稱為還原性戊醣途徑（第 33 頁）和卡爾文循環（↑）。

C_4 pathway of CO_2 fixation
CO_2 固定的 C_1 途徑

| bundle sheath cells 維管束鞘細胞 | mesophyll cells 葉肉細胞 |

ribulose-diphosphate 核酮糖二磷酸 — Calvin cycle 卡爾文循環
pyruvic acid 丙酮酸 — 3 個 carbon atoms 碳原子
phosphoenol-pyruvic acid 磷酸烯醇丙酮酸 — 3 個 carbon atoms 碳原子
carbohydrate 碳水化合物
malic acid 蘋果酸 — 4 個 carbon atoms 碳原子
oxaloacetic acid 草醯乙酸 — 4 個 carbon atoms 碳原子

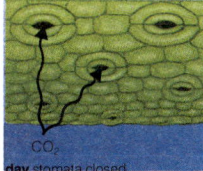

crassulacean acid metabolism 景天酸代謝 (CAM)
night stomata open, CO_2 enters 晚上氣孔張開，CO_2 進入
CO_2 fixed by phosphoenol-pyruvic acid, producing malic acid CO_2 被磷酸烯醇丙酮酸固定，產生蘋果酸

day stomata closed, preventing water loss 白天氣孔關閉，阻止水分耗散
CO_2 released from malate and fixed by ribulose-diphosphate 從蘋果酸釋放 CO_2 被二磷酸核糖固定

C_4 pathway kind of CO_2 fixation (p. 33), found especially in tropical (p. 162) monocotyledons (p. 130). In this pathway CO_2 is fixed by a compound with three carbon atoms (p. 8) (phosphoenolpyruvate) to produce a molecule (p. 9) with four carbon atoms (malate). This occurs in mesophyll (p. 95) cells in the leaf. The malate is then transported to bundle sheath (p. 106) cells, where the CO_2 is released and fixed by ribulose-diphosphate (p. 33) in the normal way. Plants which do this are called C_4 plants.

crassulacean acid metabolism CAM. A kind of CO_2 fixation (p. 33), found in many succulent (p. 99) plants, such as the family Crassulaceae. CO_2 is fixed by phosphoenolpyruvate, as in the C_4 pathway (↑), to produce malate; this occurs in the night, when the stomata (p. 96) are open. During the day, when the stomata are closed, the CO_2 is released and fixed again by ribulose-diphosphate (p. 33). This reduces the water loss by transpiration (p. 101) on hot days.

四碳（C_4）途徑 為 CO_2 固定（第 33 頁）的一種形式，特別常見於熱帶（第 162 頁）單子葉植物（第 130 頁）。在這途徑中，CO_2 是被一個含有三個碳原子（第 8 頁）的化合物（磷酸烯醇丙酮酸）固定，產生一個含有四個碳原子的分子（第 9 頁）（蘋果酸）。反應在葉片中的葉肉（第 95 頁）細胞中發生。蘋果酸隨後轉移到維管束鞘（第 106 頁）細胞。在此釋放出 CO_2 並由核酮糖二磷酸（第 33 頁）以正常方式固定下來。行這種途徑的植物稱為 C_4 植物。

景天酸代謝作用（CAM） CO_2 固定（第 33 頁）的一種形式，見於許多肉質植物（第 99 頁）。例如景天科植物。CO_2 為磷酸烯醇丙酮酸所固定，和 C_4 途徑（↑）一樣產生蘋果酸；這過程發生於植物氣孔（第 96 頁）張開的晚間。白天時氣孔關閉，釋放出 CO_2 再次為核酮糖二磷酸（第 33 頁）固定。這樣，可減少酷熱白天蒸騰作用（第 101 頁）所致的水分損耗。

light reaction the chemical reactions of photosynthesis (p. 32) which require light. These reactions, in which pigments (↓) are used to trap light energy, are the splitting of water (H_2O) molecules (p. 9) to give hydrogen and oxygen, and the production of ATP (p. 26) and $NADPH_2$ (p. 38).

photolysis of water the part of the light reaction (↑) of photosynthesis (p. 32) in which water molecules (p. 9) are split to give hydrogen and oxygen.

Hill reaction the name for the part of the light reaction (↑) of photosynthesis (p. 32), after R. Hill who first observed it in 1937. It involves the reduction (p. 11) of NADP to $NADPH_2$ (p. 38).

光反應 需要光照的光合作用（第32頁）的化學反應。在這些反應中，色素（↓）用於捕捉光能，水（H_2O）分子（第9頁）被分解，釋放出氫和氧，並產生ATP（第26頁）和$NADPH_2$（第38頁）。

水的光解 光合作用（第32頁）的光反應（↑）部分。反應中水分子（第9頁）被分解，放出氫和氧。

希爾反應 光合作用（第32頁）光反應（↑）部分的名稱，係1973年希爾首先觀察到。反應包括NADP還原（第11頁）為$NADPH_2$（第38頁）。

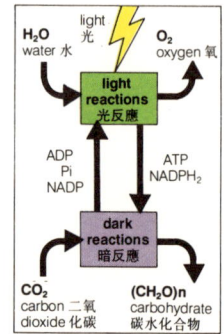

the links between the light reactions and dark reactions of photosynthesis
光合作用的光反應與暗反應間的聯繫

Hill reaction 希爾反應
(photolysis of water in chloroplast
葉綠素中水的光解)
light 光
$2NADP + 2H_2O \longrightarrow 2NADPH_2 + O_2$

pigment (n) any coloured substance present in the tissues (p. 88) of an organism. Pigments absorb energy from light. Some of them, like chlorophyll (↓), are important in photosynthesis (p. 32). Others, e.g. phytochromes (p. 116), help to control growth.

chlorophylls (n.pl.) magnesium-containing green pigments (↑), found in the chloroplasts (p. 32) of all plants, which trap light energy, of blue and red wavelengths (p. 38), for photosynthesis (p. 32). Chlorophylls give plants their green colour. The two most important chlorophylls are chlorophyll a ($C_{55}H_{72}O_5N_4Mg$) and chlorophyll b ($C_{55}H_{70}O_6N_4Mg$).

accessory pigment a pigment (↑) which is involved in photosynthesis (p. 32) but not directly in the capture of sunlight energy, e.g. carotenoids (↓).

plastocyanin (n) a blue, copper-containing protein (p. 56), which is an electron (p. 8) carrier in the light reaction of photosynthesis (p. 32).

plastoquinone (n) a non-protein (p. 56) electron (p. 8) carrier in the light reaction (↑) of photosynthesis (p. 32).

色素（名） 有機體組織中（第88頁）含有的任何有色物質。色素可吸收光能。其中有些色素如葉綠素（↓）對光合作用（第32頁）很重要。其他色素如感光色素（第116頁）有助於控制生長。

葉綠素（名、複） 一切植物的葉綠體（第32頁）都有的含鎂綠色色素（↑）。它們能為光合作用（第32頁）捕捉藍及紅波長（第38頁）的光能。各種葉綠素使植物呈現綠色。葉綠素 a ($C_{55}H_{72}O_5N_4Mg$)及葉綠素 b ($C_{55}H_{70}O_6N_4Mg$)是最重要的兩種葉綠素。

輔助色素 介入光合作用（第32頁），但不直接捕捉光能的一種色素（↑）。例如類胡蘿蔔素（↓）。

質體藍素（名） 一種藍色的含銅蛋白質（第56頁），為光合作用（第32頁）的光反應的電子（第8頁）載體。

質體醌（名） 為光合作用（第32頁）的光反應（↑）中的非蛋白質（第56頁）電子（第8頁）載體。

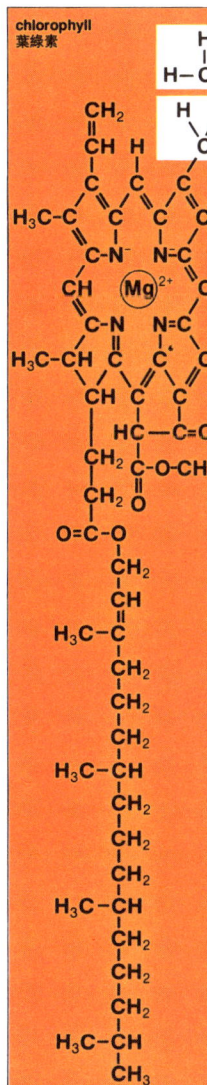

chlorophyll 葉綠素

in chlorophyll a 在葉綠素 a 中 ⟹ $C_{55}H_{72}O_5N_4Mg$

in chlorophyll b 在葉綠素 b 中 ⟹ $C_{55}H_{70}O_6N_4Mg$

porphyrin section of molecule, consisting of four pyrrole rings around a magnesium ion
分子的卟啉部分，一個鎂離子周圍有 4 個吡咯環

xanthophyll (n) a brown or yellow carotenoid (↓) pigment (↑), present in the plastids (p. 18) of many plants.

carotenoids (n.pl.) a class of non-protein (p. 56) brown, orange and yellow pigments (↑), some of which are accessory pigments (↑) in photosynthesis (p. 32).

carotene (n) an orange carotenoid (↑) accessory pigment (↑), found in chloroplasts (p. 32). $C_{40}H_{56}$.

riboflavin (n) vitamin B_2. Riboflavin is required by cells as a coenzyme (p. 15) in many oxidation (p. 11) reactions, including those of photosynthesis (p. 32).

flavoprotein (n) the group name for yellow-coloured protein (p. 56) bonded to riboflavin (↑), involved in electron transfer (p. 40) reactions.

cytochromes (n.pl.) a group of iron-containing proteins (p. 56) concerned with the electron transfer chain (p. 40) of photosynthesis (p. 32), and with the use of oxygen in aerobic respiration (p. 22). The iron atom (p. 8) in a cytochrome molecule (p. 9) is at the centre of a porphyrin (p. 38) ring, or haem (↓).

ferredoxin (n) a non-haem (↓) iron-containing protein (p. 56) in the chloroplast (p. 32), which is involved in the photosynthetic (p. 32) light reaction (p. 32).

haem (n) a porphyrin (p. 38) ring, with an atom (p. 8) of iron in its centre, e.g. in cytochromes (↑).

葉黃素(名) 是褐色或黃色類胡蘿蔔素(↓)的色素(↑)，見於許多植物的質體(第 18 頁)。

類胡蘿蔔素(名、複) 一類褐色、橙色及黃色的非蛋白質(第 56 頁)色素(↑)，其中有些是光合作用(第 32 頁)的輔助色素(↑)。

胡蘿蔔素(名) 一種橙色的類胡蘿蔔素(↑)輔助色素(↑)，見於葉綠體(第 32 頁)。$C_{40}H_{56}$。

核黃素(維生素 B_2) 核黃素在許多氧化(第 11 頁)反應，包括在光合作用(第 32 頁)的氧化反應中，許多細胞都需要它作為一種輔酶(第 15 頁)。

黃素蛋白(名) 和核黃素(↑)相連接的黃色蛋白質(第 56 頁)類，它與電子傳遞(第 40 頁)反應相關。

細胞色素(名、複) 一類與光合作用(第 32 頁)中電子傳遞鏈(第 40 頁)有關及在需氧呼吸(第 22 頁)中利用氧氣的含鐵蛋白質(第 56 頁)。細胞色素分子(第 9 頁)的鐵原子(第 8 頁)在卟啉(第 38 頁)環或血紅素(↓)的中心。

鐵氧化還原蛋白(名) 葉綠體(第 32 頁)內的一種非血紅素(↓)含鐵蛋白質(第 56 頁)，與光合作用(第 32 頁)的光反應(第 32 頁)有關。

血紅素(名) 是一種其中心含有一個鐵原子(第 8 頁)的卟啉環(第 38 頁)。例如在細胞色素中(↑)所見者。

porphyrin (*n*) type of molecular structure in which four pyrrole (↓) groups are arranged in a ring around a central metal atom (p. 8). Part of the chlorophyll (p. 36) molecule (p. 9) has this structure, the metal atom being magnesium.

pyrrole (*n*) an organic (p. 11) compound with one atom (p. 8) of nitrogen and four atoms of carbon, each bonded to a hydrogen atom, arranged in a ring. Four pyrrole groups make up the porphyrin (↑) structure in chlorophyll (p. 36) and cytochromes (p. 37).

action spectrum the wavelength (↓) or wavelengths of light which activate a biochemical process. Blue and red light are needed for photosynthesis (p. 32).

absorption spectrum the wavelengths (↓) or colours of light which are absorbed by a pigment (p. 36). Plants appear green because chlorophyll (p. 36) absorbs red and blue light and reflects green light. **spectra** (*pl.*).

wavelength (*n*) the length of a wave of light. Different wavelengths have different colours and different levels of energy.

fluorescence (*n*) the fast production of light by molecules (p. 9) of pigment (p. 36), releasing the energy trapped by the pigment from a light source. Fluorescence happens about 10^{-9} seconds after the energy was trapped, at a slightly longer wavelength (↑) than the absorption spectrum (↑). **fluoresce** (*v*).

phosphorescence (*n*) the slow production of light by molecules (p. 9) of pigment (p. 36) in a semi-stable high-energy state. Phosphorescence takes place milliseconds after the trapping of the light energy, at a longer wavelength than fluorescence (↑). **phosphoresce** (*v*).

NADP nicotinamide adenine dinucleotide phosphate. A compound which can exist in oxidized (p. 11) or reduced (p. 11) forms. The reduced form is $NADPH_2$. During the light reaction (p. 36) NADP accepts the hydrogen atoms resulting from the splitting of water molecules, giving $NADPH_2$, which is then used in the reduction of CO_2 to carbohydrate (p. 28) in the dark reaction (p. 33).

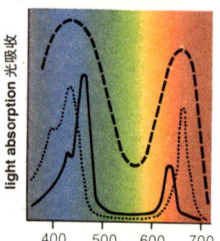

action spectra and absorption spectra in photosynthesis
光合作用的作用光譜和吸收光譜

------- action spectrum of photosynthesis
光合作用的作用光譜
········ absorption spectrum chlorophyll a
葉綠素 a 的吸收光譜
——— absorption spectrum chlorophyll b
葉綠素 b 的吸收光譜

PHOTOSYNTHESIS/PHOSPHORYLATION 光合作用/磷酸化作用 · 39

photophosphorylation (*n*) the part of the light reaction (p. 36), in which ADP (p. 26) is phosphorylated (p. 26) to ATP (p. 26) using energy from light.

光合磷酸化作用（名） 屬光反應（第36頁）的一部分，其中利用光的能量將ADP（第26頁）磷酸化（第26頁）反應成ATP（第26頁）。

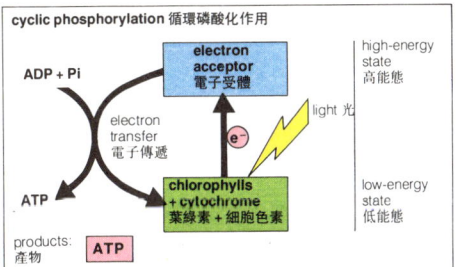

cyclic phosphorylation a photosynthetic (p. 32) reaction (p. 11) cycle in which light energy is used to produce ATP (p. 26) from ADP (p. 26) and orthophosphate (p. 13).

循環磷酸化作用 是光合（第32頁）反應（第11頁）循環，其中利用光能由ADP（第26頁）和正磷酸（第13頁）產生ATP（第26頁）。

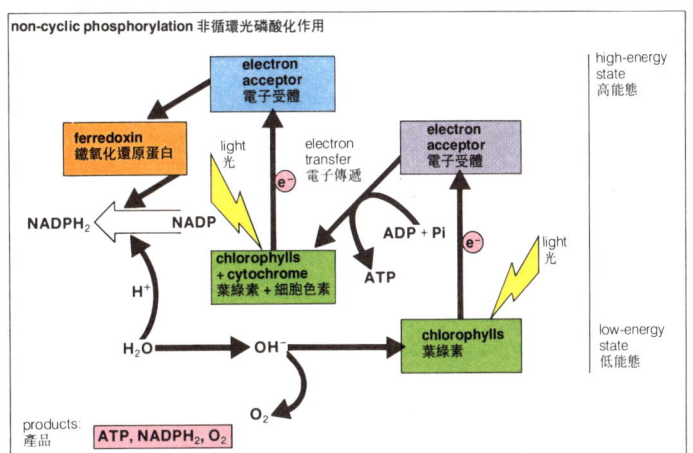

non-cyclic phosphorylation a photosynthetic (p. 32) reaction sequence in which light energy is used to produce NADPH$_2$ (↑) and oxygen from NADP and water, and ATP (p. 26) from ADP (p. 26) and orthophosphate (p. 13).

非循環磷酸化作用 光合作用（第32頁）的反應順序，其中利用光能由NADP和水產生NADPH$_2$（↑）及氧，並由ADP（第26頁）和正磷酸（第13頁）產生ATP（第26頁）。

electron transfer chain (1) a set of redox (p. 11) reactions in the light reaction (p. 36) of photosynthesis (p. 32), involving plastocyanin (p. 36), plastoquinone (p. 36) and cytochromes (p. 37), in which ATP (p. 26) is produced; (2) a set of redox reactions in aerobic respiration (p. 22), involving cytochromes and also producing ATP.

電子傳遞鏈 (1)光合作用(第32頁)的光反應(第36頁)中的一組氧化還原(第11頁)反應，和質體藍素(第36頁)、質體醌(第36頁)及細胞色素(第37頁)有關，反應產生ATP(第26頁)；(2)需氧呼吸(第22頁)中的一組氧化還原反應，和細胞色素有關，也產生ATP。

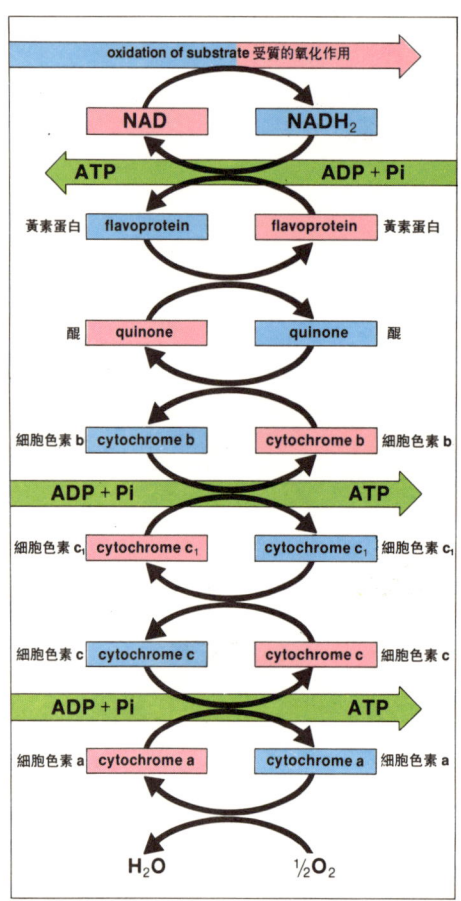

electron transfer chain
電子傳遞鏈
oxidative phosphorylation *via* the electron transfer chain produces three molecules of ATP
氧化磷酸化經由電子傳遞鏈產生三個ATP分子

oxidized compounds 氧化的化合物
reduced compounds 還原的化合物

GENETICS/GENERAL 遺傳學／概述

genetics (n) the study of the mechanism of inheritance (↓), and the control of the characteristics of an organism by its genes (↓). **geneticist** (n).

gene (n) a length of DNA (p. 51) in a chromosome (p. 46). A gene codes for a particular character of the cell or organism, and can be regarded as a unit of inheritance (↓). **genetic** (adj).

pleiotropic (adj) of genes (↑) which appear to control several different characters in an organism.

genome (n) the genetic (↑) material on the sets of chromosomes (p. 46) in a cell. The smallest genome consists of all the genes (↑) on a haploid (p. 50) set of chromosomes. A diploid (p. 50) cell, with two sets of chromosomes, is said to have a diploid genome.

clone (n) a set of cells or individuals (p. 135) reproduced (p. 59) vegetatively (p. 60) from the same original cell or organism. All members of a clone have exactly the same genome (↑) or genetic (↑) material.

genotype (n) the allelic (p. 43) composition of a single locus (p. 44), several loci, or the entire genome (↑) of an individual (p. 135). **genotypic** (adj).

phenotype (n) the observable characteristics of an organism, regarded as the result of the interaction (p. 144) between the genotype (↑) and the environment (p. 149). **phenotypic** (adj).

aberration (n) an unusual phenotype (↑), resulting from genetic (↑) abnormality or mutation (p. 54). **aberrant** (adj).

genecology (n) the study of the distribution of genes (↑) in a population (p. 135) of organisms, in relation to their habitat (p. 149).

heredity (n) the passing of characters from one generation (p. 63) to the next. **hereditary** (adj).

inherit (v) to receive genetic (↑) material or characters from parents and ancestors. **inheritance** (n).

trait (n) a character or set of characters.

wild type the phenotype (↑) which most of the individuals (p. 135) in a population (p. 135) have in their natural habitat (p. 149).

遺傳學（名） 研究遺傳（↓）機制及基因（↓）控制生物體特徵（或性狀）的學科。（名詞 geneticist 意為遺傳學家）

基因（名） 染色體（第 46 頁）上 DNA（第 51 頁）分子的一個片段。基因是細胞或生物體特定性狀的密碼，也可以看作是一個遺傳（↓）單位。（形容詞 genetic 意為遺傳的）

多效性的（形） 指基因（↑）而言，在一個生物體內，基因似乎是支配著幾個不同的性狀。

基因組（名） 細胞內染色體（第 46 頁）組的遺傳（↑）物質。最小的基因組是由單倍體（第 50 頁）染色體組的全部基因（↑）組成。一個二倍體（第 50 頁）細胞有兩組染色體，稱為有二倍體基因組。

無性繁殖系（名） 從同源細胞或生物體以營養（第 60 頁）繁殖（第 59 頁）方式產生的一群細胞或一組個體（第 135 頁）。一個無性繁殖系的全部成員具有完全相同的基因（↑）或遺傳（↑）物質。

基因型（名） 一個基因座位（第 44 頁），幾個基因座位，或一個個體（第 135 頁）的整個基因組（↑）的等位基因（第 43 頁）組成。（形容詞為 genotypic）

表現型（名） 人們認為一個生物體的可觀察特徵是基因型（↑）與環境（第 149 頁）間相互作用（第 144 頁）的結果。（形容詞為 phenotypic）

畸變（名） 遺傳（↑）反常或突變（第 54 頁）所造成的異常表現型（↑）。（形容詞為 aberrant）

基因生態學（名） 研究一個生物種群（第 135 頁）中的基因（↑）分佈與其生境（第 149 頁）關係的學科。

遺傳（名） 從一代（第 63 頁）到下一代的性狀傳遞。（形容詞為 hereditary）

遺傳（動） 從親代及祖先接受遺傳（↑）物質或遺傳性狀。（名詞為 inheritance）

特質（名） 特徵或特徵組。

野生型 在一個種群（第 135 頁）內大多數個體（第 135 頁）都有其自然生境（第 149 頁）的表現型（↑）。

42 · GENETICS/MENDEL'S LAWS 遺傳學／孟德爾定律

Mendel's Laws the laws of inheritance (p. 41) worked out by an Austrian, Gregor Mendel, (1822–1884) in 1866. Mendel's first law is the law of segregation (↓) and the second is the law of independent assortment (↓). **Mendelian inheritance**.

segregation (*n*) the separation of each of a pair of alleles (↓) into different gametes (p. 61), as a result of meiosis (p. 49). This is the mechanism behind Mendel's first law (↑), which states that alleles brought together in the F_1 generation (↓) can be segregated in the F_2 generation (↓).

孟德爾定律　奧國人孟德爾(1822-1884)在1866年提出的遺傳(第41頁)定律。孟德爾第一定律為分離律(↓)，第二定律為獨立分配律(↓)。亦稱**孟德爾遺傳**。

分離律(名)　一對等位基因(↓)中的二個成員分離入不同的配子(第61頁)中，猶如減數分裂(第49頁)的結果。這是孟德爾第一定律(↑)的機制，該定律表明子一代F_1(↓)滙集在一起的對偶基因可在子二代F_2(↓)分離。

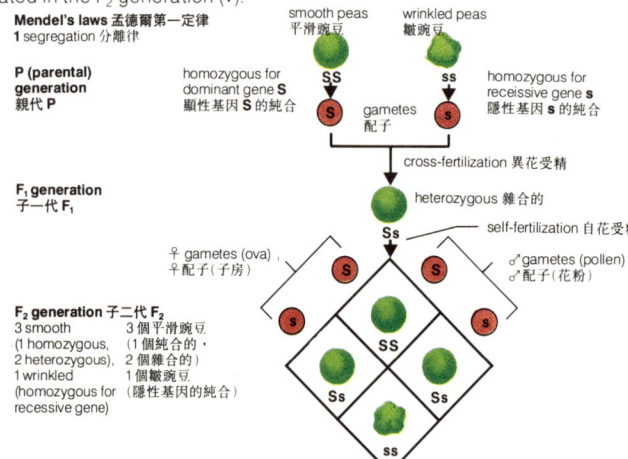

Mendel's laws 孟德爾第一定律
1 segregation 分離律

P (parental) generation 親代 P
smooth peas 平滑豌豆 — homozygous for dominant gene **S** 顯性基因 **S** 的純合
wrinkled peas 皺豌豆 — homozygous for recessive gene **s** 隱性基因 **s** 的純合
gametes 配子
cross-fertilization 異花受精

F_1 generation 子一代 F_1
heterozygous 雜合的
self-fertilization 自花受精
♀ gametes (ova) ♀配子(子房)　♂ gametes (pollen) ♂配子(花粉)

F_2 generation 子二代 F_2
3 smooth (1 homozygous, 2 heterozygous), 1 wrinkled (homozygous for recessive gene)
3個平滑豌豆(1個純合的，2個雜合的)，1個皺豌豆(隱性基因的純合)

linkage (*n*) the usual inheritance (p. 41) of two or more characters together, which happens when the genes (p. 41) controlling these characters are on the same chromosome (p. 46). Linked genes can only be separated by crossing-over (p. 47) during meiosis (p. 49). Genes on the same chromosome form a linkage group.

F_1 **generation** the first filial generation. The offspring (p. 44) of the parental generation at the beginning of a genetic (p. 41) experiment.

F_2 **generation** the second filial generation. The offspring (p. 44) resulting from sexual (p. 59) reproduction (p. 59) in the F_1 generation (↑).

連鎖(名)　兩個或多個特徵結合在一起的常見遺傳(第41頁)，見於支配這些性狀的基因(第41頁)是在同一個染色體(第46頁)上時，連鎖的基因只在減數分裂(第49頁)時由於交換(第47頁)而被分開。在同一個染色體上的基因形成一個連鎖群。

F_1 子一代　即雜交的子一代，遺傳(第41頁)實驗開始時親代的後代(第44頁)。

F_2 子二代　第二代。F_1子代(↑)有性(第59頁)生殖(第59頁)所產生的後代(第44頁)。

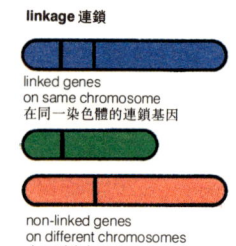

linkage 連鎖

linked genes on same chromosome 在同一染色體的連鎖基因

non-linked genes on different chromosomes 在不同染色體的非連鎖基因

GENETICS/MENDEL'S LAWS 遺傳學／孟德爾定律・43

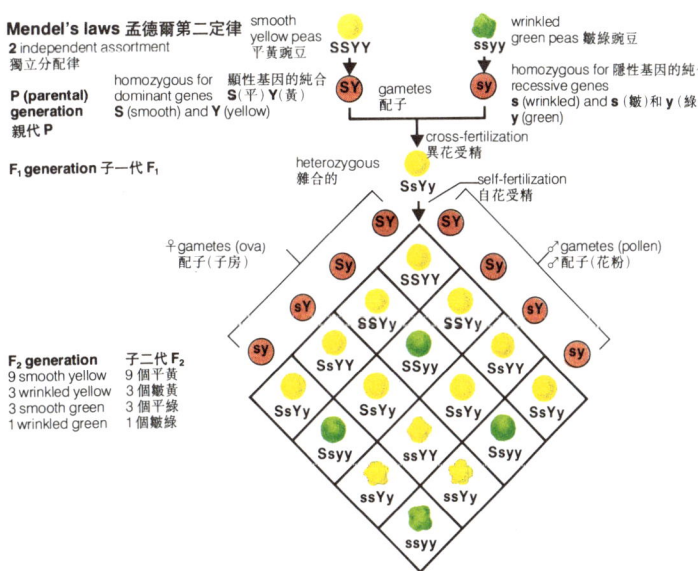

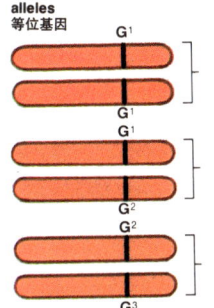

some possible combinations of 3 alleles on a chromosome pair
在一個染色體對內三個等位基因的可能組合

independent assortment Mendel's second law (↑), which states that most of the characters of parents can appear in any combination in their offspring (p. 44).

alleles (n.pl.) two genes (p. 41), each occupying the same position or locus (p. 44) on two homologous (p. 46) chromosomes (p. 46). Alleles may have small differences in the sequence of bases (p. 52) in their DNA (p. 51).

cytoplasmic inheritance inheritance (p. 41) of characters coded (p. 53) by DNA (p. 51) in the mitochondria (p. 21), chloroplasts (p. 32) or in other parts of the cytoplasm (p. 18).

plasmagene (n) any gene (p. 41) that is not contained in the nucleus (p. 19), i.e. genes that are found in the cytoplasm (p. 18). These genes are passed from one generation (p. 63) to the next by cytoplasmic inheritance (↑). The inheritance of plasmagenes does not usually obey Mendel's laws, since they are not organized into chromosomes (p. 46).

獨立分配律 為孟德爾第二定律(↑)，該定律表明親代的大多數性狀可在其後代(第44頁)的任何組合表現出來。

對偶基因；等位基因(名、複) 指兩個基因(第41頁)各自在兩個同源(第46頁)染色體(第46頁)上佔據相同位置或座位(第44頁)。對偶基因在其DNA(第51頁)鹼基(第52頁)順序上可有輕微差別。

細胞質遺傳 在粒線體(第21頁)、葉綠體(第32頁)或細胞質(第18頁)的其它部分上由DNA(第51頁)編碼(第53頁)的性狀遺傳(第41頁)。

細胞質基因(名) 指細胞核(第19頁)不包含的任何基因(第41頁)，即在細胞質(第18頁)中存在的基因。這些基因是靠細胞質遺傳(↑)從一代(第63頁)傳到下一代。細胞質基因遺傳通常不遵守孟德爾定律，因為它們編入染色體(第46頁)中。

44 · GENETICS/LOCL, DOMINANCE, INTERITANCE 遺傳學/基因座位、顯性、遺傳

locus (n) a position on a chromosome (p. 46).
loci (pl.).
homozygous (adj) having identical alleles (p. 43) at the same loci (↑) on two homologous (p. 46) chromosomes (p. 46). **homozygosity** (n).
heterozygous (adj) having non-identical alleles (p. 43) at the same loci (↑) on two homologous (p. 46) chromosomes (p. 46). **heterozygosity** (n).
dominant[1] (adj) of alleles (p. 43) which have the same effects in the heterozygous (↑) and homozygous (↑) conditions. **dominance** (n).
recessive (adj) of alleles (p. 43) whose effects can only be seen in the homozygous (↑) condition. When in the heterozygous (↑) condition, it is the dominant (↑) allele and not the recessive allele which controls the phenotype (p. 41).
isolation (n) the separation of one object from another, or the inability of two substances or organisms to mix with each other. Reproductive (p. 59) isolation is the inability of two or more populations (p. 135) to breed with each other, e.g. because they live in different places or different habitats (p. 149), because they flower at different times of year, or because they have different genomes (p. 41). **isolated** (adj).
deme (n) any population (p. 135) of organisms that is genetically isolated (↑) from other populations. Individuals (p. 135) in a deme breed amongst themselves, and there is no input of genetic material from other demes.
gene pool all the different genes (p. 41) present in a population (p. 135).
monohybrid inheritance the inheritance (p. 41) of one pair of genes (p. 41).
dihybrid inheritance the inheritance (p. 41) of two pairs of genes (p. 41).
pure line a series of generations (p. 63) that are homozygous (↑) for all characters.
offspring (n) = progeny (p. 59).
chimaera (n) a plant whose tissues (p. 88) are of more than one genetic (p. 41) kind. This can happen due to mutations (p. 54) in a cell of a very young plant, or can be caused by grafting (p. 69).

基因座位 (名) 指基因在染色體(第 46 頁)上的位置。(複數為 loci)。
純合的 (形) 在兩個同源(第 46 頁)染色體(第 46 頁)的相同基因座位(↑)上具有相同的對偶基因(第 43 頁)。(名詞為 homozygosity)
雜合的 (形) 在兩個同源(第 46 頁)染色體(第 46 頁)的相同基因座位(↑)上具有不同的對偶基因(第 43 頁)。(名詞為 heterozygosity)
顯性的 (形) 指對偶基因(第 43 頁)在雜合(↑)及純合(↑)情況下，都具有相同的影響。(名詞為 dominance)
隱性的 (形) 僅在純合的(↑)情況下才能看見等位基因(第 43 頁)所具有的影響。在雜合的(↑)情況下支配表現型(第 41 頁)的只有顯性(↑)對偶基因，而不是隱性對偶基因。
隔離 (名) 一個物體和另一個隔開，或者兩種物質或生物體無能力混和在一起。生殖(第 59 頁)隔離是指兩個或多個種群(第 135 頁)由於它們居住在不同的地方或不同的生境(第 149 頁)；或由於在一年中的不同時間開花；或者由於它們有不同的基因組(第 41 頁)而無能力彼此繁殖。(形容詞為 isolated)
同類群 (名) 指遺傳上與其他種群隔離(↑)的任何一種生物種群(第 135 頁)。在一個同類群中的個體，它們之間彼此繁育，沒有從其他類群輸入遺傳物質。
基因庫 在一個種群(第 135 頁)內存在的所有不同基因(第 41 頁)。
單基因雜種遺傳 一對基因(第 41 頁)的遺傳(第 41 頁)。
雙基因雜種遺傳 兩對基因(第 41 頁)的遺傳(第 41 頁)。
純系 全部性狀均為純合的(↑)一系列世代(第
後裔 (名) 同後代(第 59 頁)。
嵌合體 (名) 指其組織(第 88 頁)屬於一個以上遺傳(第 41 頁)類型的一個植物體。極幼嫩植物的細胞突變(第 54 頁)或者嫁接(第 69 頁)都可產生嵌合體。

locus 基因座位
homologous chromosomes
同源染色體

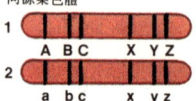

alleles **A, B, C, X, Y, Z**, occupy same loci (positions) on chromosome 1 as alleles **a, b, c, x, y, z**, on chromosome 2
等位基因 **A、B、C、X、Y、Z** 在染色體 1 和染色體 2 上
a、b、c、x、y、z 都佔有相同的基因座

homozygosity and heterozygosity
純性與雜合性

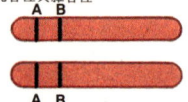

chromosome pair homozygous for gene **A** and gene **B**
染色體對
基因 A 及 B 的純合

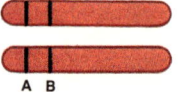

chromosome pair homozygous for gene **A**, heterozygous for gene **B**
染色體對
對基因 A 純合，
對基因 B 雜合

對頁①、②、③、④的譯文

①前期能看見核內的染色體，每個染色體分裂為兩個染色分體，染色分體在著絲點相連

②中期核膜及核仁解體，紡錘絲形成，染色體變粗短，排列在紡錘體兩極之間的中部

③後期染色分體在著絲點分離，姊妹染色分體被拉向紡錘體兩極

④末期核膜及核仁重新出現，染色體開始失去緊湊結構，出現新細胞壁

CELL DIVISION/MITOSIS 細胞分裂／有絲分裂 · 45

mitosis
有絲分裂
(only two pairs of homologous chromosomes shown for clarity)
(為清楚只顯示兩對同源染色體)

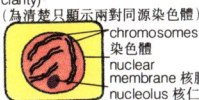
chromosomes 染色體
nuclear membrane 核膜
nucleolus 核仁

①**prophase** chromosomes become visible in the nucleus, each one split into two chromatids, joined at the centromere.

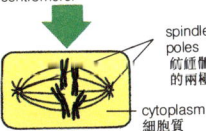
spindle poles 紡錘體的兩極
cytoplasm 細胞質

②**metaphase** nuclear membrane and nucleolus have disintegrated. Spindle fibres form. Chromosomes shorter and thicker, arranged midway between the spindle poles.

③**anaphase** chromatids separate at centromeres. Sister chromatids drawn to opposite poles of the spindle.

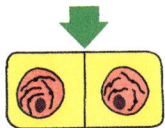

④**telophase** nuclear membranes and nucleoli reform. Chromosomes begin to lose their compact structure. The new cell wall is laid down.

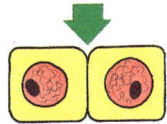

interphase chromosomes no longer visible.
分裂間期已看不到染色體

註：①、②、③、④的譯文見對頁

cell division the process in which a cell divides to form two new cells, each containing a nucleus (p. 19). Cell division is either mitotic (↓) or meiotic (p. 49).

cytokinesis (n) the division of a cell into two cells.

binary fission the division of a cell into two identical cells.

mitosis (n) vegetative (p. 60) or somatic (↓) cell division (↑), in which the chromosomes (p. 46) in the nucleus (p. 19) are duplicated into two chromatids (p. 46). The nuclear membrane (p. 19) breaks down, the centromeres (p. 46) divide, and the chromatids move to either end of the cell on the spindle (p. 46). Nuclear membranes re-form around each group of chromatids, and a new cell wall (p. 17) is laid down between them. In this way each new cell gains exactly the same chromosomes and genetic (p. 41) material. The four stages of mitosis are prophase (↓), metaphase (↓), anaphase (↓) and telophase (↓). **mitotic** (adj)

somatic (adj) of any process, or part of an organism, that is not connected with sexual reproduction (p. 59), e.g. mitosis (↑) is somatic cell division (↑).

prophase (n) the first stage in cell division (↑), when the chromosomes (p. 46) in the nucleus (p. 19), when stained (p. 171), become visible as short, thick, helically (p. 51) coiled threads.

metaphase (n) the second stage in cell division (↑), in which the nuclear membrane (p. 19) breaks down, and the centromeres (p. 46) of the chromosomes (p. 46) position themselves at the centre of the spindle (p. 46), forming the metaphase plate.

anaphase (n) the third stage of cell division (↑), in which the chromatids (p. 46) separate from each other and move to the ends, or poles, of the spindle (p. 46).

telophase (n) the last stage in cell division (↑), when the chromatids (p. 46) are at either end of the spindle (p. 46) and the new nuclear membranes (p. 19) are being synthesized (p. 13).

interphase (n) the period between one cell division (↑) and the next.

細胞分裂　一個細胞分裂成兩個新細胞的過程，這兩個細胞各有一個細胞核（第19頁）。細胞分裂方式分為有絲分裂（↓）和減數分裂（第49頁）。

胞質分裂（名）　一個細胞分裂為兩個細胞。

二分裂　一個細胞分裂為兩個相等的細胞。

有絲分裂（名）　為營養（第60頁）細胞或體細胞（↓）分裂（↑），其中細胞核（第19頁）內的染色體（第46頁）複製加倍形成兩個染色分體（第46頁），核膜（第19頁）破裂，著絲點（第46頁）分裂，染色分體移向紡錘體上（第46頁）的細胞兩端，圍繞每組染色分體重新形成核膜，在它們之間出現新的細胞壁（第17頁）。在這過程中，每個新細胞獲得完全相等的染色體及遺傳（第41頁）物質。有絲分裂的過程分為四期，即前期（↓）、中期（↓）、後期（↓）及末期（↓）。（形容詞為 mitotic）

體細胞的（形）　描述和有性生殖（第59頁）不發生聯繫的任何過程或有機體的部分。例如有絲分裂（↑）是體細胞分裂（↑）。

前期（名）　細胞分裂（↑）的早期，此時如將細胞核（第19頁）內的染色體（第46頁）著色（第171頁）可看見染色體變成粗短的螺旋狀（第51頁）捲絲。

中期（名）　細胞分裂（↑）的第二期，此時期核膜（第19頁）破裂，染色體（第46頁）的著絲點（第46頁）本身的位置在紡錘體（第46頁）兩極中心，形成中期板。

後期（名）　細胞分裂（↑）的第三期，這時期著絲點（第46頁）彼此分離，向紡錘體（第46頁）兩端即兩極移動。

末期（名）　細胞分裂（↑）的最後期。這時期著絲點（第46頁）位於紡錘體（第46頁）的兩端，合成（第13頁）新的核膜（第19頁）。

分裂間期（名）　一個細胞分裂（↑）開始至下一次分裂完為止的周期。

CELL DIVISION/CHROMOSOMES 細胞分裂／染色體

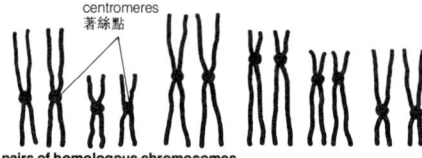

centromeres 著絲點

pairs of homologous chromosomes 同源的染色體對

chromosome (*n*) threadlike bodies containing DNA (p. 51), RNA (p. 51), and protein (p. 56), found in the nuclei (p. 19) of all cells. They are usually only visible during cell division (p. 45), during which they become shorter and thicker. All the vegetative (p. 60) cells in a plant, and in a species (p. 134), have the same number of chromosomes.

homologous (*adj*) of two similar chromosomes (↑) which pair with each other during meiosis (p. 49). Homologous chromosomes have identical sequences of loci (p. 44). Members of a pair of homologous chromosomes have centromeres (↓) in the same position and arms of the same length as each other.

chromatid (*n*) one of the pair of strands which result from the duplication of a chromosome (↑) during prophase (p. 45) and metaphase (p. 45).

centromere (*n*) the point of attachment of a chromosome (↑) to the spindle (↓) during cell division (p. 45).

spindle (*n*) the minute curved threads of protein (p. 56) which appear during cell division (p. 45), spreading across the cell from either end. The movement of chromosomes (↑) during cell division is organized on the spindle. The protein threads of the spindle are microtubules (p. 21), which are formed during metaphase (p. 45).

centriole (*n*) a small granule outside the nuclear membrane (p. 19), which divides at mitosis (p. 45), forming the two ends of the spindle (↑). Centrioles are found in all animal cells, but in plants they can only be seen in motile (p. 121) male gametes (p. 61).

centrosome (*n*) the region in the cytoplasm (p. 18) that gives rise to centrioles (↑) during cell division (p. 45).

染色體（名） 存在於一切細胞的細胞核（第 19 頁）的絲狀體，其中含有 DNA（第 51 頁）、RNA（第 51 頁）及蛋白質（第 56 頁）。染色體僅在細胞分裂（第 45 頁）時可見到。這時它們變成更短、更粗。一種植物及一個物種（第 134 頁）的一切營養（第 60 頁）細胞都有相同數目的染色體。

同源的（形） 指減數分裂（第 49 頁）時，兩個相似染色體（↑）彼此成對。同源染色體具有相同的基因座位（第 44 頁）序。成對同源染色體成員的著絲點（↓）位置相同及彼此的臂長度相同。

染色分體（名） 細胞分裂前期（第 45 頁）及中期（第 45 頁），一個染色體（↑）加倍複製所成的兩個子染色體之一。

著絲點（名） 細胞分裂（第 45 頁）時染色體（↑）與紡錘體（↓）相連接的點。

紡錘體（名） 細胞分裂（第 45 頁）時出現的一些彎曲的蛋白質（第 56 頁）纖細絲狀物。在細胞分裂時，從兩端散開越過細胞。細胞分裂時染色體（↑）的運動是紡錘體策動的。紡錘體的蛋白質絲是在分裂中期（第 45 頁）形成的許多微管（第 21 頁）。

中心粒（名） 是核膜（第 19 頁）外側的小粒，它在有絲分裂（第 45 頁）分開組成紡錘體（↑）的兩端。所有動物的細胞中都有中心粒，而在植物它只見於能動的（第 121 頁）雄性配子（第 61 頁）。

中心體（名） 在細胞分裂（第 45 頁）時，細胞質（第 18 頁）部位形成的中心粒（↑）。

CELL DIVISION/MEIOSIS 細胞分裂/減數分裂 · 47

the effect of colchicine in mitosis
秋水仙素對有絲分裂的影響

prophase nucleus
前期的細胞核

chromosomes divide normally into chromatids, no spindle formed
染色體正常地分裂為染色分體，不形成紡錘體

nuclear membrane reforms, containing twice as many chromosomes as original nucleus
重新形成核膜，比原先的細胞核含多達兩倍的染色體

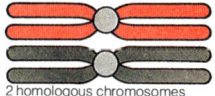

crossing-over during first meiotic division
初次減數分裂時的交換

2 homologous chromosomes
兩個同源染色體

chiasma
交叉

chiasmata formed 形成交叉

bivalent or tetrad 二價染色體或四分體

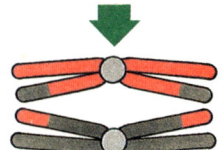

genetic material exchanged, chromosomes separate
遺傳物質交換染色體分離

colchicine (n) an alkaloid (p. 148) which inhibits (p. 14) the synthesis (p. 13) of the spindle (↑) in the metaphase (p. 45) stage of mitosis (p. 45). It is used in many genetic (p. 41) experiments, as its effect is to produce tetraploid (p. 50) cells when the new nuclear membrane (p. 19) forms at the end of mitosis.

crossing-over the exchange of some of the corresponding parts, each with the same set of loci (p. 44), of homologous (↑) chromosomes (↑) at the first meiotic (p. 49) division, during which chiasmata (↓) are formed. The result of crossing-over is recombination (↓).

synapsis (n) the pairing of homologous (↑) chromosomes (↑) during zygotene (p. 49) in the first meiotic (p. 49) prophase (p. 45).

chiasmata (n.pl.) the points in a bivalent (↓) where the two chromosomes (↑) appear to be joined and crossed over. Chiasmata can be observed during diplotene (p. 49). **chiasma** (sing.).

bivalent (n) a pair of joined homologous (↑) chromosomes (↑) during the first meiotic (p. 49) prophase (p. 45).

tetrad[1] (n) another name for a bivalent (↑) in meiosis (p. 49), so called because it consists of four chromatids (↑) in two pairs.

recombination (n) the process by which offspring (p. 44) can gain combinations of genes (p. 41) different from the combinations in either of their parents. This happens as a result of crossing-over (↑) of chromosomes (↑).

秋水仙素（名） 在有絲分裂（第 45 頁）中期（第 45 頁）抑制（第 14 頁）紡錘體（↑）合成（第 13 頁）的一種生物鹼（第 148 頁），可用於許多遺傳學（第 41 頁）實驗，其作用是在有絲分裂之末形成新核膜（第 19 頁）之際，產生四倍體（第 50 頁）細胞。

交換 同源（↑）染色體（↑）在初次減數（第 49 頁）分裂時，某些相對應部分各以同組基因座位（第 44 頁）交換，此時形成交叉（↓）。交換的結果是重組（↓）。

聯會（名） 成對同源（↑）染色體（↑）在首次減數分裂的（第 49 頁）前期（第 45 頁）的合絲期（第 49 頁）。

交叉（名，複） 兩個染色體（↑）出現聯合和交叉時的一個二價染色體（↓）的點，在雙絲期（第 49 頁）可觀察到交叉。（單數為 chiasma）

二價染色體（名） 在首次減數分裂（第 49 頁）前期（第 45 頁）的一對連結的同源（↑）染色體（↑）。

四分體（名） 為二價染色體（↑）減數（第 49 頁）分裂時另一稱謂，這是因為它含有四個成兩對的染色分體（↑）。

重組（名） 後代（第 44 頁）能獲得與其親代任何一方不同的基因（第 41 頁）組合的過程。這是染色體（↑）出現交換（↑）的結果。

48 · CELL DIVISION/MEIOSIS 細胞分裂／減數分裂

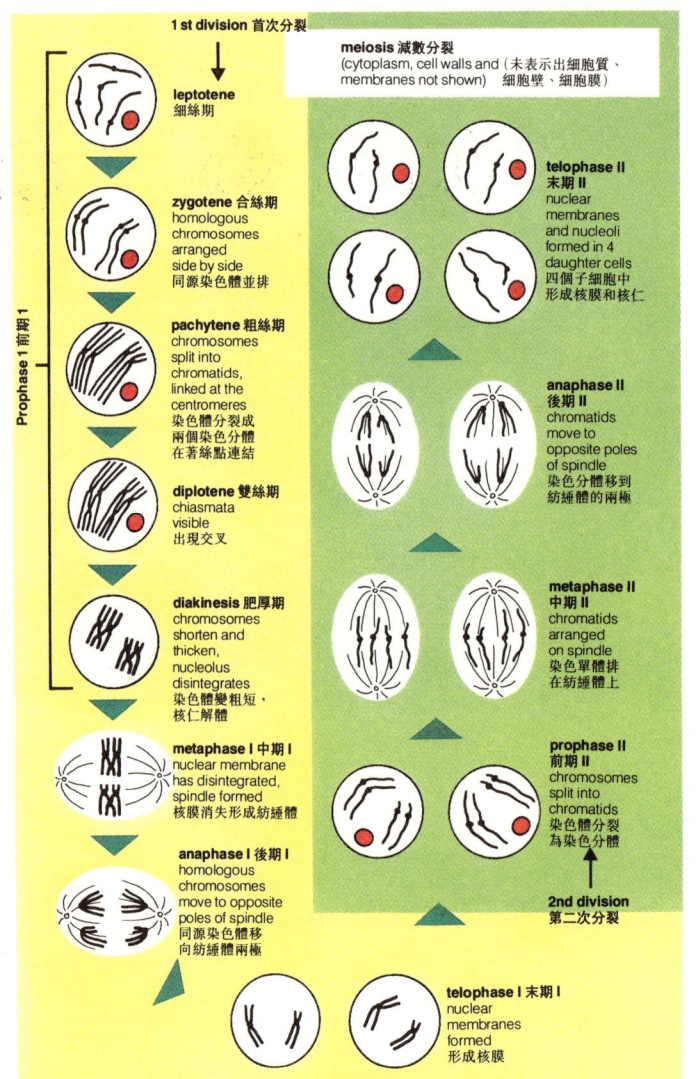

CELL DIVISION/MEIOSIS 細胞分裂／減數分裂

mitosis 有絲分裂	meiosis 減數分裂
no pairing of homologous chromosomes 沒有同源染色體配對 splitting of chromatids at centromere 染色分體在著絲點裂開 daughter nuclei have *same* number of chromosomes as parent nucleus 子核和親核都含同數的染色體 2 daughter nuclei produced 產生兩個子核	pairing of homologous chromosomes 同源染色體配對 no splitting of chromatids at centromere until 2nd prophase 第二次分裂前期之前沒有染色單體在著絲點分裂 daughter nuclei have *half* the number of chromosomes as parent nuclei 子核只具親核半數的染色體 4 daughter nuclei produced 產生四個子核

differences between mitosis 2nd meiosis
有絲分裂與減數分裂的差別

meiosis (*n*) cell division (p. 45) that produces haploid (p. 50) sex cells (p. 61) from diploid (p. 50) cells. Meiosis involves two cell divisions: (1) the replicated homologous (p. 46) chromosomes (p. 46) pair with each other on the spindle (p. 46); crossing-over (p. 47) happens at this stage. The chromosomes then separate to either end of the spindle. (2) The chromatids of each chromosome come apart at the centromere (p. 46), and separate to each end of the second spindle. There is usually no interphase (p. 45) between the two divisions. Meiosis occurs in all organisms which reproduce (p. 59) sexually (p. 59). **meiotic** (*adj*).

reduction division a name sometimes given to meiosis (↑), because the daughter cells receive a haploid (p. 50) set of chromosomes (p. 46) from the diploid (p. 50) parent cell.

leptotene (*n*) the first stage in the first meiotic (↑) prophase (p. 45), in which the chromosomes (p. 46) appear as thin threads.

zygotene (*n*) the stage in the first meiotic (↑) prophase (p. 45) when the homologous (p. 46) chromosomes (p. 46) come together to form bivalents (p. 47).

pachytene (*n*) the stage of the first meiotic (↑) prophase (p. 45), when the chromosomes (p. 46) become shorter and thicker, and can clearly be seen to have replicated (p. 54) into chromatids (p. 46).

diplotene (*n*) the stage in the first meiotic (↑) prophase (p. 45), when the centromeres (p. 46) of paired chromosomes (p. 46) move away from each other and crossing-over (p. 47) can be seen.

diakinesis (*n*) the last stage in the first meiotic (↑) prophase (p. 45), when the chromosomes (p. 46) are shortest and thickest, and the nuclear membrane (p. 19) disappears.

減數分裂；成熟分裂（名） 從二倍體（第50頁）細胞產生單倍（第50頁）體性細胞（第61頁）的細胞分裂（第45頁）。減數分裂包括兩次細胞分裂：（1）同源（第46頁）染色體（第46頁）對彼此在紡縛體（第46頁）上複製：這階段出現交換（第47頁），然後染色體分離，分別移到紡縛體兩端；（2）每個染色體的染色分體離開著絲點（第46頁），並分開移到第二個紡縛體的一端。兩次分裂之間通常沒有分裂間期（第45頁）。全部能進行有性（第59頁）生殖（第59頁）的生物體都有減數分裂。（形容詞為 meiotic）

減數分裂（↑）英文有時亦稱 reduction division，因子細胞從二倍體（第50頁）母細胞接受單倍體（第50頁）的染色體（第46頁）組而得名。

細絲期（名） 減數分裂（↑）首次分裂前期（第45頁）的第一階段，此階段染色體（第46頁）呈細絲狀。

偶絲期（名） 減數分裂（↑）首次分裂前期（第45頁）的一個階段，此階段同源（第46頁）染色體（第46頁）滙合一起，形成二價染色體（第47頁）。

粗絲期（名） 減數分裂（↑）首次分裂前期（第45頁）的一個階段，此時染色體（第46頁）變粗短，並可清楚看出已複製（第54頁）成染色分體（第46頁）。

雙絲期（名） 減數分裂（↑）首次分裂前期（第45頁）的一個階段，此時成對的染色體（第46頁）著絲點（第46頁）彼此離開，並可看到交換（第47頁）。

肥厚期（名） 減數分裂（↑）首次分裂前期（第45頁）的最後階段，此時期染色體（第46頁）變粗短，核膜消失（第19頁）。

50 · CELL DIVISION/HAPLOID, DIPLOID, POLYPLOID 細胞分裂／單倍體、二倍體、多倍體

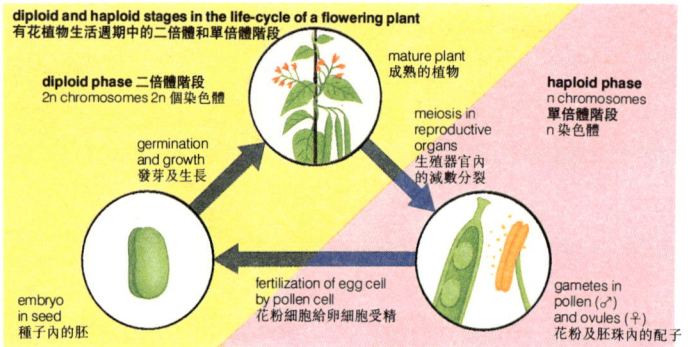

haploid (adj) of cells with one set of chromosomes (p. 46) in their nuclei (p. 19).
diploid (adj) of cells with two sets of chromosomes (p. 46) in their nuclei (p. 19). The sets are said to be homologous (p. 46).
triploid (adj) of cells which have three sets of homologous (p. 46) chromosomes (p. 46) in their nuclei (p. 19).
tetraploid (adj) of cells which have four sets of homologous (p. 46) chromosomes (p. 46) in their nuclei (p. 19).

單倍體的（形） 指細胞核（第19頁）內有一組染色體（第46頁）的細胞。
二倍體的（形） 指細胞核（第19頁）內有兩組染色體（第46頁）的細胞。這成組的染色體是同源的（第46頁）。
三倍體的（形） 指細胞核（第19頁）內有三組同源（第46頁）染色體（第46頁）的細胞。
四倍體的（形） 指細胞核（第19頁）內有四組同源（第46頁）染色體（第46頁）的細胞。

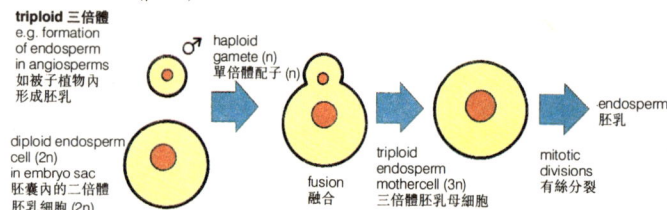

polyploid (adj) of cells which have three or more sets of homologous (p. 46) chromosomes (p. 46) in their nuclei (p. 19).
allopolyploid (n) a polyploid (↑) species (p. 134) with sets of chromosomes (p. 46) from two or more different species. This can be the result of hybridization (p. 63) between species.
autopolyploid (n) a polyploid (↑) species (p. 134) with all sets of chromosomes (p. 46) coming from the same species.

多倍體的（形） 指細胞核（第19頁）內有三組或多組同源（第46頁）染色體（第46頁）的細胞。
異源多倍體（名） 含有來自兩個或多個不同種染色體（第46頁）組的多倍體（↑）種（第134頁）。這是種與種之間雜交（第63頁）得到的結果。
同源多倍體（名） 全部染色體（第46頁）組都來自同一種的多倍體（↑）種（第134頁）。

nucleic acid a long chain polymer (p. 10) consisting of nucleotide (p. 52) units. There are two kinds of nucleic acid, DNA (↓) and RNA (↓), which are found in the cells of all living organisms.

DNA deoxyribonucleic acid. The main nucleic acid (↑) in the chromosomes (p. 46) of the nucleus (p. 19) of a cell. The DNA molecule consists of two chains of nucleotide (p. 52) polymer (p. 10), arranged in a double helix (↓). The sugar in the nucleotides of DNA is deoxyribose. DNA controls protein synthesis (p. 57) by the processes of transcription (p. 56) and translation (p. 56). It is replicated (p. 54) by a self-copying process, and it is the hereditary (p. 41) material of all cellular (p. 169) organisms and some viruses (p. 118).

核酸 由核苷酸(第52頁)結構單元組成的長鏈聚合物(第10頁)。核酸有兩種，即DNA(↓)和RNA(↓)，它們存在於一切活生物體的細胞內。

去氧核糖核酸(DNA) 細胞核(第19頁)內染色體(第46頁)的主要核酸(↑)。DNA分子由兩條核苷酸(第52頁)聚合物(第10頁)鏈構成，兩條鏈排成雙螺旋(↓)。DNA的核苷酸中的糖是去氧核糖。DNA通過轉錄(第56頁)和轉譯(第56頁)過程控制蛋白質的合成(第57頁)，係由自體複製過程所複製(第54頁)的。並且是一切具細胞(第169頁)的有機體及某些病毒(第118頁)的遺傳(第41頁)物質。

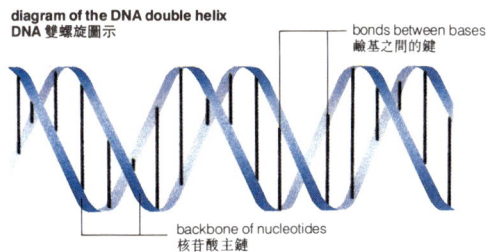

diagram of the DNA double helix
DNA 雙螺旋圖示

bonds between bases
鹼基之間的鍵

backbone of nucleotides
核苷酸主鏈

helix (*n*) a thread or line coiled like a screw. Molecules of DNA (↑) have this shape, with two helices (*pl.*) coiled together. **double helix**.

RNA ribonucleic acid. The nucleic acid (↑) directly involved in protein synthesis (p. 57). RNA differs from DNA (↑) by having uracil (p. 53) instead of thymine (p. 53) and ribose instead of deoxyribose in its nucleotides (p. 52). The RNA polymer (p. 10) is usually a single strand. There are three main kinds of RNA: *messenger RNA* (mRNA), which carries the genetic code (p. 54) from the nucleus to the cytoplasm (p. 18); *transfer RNA* (tRNA), to which amino acids (p. 56) are attached before protein synthesis (p. 57); and *ribosomal RNA* (rRNA), which is a structural part of the ribosomes (p. 56).

螺旋結構(名) 盤捲如螺旋的一條絲或線。DNA(↑)分子具有這種形態，其兩條螺旋盤捲在一起成為**雙螺旋**。

核糖核酸(RNA) 直接參與蛋白質合成(第57頁)核酸(↑)。RNA與DNA(↑)的差別在於：RNA的核苷酸(第52頁)分子有尿嘧啶(第53頁)而沒有胸腺嘧啶(第53頁)，含有核糖而不含去氧核糖。RNA聚合物(第10頁)常為單股。RNA主要有三種：信使RNA(mRNA)，它從細胞核携帶遺傳密碼(第54頁)到細胞質(第18頁)；轉運RNA(tRNA)，在蛋白質合成(第57頁)之前有氨基酸(第56頁)與之相連；核蛋白體RNA(rRNA)，它是各種核蛋白體(第56頁)的構造部分。

NUCLEIC ACIDS/NUCLEOTIDES 核酸／核苷酸

the common bases in the nucleotides of DNA and RNA
DNA 和 RNA 的核苷酸的共同鹼基

	purines 嘌呤	pyrimidines 嘧啶
DNA only 單獨 DNA		thymine 胸腺嘧啶
DNA and RNA DNA 和 RNA	adenine 腺嘌呤 / guanine 鳥嘌呤	cytosine 胞嘧啶
RNA only 單獨 RNA		uracil 尿嘧啶

nucleotide (n) a molecule with a pentose (p.28) sugar, a phosphate (p.13) group, and a purine (↓) or pyrimidine (↓) base (↓) containing nitrogen. Nucleotides are the units which form the long chain polymers (p.10), nucleic acids (p.51).
base² (n) a purine (↓) or pyrimidine (↓) unit.

核苷酸（名） 含有一個戊醣（第 28 頁）、一個磷酸鹽（第 13 頁）基團、一個嘌呤（↓）或一個含氮的嘧啶（↓）鹼基（↓）的分子。核苷酸是形成長鏈聚合物（第 10 頁）核酸（第 51 頁）的結構單元。
鹼基（名） 一個嘌呤（↓）或嘧啶（↓）的單元。

a chain of nucleotides 核苷酸鏈

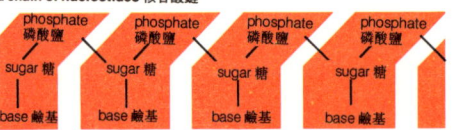

purine (n) one of the two kinds of nitrogen-containing base (↑) in nucleic acids (p.51). A purine molecule (p.9) consists of two rings of carbon and nitrogen atoms (p.8). The main purines in nucleic acids are adenine (↓) and guanine (↓).

嘌呤（名） 核酸（第 51 頁）中兩種含氮鹼基（↑）的一種。一個嘌呤分子（第 9 頁）由兩個含碳及氮原子（第 8 頁）的環構成，核酸中主要的嘌呤是腺嘌呤（↓）和鳥嘌呤（↓）。

nucleotide basic structure
核苷酸的基本結構
phosphate 磷酸鹽
sugar 糖
base 鹼基

different nucleotides have different sugars and different bases
不同的核苷酸有不同的糖和不同的鹼基

adenine (*n*) a purine (↑) base (↑) which pairs with thymine (↓) in DNA (p. 51) and with uracil (↓) in RNA (p. 51).

guanine (*n*) a purine (↑) base (↑) which pairs with cytosine (↓) in DNA (p. 51) and RNA (p. 51).

pyrimidine (*n*) one of the two kinds of nitrogen-containing base (↑) found in nucleic acids (p. 51). The molecules (p. 9) consist of a single ring of carbon and nitrogen atoms (p. 8). The main pyrimidines in nucleic acids are thymine (↓), cytosine (↓), and uracil (↓).

cytosine (*n*) a pyrimidine (↑) base (↑), which pairs with guanine (↑) in DNA (p. 51) and RNA (p. 51).

thymine (*n*) a pyrimidine (↑) base (↑), which pairs with adenine (↑) in DNA (p. 51).

uracil (*n*) a pyrimidine (↑) base (↑) found in RNA (p. 51), which pairs with adenine (↑) during transcription (p. 56) and translation (p. 56).

codon (*n*) a sequence of three nitrogen-containing bases (↑) in the triplet code (p. 54) on a molecule (p. 9) of messenger RNA (p. 51), which pairs with an anticodon (↓) on a molecule of transfer RNA (p. 51) during translation (p. 56). Because each base only pairs with one other base, each codon has its own anticodon, e.g. a codon consisting of adenine (↑), guanine (↑) and cytosine (↑), or AGC, will pair with an anticodon of uracil (↑), cytosine and guanine, or UCG.

anticodon (*n*) a sequence of three nitrogen-containing bases (↑) in a molecule (p. 9) of transfer RNA (p. 51), which pairs with a codon (↑) on a molecule of messenger RNA (p. 51) during translation (p. 56). Each molecule of transfer RNA has only one anticodon, corresponding to the particular amino acid (p. 56) to which it is attached during protein synthesis (p. 57), e.g. one of the anticodons for the amino acid serine is a sequence consisting of uracil (↑), cytosine (↑) and guanine (↑), or UCG.

nonsense codon a codon (↑) that does not code for any amino acid (p. 56). Only three of the 64 codons in the genetic code (p. 54) are nonsense, and their function is to code for the ends of polypeptide (p. 56) chains.

腺嘌呤（名） 在DNA（第51頁）中與胸腺嘧啶（↓）配對，在RNA中與尿嘧啶（↓）配對的一種嘌呤（↑）鹼（↑）。

鳥嘌呤（名） 在DNA（第51頁）和RNA（第51頁）中與胞嘧啶（↓）配對的一種嘌呤（↑）鹼（↑）。

嘧啶（名） 存在於核酸（第51頁）中的兩種含氮鹼基（↑）之一。嘧啶分子（第9頁）由碳原子和氮原子（第8頁）的單環構成。核酸內的主要嘧啶是胸腺嘧啶（↓）、胞嘧啶（↓）及尿嘧啶（↓）。

胞嘧啶（名） 在DNA（第51頁）中RNA（第51頁）中與鳥嘌呤（↑）配對的 種嘧啶(↑)鹼基(↑)。

胸腺嘧啶（名） 在DNA（第51頁）中和腺嘌呤（↑）配對的一種嘧啶（↑）鹼基（↑）。

尿嘧啶（名） 存在於RNA（第51頁）中，在轉錄（第56頁）及轉譯（第56頁）時與腺嘌呤（↑）配對的一種嘧啶（↑）鹼基（↑）。

密碼子（名） 一個信息RNA（第51頁）分子（第9頁）上的三聯體密碼（第54頁）的三個含氮鹼基（↑）順序，它在轉譯（第56頁）時在一個轉運RNA（第51頁）分子上和一個反密碼子（↓）配對。由於每個鹼基只和另一個鹼基配對，因此每個密碼子都有本身的反密碼子。例如一個由腺嘌呤（↑）、鳥嘌呤（↑）及胞嘧啶（↑）組成的密碼子（即 **AGC**，與由尿嘧啶（↑）、胞嘧啶（↑）及鳥嘌呤組成的一個反密碼子（即 **UCG**）配對。

反密碼子（名） 在轉運RNA（第51頁）一個分子（第9頁）內三個含氮鹼基（↑）的順序，它在轉譯（第56頁）時，在一個信息RNA（第51頁）分子上和一個密碼子（↑）配對。每個轉運RNA分子只有一個反密碼子，對應於蛋白質合成（第57頁）時被連接的特定氨基酸（第56頁）。例如氨基酸絲氨酸的其中一個反密碼子是一種由尿嘧啶（↑）、胞嘧啶（↑）及鳥嘌呤（↑）（即 **UCG**）組成的順序。

無義密碼子 指不為任何氨基酸（第56頁）編碼的一種密碼子（↑）。遺傳密碼（第54頁）中的64個密碼子中，僅有三個是無意義的密碼子，它們的功能是為多肽（第56頁）鏈各末端編碼。

genetic code the name given to the 64 possible sequences in which any three of the four nitrogen-containing bases (p. 52) of RNA (p. 51), adenine (p. 53), uracil (p. 53), guanine (p. 53) and cytosine (p. 53), can be arranged. Each group of three, or triplet, codes for a particular amino acid (p. 56) in protein synthesis (p. 57). Because there are only 20 amino acids, most of them are coded for by more than one triplet of bases.

triplet code a name for the genetic code (↑), so called because the code consists of nitrogen-containing bases (p. 52) in groups of three.

replication (n) the process by which new DNA (p. 51) is made. The two strands in the DNA double helix (p. 51) separate, and a new strand of nucleotide (p. 52) polymer (p. 10) is synthesized (p. 13) on each one. Because each nitrogen-containing base (p. 52) in the nucleotide units of the polymer will only pair with one other base, the new DNA has exactly the same sequence of bases as the old. This self-copying process is the basis of heredity (p. 41). **replicate** (v).

mutation (n) the general term for a change in the sequence of nucleotides (p. 52) in the DNA (p. 51) of a cell e.g. the replacement of a pair of nitrogen-containing bases (p. 52) in the DNA chain by another pair, the turning round of a sequence of nucleotides in the chromosome (p. 46), or the loss of a whole chromosome or piece of DNA. Mutations can be inherited (p. 41) if they are present in the gametes (p. 61). They can be harmless, useful, or can cause death, depending on where they occur in a chromosome. They occur rarely, but this rate can be speeded up by mutagens (↓). Mutations result in variation (p. 135) between individuals (p. 135), and the natural selection (p. 140) of this variation leads to evolution (p. 139). **mutate** (v).

mutagen (n) a factor that causes mutations (↑), e.g. X-rays, gamma rays or certain chemicals. **mutagenic** (adj).

mutant (n) an individual (p. 135) which shows the effects of a mutation (↑) phenotypically (p. 41).

遺傳密碼　給 RNA（第 51 頁）的四個含氮鹼基（第 52 頁），即腺嘌呤（第 53 頁）、尿嘧啶（第 53 頁）、鳥嘌呤（第 53 頁）及胞嘧啶，其中任何三個鹼基可以排列成 64 個可能順序所取的名稱。每三個鹼基成一組（即三聯體），在蛋白質合成（第 57 頁）中都為一個特定的氨基酸（第 56 頁）編碼。由於只有 20 個氨基酸，所以其中多數都會為一個以上的鹼基三聯體所編碼。

三聯體密碼　遺傳密碼（↑）的一種名稱，因密碼是由含氮鹼基（第 52 頁）三個成一組組成而得名。

複製（名）　製成新的 DNA（第 51 頁）的過程。DNA 雙螺旋（第 51 頁）的兩股分開，在每一股上合成（第 13 頁）新的一股核苷酸（第 52 頁）聚合體（第 10 頁）。由於聚合體核苷酸單元中的各個含氮鹼基（第 52 頁）僅和其他一個鹼基配對，所以新的 DNA 正好和原先的鹼基相同順序。這種自體複製過程就是遺傳（第 41 頁）的基礎。（動詞為 replicate）

突變（名）　細胞中 DNA（第 51 頁）的核苷酸（第 52 頁）順序起變化的通稱。例如在 DNA 鏈的一對含氮鹼基（第 52 頁）被另一對取代，結果改變了染色體（第 46 頁）中核苷酸的順序，或者失去整個染色體或一段 DNA。如果突變是存在配子（第 61 頁）之中則可以遺傳（第 41 頁）。視乎突變在染色體中發生的位置，它可能是無害的，也可能是有利的，或者致死的。突變很少發生，但可因誘變劑（↓）而加速。突變會使個體（第 135 頁）之間出現變異（第 135 頁），這種變異的自然選擇（第 140 頁）將推動演化（第 139 頁）。（動詞為 mutate）

誘變劑（名）　能引致突變（↑）的因素。例如 X 射線、γ 射線和某些化學品。（形容詞為 mutagenic）

突變體（名）　表現型（第 41 頁）上顯現突變（↑）影響的一個個體（第 135 頁）。

NUCLEIC ACIDS/GENETIC CODE 核酸／遺傳密碼

amino acids and the genetic code
氨基酸和遺傳密碼

amino acid general formula 氨基酸的通式

$$\text{NH}_3^+ - \underset{\underset{R}{|}}{\overset{\overset{COO^-}{|}}{C}} - H$$

R = side group 側基

codon 密碼子	amino acid 氨基酸	side group 側基 (R)	side group 側基 (R)	amino acid 氨基酸	codon 密碼子
AAA, AAG	lysine 賴胺酸	$-CH_2CH_2CH_2CH_2NH_3^+$	$-H$	glycine 甘胺酸	GGU, GGC, GGA, GGG
AAU, AAC	asparagine 天冬醯胺	$-CH_2CONH_2$	$-CH_2COO^-$	aspartic acid 天冬胺酸	GAU, GAC
ACU, ACC, ACA, ACG	threonine 蘇胺酸	$-CHOHCH_3$	$-CH_2CH_2COO^-$	glutamic acid 穀胺酸	GAA, GAG
AGU, AGC	serine 絲胺酸	$-CH_2OH$	$-CH_3$	alanine 丙胺酸	GCU, GCC, GCA, GCG
AGA, AGG	arginine 精胺酸	$-CH_2CH_2CH_2NHC(NH_2)(N^+H_2)$			
AUU, AUC, AUA	isoleucine 異亮胺酸	$CH_3CH_2CHCH_3$	CH_3CHCH_3	valine 纈胺酸	GUU, GUC, GUA, GUG
AUG	methionine 甲硫胺酸	$-CH_2CH_2SCH_3$	$-CH_2-C_6H_5$	phenylalanine 苯丙胺酸	UUU, UUC
CCU, CCC, CCA, CCG	proline 脯胺酸	(ring structure)	$-CH(CH_3)_2-CH_3$	leucine 亮胺酸	UUA, UUG
CAU, CAC	histidine 組織胺酸	$-CH_2-$(imidazole)	$-CH_2-C_6H_4-OH$	tyrosine 酪胺酸	UAU, UAC
				NONSENSE 無意義	UAA, UAG
CAA, CAG	glutamine 穀氨醯胺	$-CH_2CH_2CONH_2$	$-CH_2SH$	cysteine 半胱胺酸	UGU, UGC
CGU, CGC, CGA, CGG	arginine 精胺酸	$-CH_2CH_2CH_2NHC(N^+H_2)(NH_2)$	$-CH_2-$(indole)	tryptophan 色胺酸	UGG
CUU, CUC, CUA, CUG	leucine 亮胺酸	$-CH_2CH(CH_3)_2$		NONSENSE 無意義	UGA
			$-CH_2OH$	serine 絲胺酸	UCU, UCC, UCA, UCG

PROTEINS/GENERAL 蛋白質/概述

protein (n) a substance made of one or more polypeptides (↓), which themselves are made of amino acids (↓). There are very many different kinds of protein, each with its own sequence of amino acids. Some are structural, e.g. in membranes (p. 18), and others are enzymes (p. 15) which catalyze (p. 15) reactions in cells.

amino acid any one of a class of organic (p. 11) compounds with a carboxyl group (—COOH), an amino group (NH₂) and a 'side-group' all attached to a central carbon atom (p. 8). Different amino acids have different side-groups. There are about 20 different amino acids found in proteins (↑), which they form when linked together in a chain or polymer (p. 10).

peptide (n) a compound made of two or more amino acids (↑) joined in a polymer (p. 10).

polypeptide (n) a peptide (↑) with a large number of amino acids (↑). Polypeptide chains become folded to form proteins (↑).

transcription (n) the process in which mRNA (p. 51) is produced in the nucleus (p. 19) of a cell, carrying in its sequence of nitrogen-containing bases (p. 52) the genetic code (p. 54) of the DNA (p. 51) in the nucleus.

translation (n) the part of protein synthesis (↓) in which molecules of tRNA (p. 51) carrying amino acids (↑) are matched with the genetic code (p. 54) carried by mRNA (p. 51), so that the amino acids are joined together in the correct sequence to make a polypeptide (↑). This takes place on the ribosomes (↓).

endoplasmic reticulum a system of membranes (p. 18) in the cytoplasm (p. 18), where much of protein synthesis (↓) takes place. Endoplasmic reticulum may be rough (with ribosomes (↓)), or smooth (without ribosomes).

ribosome (n) small body made of rRNA (p. 51) and protein (↑). It is the site of protein synthesis (↓) and the process of translation (↑). Cells can contain many thousands of ribosomes, which are found either on the endoplasmic reticulum (↑) or as polysomes (↓).

polysome (n) a group of ribosomes (↑) joined together by a strand of mRNA (p. 51).

蛋白質（名）　由一個或多個多肽（↓）組成的物質，多肽則由氨基酸（↓）組成。蛋白質有許多不同種類，每種都各有本身的氨基酸排列順序。有些是結構蛋白質，例如膜（第18頁）；有些則是酶（第15頁），酶在細胞內起催化（第15頁）反應。

氨基酸　任何一類分子中含有一個羧基（—COOH）、一個氨基（—NH₂）及一個側鏈基，而全部基團都連接於一個中心碳原子（第8頁）的有機（第11頁）化合物。不同的氨基酸有不同的側鏈基團。蛋白質（↑）中含有約20種不同的氨基酸，這些氨基酸連在一起形成一條鏈或聚合體（第10頁），組成蛋白質。

肽（名）　由兩個或多個氨基酸（↑）加入一個聚體（第10頁）所構成的一種化合物。

多肽（名）　含有大量氨基酸（↑）的肽（↑）。多肽鏈摺疊就形成蛋白質（↑）。

轉錄（名）　細胞的細胞核（第19頁）中產生mRNA（第51頁）的過程，其含氮鹼基（第52頁）的順序攜帶有細胞核DNA（第51頁）的遺傳密碼（第54頁）。

轉譯（名）　指蛋白質合成（↓）部分，其中攜帶氨基酸（↑）的tRNA（第51頁）分子和mRNA（第51頁）所攜帶的遺傳密碼（第54頁）相配稱，從而使氨基酸以正確順序連接一起，組成一個多肽（↑），轉譯發生在核糖體（↓）內。

內質網　在細胞質（第18頁）內的膜（第18頁）系統，在此合成（↓）大量蛋白質。內質網可以是粗糙的（具有核糖體（↓）），或者是平滑的（不具核糖體）。

核糖體（名）　由rRNA（第51頁）與蛋白質（↑）組成的小體。這是蛋白質合成（↓）的部位及轉譯（↑）的過程。細胞可含有數千個核糖體，它們或者見於內質網（↑），或者作為多核蛋白體（↓）。

聚核糖體（名）　一群由一股mRNA（第51頁）連結在一起的一組核糖體（↑）。

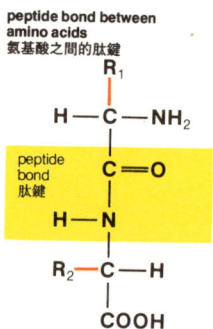

peptide bond between amino acids
氨基酸之間的肽鍵

R_1 and R_2 are side groups
R_1、R_2 為側基

protein synthesis the building of polymers (p.10) of amino acids (↑), which takes place on the ribosomes (↑) of a cell. Each amino acid is attached to a molecule (p. 9) of tRNA (p. 51) before synthesis starts. The anticodon (p. 53) on the tRNA must be matched with the codon (p. 53) on a molecule of mRNA (p. 51) which runs through the ribosome, before its amino acid can be joined onto the polypeptide (↑) chain which will become the protein (↑).

蛋白質合成　指氨基酸(↑)聚合體(第 10 頁)的建成。這是在一個細胞的核糖體(↑)上發生的。各個氨基酸連接到一個 tRNA(第 51 頁)分子(第 9 頁)上而開始合成。tRNA 上的反密碼子(第 53 頁)必須與一個 mRNA(第 51 頁)分子的密碼子(第 53 頁)相配稱，mRNA 進入核糖體，其氨基酸連結到多肽(↑)鏈上而成為蛋白質(↑)。

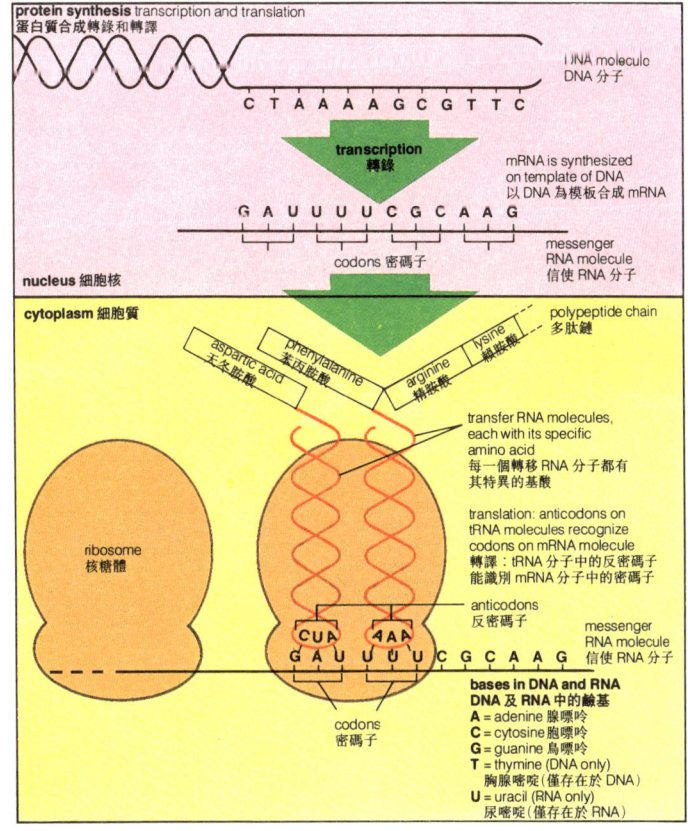

58 · PROTEINS/STRUCTURE 蛋白質/結構

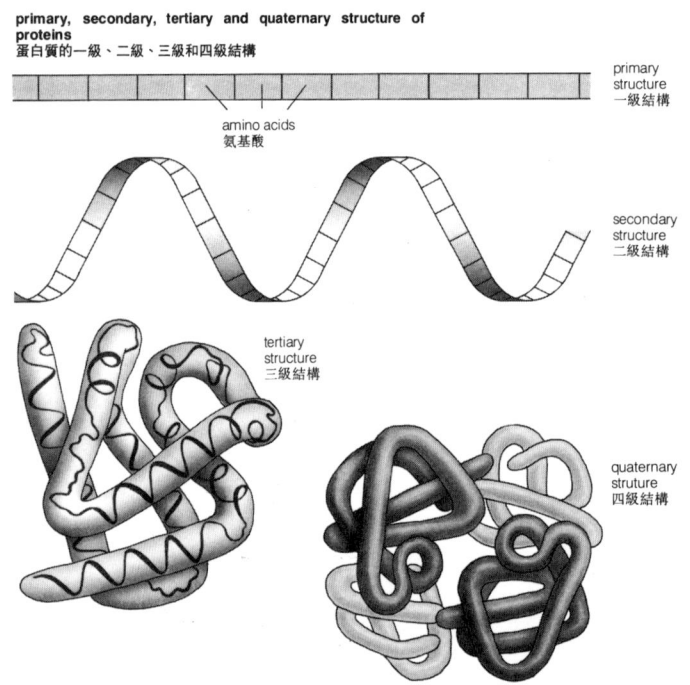

protein structure the structure of proteins (p. 56) can be studied at four levels – primary, secondary, tertiary and quaternary. The *primary structure of a protein* is the sequence of amino acids (p. 56) in a polypeptide (p. 56). The *secondary structure of a protein* is the coiling of the polypeptide into a helix (p. 51) or pleated sheet. The *tertiary structure of a protein* is the twisting and folding of the polypeptide helix or pleated sheet to form a three-dimensional protein molecule (p. 9). The *quaternary structure of a protein* is the structure of several protein molecules when bonded together.

glycoprotein (*n*) a protein (p. 56) bonded to a sugar.

蛋白質結構 蛋白質(第56頁)的結構可從一級、二級、三級和四級這四級結構進行研究。蛋白質的一級結構是氨基酸(第56頁)在一個多肽(第56頁)鏈的順序。二級結構是多肽鏈盤繞成螺旋(第51頁)型式或是摺疊結構。三級結構是多肽螺旋或摺疊薄片纏繞或摺疊形成三維的蛋白質分子(第9頁)。四級結構則是幾個蛋白質分子鏈結合在一起的結構。

糖蛋白(名) 與糖結合的蛋白質(第56頁)。

reproduction (n) the process in which an organism produces offspring (p. 44) like itself. Reproduction can be sexual (↓) or asexual (↓), and is one of the most important characteristics of living organisms. **reproduce** (v), **reproductive** (adj).

progeny (n) the offspring (p. 44) or young produced by an organism during reproduction (↑).

sexual (adj) of reproduction (↑) which involves the fusion (p. 61) of two cells and their nuclei (p. 19) from two parent individuals (p. 135), so that the offspring (p. 44) receives genetic (p. 41) material from both parents. Sexual reproduction occurs in all the divisions (p. 134) of the plant kingdom (p. 134). **sex** (n).

breed (v) to reproduce (↑) sexually (↑).

asexual (adj) of reproduction (↑) from one individual (p. 135), without the fusion (p. 61) of sex cells (p. 61) from two different parents. Asexual reproduction is common in the plant kingdom (p. 134). Many plant species can reproduce both sexually (↑) and asexually.

apomixis (n) the production of propagules (↓) by the female reproductive organs (p. 88) of a plant, without the sexual (↑) fusion (p. 61) of cells. In one type of apomixis, the embryo (p. 85) develops from the unfertilized haploid (p. 50) egg-cell (p. 61), in which case the offspring (p. 44) are usually sterile (p. 62). In others, the embryo develops from diploid (p. 50) tissue (p. 88) in the ovule (p. 78), in which case the offspring are fertile (p. 62). **apomictic** (adj).

propagule (n) any reproductive (↑) unit which gives rise to a new individual (p. 135), e.g. a seed, a spore (p. 66).

agamospermy (n) asexual (↑) production of embryos (p. 85) and seeds in flowering plants (p. 130).

apogamy (n) asexual (↑) reproduction (↑) in which embryos (p. 85) and propagules (↑) are produced without meiosis (p. 49) occurring.

apospory (n) production of a diploid (p. 50) gametophyte (p. 65) from vegetative (p. 60) cells of the sporophyte (p. 65), that is, without the production of spores (p. 66).

生殖(名) 生物體產生和自身相似的後代(第44頁)的過程，是活生物體最重要的特徵之一。生殖方式分成有性(↓)生殖和無性(↓)生殖兩類。(動詞為 reproduce，形容詞為 reproductive)

後代(名) 生物體生殖(↑)所產生的後裔(第44頁)或幼體。

有性的(形) 指兩個親代個體(第135頁)的兩個細胞及其細胞核(第19頁)融合(第61頁)的生殖(↑)方式，所產生後代(第44頁)從其雙親接受遺傳(第41頁)物質。有性生殖存在於植物界(第134頁)的各門(第134頁)植物。(名詞為 sex)

繁育(動) 進行有性(↑)生殖(↑)。

無性的(形) 指從一個個體(第135頁)的生殖方式(↑)，這種生殖方式沒有兩個不同親體性細胞(第61頁)的融合(第61頁)。無性生殖常見於植物界(第134頁)。許多植物種既能行有性(↑)生殖也能行無性生殖。

無融合生殖(名) 植物的雌性生殖器官(第88頁)產生繁殖體(↓)，不出現細胞的有性(↑)融合(第61頁)。在無融合生殖的一種形式中不受精的單倍體(第50頁)卵細胞(第61頁)發育成胚(第85頁)。在此情況下，其後代(第44頁)通常是不育的(第62頁)。在另一種形式中，是由胚珠(第78頁)內的二倍體(第50頁)組織(第88頁)發育成胚，在此情況下其後代能育(第62頁)。(形容詞為 apomictic)

繁殖體(名) 任何能產生新個體(第135頁)的生殖(↑)單元。例如種子、孢子(第66頁)。

無融合結籽(名) 有花植物(第130頁)的胚(第85頁)及種子的無性(↑)生殖。

無配子生殖(名) 沒有發生成熟分裂(第49頁)而產生胚(第85頁)及繁殖體(↑)的無性(↑)生殖(↑)。

無孢子生殖(名) 從孢子體(第65頁)的營養(第60頁)細胞產生二倍體的(第50頁)配子體(第65頁)，亦即沒有產生孢子(第66頁)。

REPRODUCTION/VEGETATIVE REPRODUCTION 生殖／營養生殖

vegetative reproduction a type of asexual (p. 59) reproduction (p. 59) in which a whole new plant is produced from an organ (p. 88), e.g. a rhizome (↓), bulb (↓), or tuber (↓), which is not involved in sexual (p. 59) reproduction.

vegetative (adj) of any part of a plant which is not involved in sexual (p. 59) reproduction (p. 59). Stems, leaves and roots are vegetative organs (p. 88).

bulb (n) an organ (p. 88) of perennation (p. 117) and vegetative reproduction (↑) in many monocotyledons. Bulbs are usually underground, and consist of a short axis (p. 92), with many overlapping thick leaves. These leaves generally lack chlorophyll (p. 36) and contain stored food.

bulbil (n) a small bulb (↑).

corm (n) a thickened stem base, usually underground, with buds (p. 110) in the axils (p. 96) of dead leaf bases. Corms are organs (p. 88) of vegetative reproduction (↑) and perennation (p. 117).

rhizome (n) a stem which grows along under the ground, bearing buds (p. 110) which produce shoots. This is a way of vegetative reproduction (↑) and perennation (p. 117), as the shoots grow into whole new plants. **rhizomatous** (adj).

stolon (n) a stem which grows along the ground, producing at its nodes (p. 90) new plants with roots and upright stems. **stoloniferous** (adj).

runner (n) a stolon (↑) which produces roots and a new plant at its apex (p. 90). This is a way of vegetative reproduction (↑). After the new plant has started to grow, the runner dies and decays.

sucker (n) a new shoot which develops from the base of a plant or from its roots. This is a way of vegetative reproduction (↑).

tiller (n) a single new plant growing from the base of an old plant, especially in grasses.

tuber (n) a thick underground stem in which food is stored. Tubers have buds (p. 110) in modified leaf axils (p. 96), from which new plants can grow, e.g. potato. Tubers are organs (p. 88) of perennation (p. 117) and vegetative reproduction (↑).

營養體生殖　為無性（第59頁）生殖（第59頁）的一種方式，指從一個器官（第88頁），例如根狀莖（↓）、鱗莖（↓）、塊莖（↓）長成一株全新植物，這種生殖方式和有性（第59頁）生殖無聯繫。

營養體的（形）指不牽涉有性（第59頁）生殖（第59頁）的植物的任何部分。莖、葉及根都是營養器官（第88頁）。

鱗莖（名）　許多單子葉植物的多年生（第117頁）和營養體生殖（↑）器官（第88頁）。鱗莖通常長於地下，有一短軸（第92頁）和許多重疊的肥厚多肉鱗葉。這些鱗葉通常缺乏葉綠素（第36頁），而含有貯藏的食物。

珠芽（名）　細小的鱗莖（↑）。

球莖（名）　肥厚的莖基，通常長於地下，在枯死葉片基部的葉腋（第96頁）長芽（第110頁）。球莖是營養體生殖（↑）和多年生（第117頁）的器官（第88頁）。

根狀莖（名）　在地面下橫向生長的莖，莖上有芽（第110頁）能長出嫩枝，這是營養體生殖（↑）的一種方式，且為多年生（第117頁），嫩枝長成全新植物。（形容詞為 rhizomatous）

匍匐莖（名）　沿着地面生長的莖，在節（第90頁）上長有根和向上生長的莖的新植物。（形容詞為 stoloniferous）

纖匐枝；走莖（名）　能產生根和在頂端（第90頁）形成新植物的一種匍匐莖（↑），這是營養生殖（↑）的一種方式。新植物開始長出後，纖匐枝死亡而腐化。

根出條（名）　從植物的基部或根部發出來的新枝條。這是營養體生殖（↑）的一種方式。

分蘖（名）　從老植物基部長出一個單株新植物，特別見於禾草類。

塊莖（名）　一種貯藏養分的肥厚地下莖，塊莖的變態葉腋（第96頁）長芽（第110頁），芽可長成新植物，例如馬鈴薯。塊莖是多年生（第117頁）和營養體生殖（↑）的器官（第88頁）。

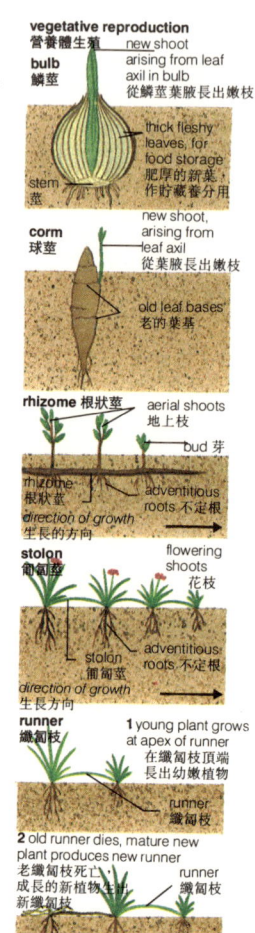

REPRODUCTION/GAMETES, ZYGOTES 生殖／配子、合子 · 61

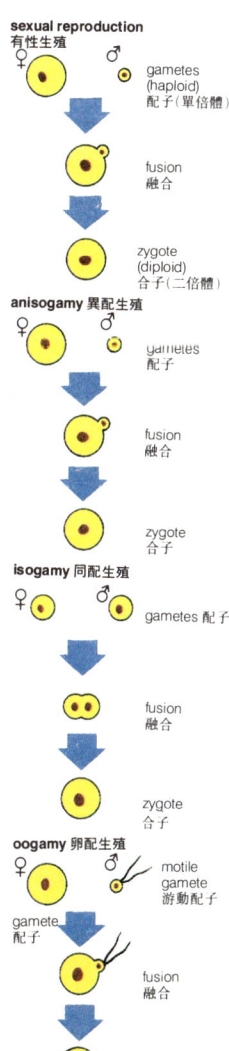

gamete (n) a haploid (p. 50) sex cell (↓), whose function is to join with a gamete of the opposite sex, to form a diploid (p. 50) zygote (↓). In plants, gametes are produced by the gametophyte (p. 65).

sex cell = a gamete (↑).

ovum (n) an egg-cell (↓) or female gamete (↑).

egg-cell (n) the female gamete (↑) or ovum (↑).

anisogamous (adj) of plants which produce gametes (↑) of different sizes, sometimes known as microgametes (male) and megagametes (female). All plants growing on land are anisogamous. **anisogamy** (n).

heterogamous (adj) having male and female gametes (↑) of different sizes, i.e. anisogamous (↑). **heterogamy** (n).

isogamous (adj) having male and female gametes (↑) of the same size. This is characteristic of some algae (p. 119). **isogamy** (n).

oogamous (adj) having a small motile (p. 121) male gamete (↑) and a large non-motile female gamete, as in bryophytes (p. 122) and pteridophytes (p. 126). **oogamy** (n).

zygote (n) a diploid (p. 50) cell which is produced by the fusion (↓) of two haploid (p. 50) gametes (↑). A fertilized (p. 62) ovum (↑) is a zygote. In plants, the zygote develops first into an embryo (p. 85) and then into a sporophyte (p. 65).

fusion (n) the joining together of two gametes (↑) to form a zygote (↑). Fusion can mean the joining of the cells, the joining of the nuclei (p. 19), or both. **fuse** (v).

conjugation (n) the joining together of two similar cells, usually male and female, in some algae (p. 119). **conjugate** (v).

gender (n) the sex of an individual (p. 135). Gender can be male, female, or neuter (↓).

female (adj) of individuals (p. 135), tissues (p. 88), organs (p. 88), etc. producing egg-cells (↑). **female** (n).

male (adj) of individuals (p. 135), organs (p. 88), tissues (p. 88), etc. producing the gametes (↑) which fertilize (p. 62) egg-cells (↑) produced by females. **male** (n).

neuter (adj) neither male nor female.

配子（名） 植物的單倍體（第 50 頁）性細胞（↓），其功能是和一個異性配子接合形成一個二倍體（第 50 頁）合子（↓），植物的配子是由配子體（第 65 頁）產生的。

性細胞 同配子（↑）。

卵（名） 指卵細胞（↓）或雌配子（↑）。

卵細胞（名） 即雌配子（↑）或卵（↑）。

異配生殖的（形） 指產生不同大小的配子（↑）的植物。有時稱為小配子（雄性）和大配子（雌性）的生殖方式。一切陸生植物都是異配生殖的。（名詞為 anisogamy）

具異型配子的（形） 有大小不同的雄配子及雌配子（↑），即異配生殖的（↑）。（名詞為 heterogamy）

同配生殖的（形） 具相同大小的雄配子及雌配子（↑），這是某些藻類植物（第 119 頁）的特徵。（名詞為 isogamy）

卵配生殖的（形） 具有一個小而能游動（第 121 頁）的雄配子（↑）和一個大而不能動的雌配子。例如苔蘚植物（第 122 頁）和蕨類植物（第 126 頁）的生殖方式。（名詞為 oogamy）

合子（名） 兩個單倍體（第 50 頁）配子（↑）融合（↓）產生的二倍體（第 50 頁）細胞，受精（第 62 頁）卵（↑）就是合子。就植物而言，合子先發育成胚（第 85 頁），繼而發育成孢子體（第 65 頁）。

融合（名） 兩個配子（↑）併合一起形成一個合子（↑）。融合是指細胞的併合，也指細胞核（第 19 頁）的併合，或指細胞和核兩者都併合。（動詞為 fuse）

接合生殖（名） 在某些藻類植物（第 119 頁）中兩個相類似的細胞，通常是雄細胞和雌細胞併合在一起。（動詞為 conjugate）

性別（名） 一個個體（第 135 頁）的性。性別可為雄的、雌的或中性的（↓）。

雌性的（形） 指能產生卵細胞（↑）的個體（第 135 頁）、組織（第 88 頁）和器官（第 88 頁）等。（名詞為 female）

雄性的（形） 指能產生配子（↑），使雌體產生的卵細胞（↑）受精（第 62 頁）的個體（第 135 頁）、器官（第 88 頁）和組織（第 88 頁）等。（名詞為 male）

中性的（形） 既不是雄性也不是雌性的。

bisexual (*adj*) of organisms with male and female reproductive (p. 59) organs (p. 88) on the same individual (p. 135).

fertile (*adj*) of organisms which produce offspring, or of reproductive (p. 59) organs (p. 88) which produce viable (↓) gametes (p. 61). **fertility** (*n*).

viable (*adj*) able to carry out its function, e.g. the ability of a dormant (p. 117) seed to germinate (p. 87) when conditions become suitable. **viability** (*n*).

sterile (*adj*) of organisms which cannot produce offspring (p. 44), or of reproductive (p. 59) organs (p. 88) which do not produce gametes (p. 61), e.g. staminodes (p. 73). **sterility** (*n*).

兩性的(形) 指在同一個體(第135頁)上具有雄性及雌性生殖(第59頁)器官(第88頁)的生物體。

能育的(形) 指能產生後代的生物體，或能產生有生活力(↓)配子(第61頁)的生殖(第59頁)器官(第88頁)。(名詞為 fertility)

有生活力的(形) 能實現其功能的。例如休眠的(第117頁)種子在條件適宜時有萌發(第87頁)的能力。(名詞為 viability)

不育的(形) 指不能產生後代(第44頁)的生物體，或者不能產生配子(第61頁)的生殖(第59頁)器官(第88頁)。例如退化雄蕊(第73頁)。(名詞為 sterility)

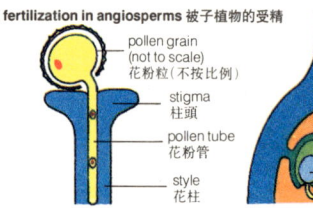
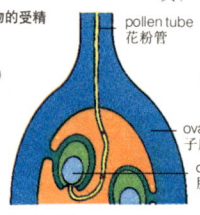
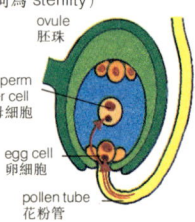

fertilization in angiosperms 被子植物的受精

pollen grain (not to scale) 花粉粒(不按比例)
stigma 柱頭
pollen tube 花粉管
style 花柱
ovary 子房
ovule 胚珠
endosperm mother cell 胚乳母細胞
egg cell 卵細胞
pollen tube 花粉管

1 pollen grain lands on stigma, pollen tube grows through tissues of style carrying the male gametes 花粉粒落在柱頭上，花粉管長過帶雄配子的花柱組織

2 pollen tube grows through ovary wall and into micropyle of ovule 花粉管長通過子房壁進入胚珠的珠孔

3 one male gamete fertilizes egg cell, the other fertilizes the endosperm nucleus forming endosperm mother cell 一個雄配子使卵細胞受精，另一雄配子使胚乳核受精，形成胚乳母細胞

fertilization (*n*) the fusion (p. 61) of a male gamete (p. 61) with a female gamete to form a zygote (p. 61). **fertilize** (*v*).

self-fertilization (*n*) the fertilization (↑) of a female gamete (p. 61) by a male gamete from the same individual (p. 135). This is sometimes called selfing.

autogamy (*n*) = self-fertilization (↑). **autogamous** (*adj*).

cleistogamy (*n*) self-fertilization (↑) before a flower opens. The flowers of some species (p. 134) never fully open, and such species are habitually cleistogamous.

cross-fertilization (*n*) fertilization (↑) of a female gamete (p. 61) of one plant by a male gamete from another.

allogamy (*n*) the production of zygotes (p. 61) by cross-fertilization (↑). **allogamous** (*adj*).

受精(名) 一個雄配子(第61頁)和一個雌配子融合(第61頁)形成合子(第61頁)。(動詞為 fertilize)

自體受精(名) 雌配子(第61頁)被同一個體(第135頁)的雄配子受精(↑)。有時稱為自花受精。

自花受粉(名) 同自體受精(↑)。(形容詞為 autogamous)

閉花受精(名) 開花前的自體受精(↑)。有些種屬(第134頁)的花永不完全開放，這種花有自花受粉的習性。

異花受精(名) 一種植物的雌配子(第61頁)被另一植物的雄配子受精(↑)。

異體受精(名) 由異花受精(↑)產生合子(第61頁)。(形容詞為 allogamous)

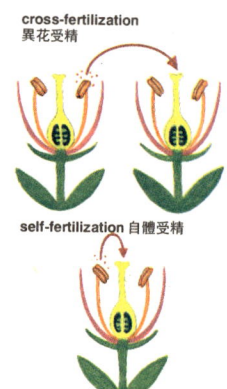

cross-fertilization 異花受精

self-fertilization 自體受精

inbreeding (n) breeding (p. 59) over many generations (↓) between closely-related individuals (p. 135) of a species (p. 134).

outbreeding (n) breeding (p. 59) between individuals (p. 135) that are not closely related.

compatible (adj) of two plants which are able to breed (p. 59) with each other. **compatibility** (n).

self-compatible (adj) of an individual (p. 135) plant which can fertilize (↑) its female gametes (p. 61) with its own male gametes.

incompatible (adj) of two plants which cannot breed (p. 59) with each other. **incompatibility** (n).

self-incompatible (adj) of an individual (p. 135) plant which cannot fertilize (↑) its female gametes (p. 61) with its own male gametes.

hybrid (n) a plant which results from the cross-fertilization (↑) of two different species (p. 134), subspecies (p. 135), varieties (p. 135), strains (p. 135), etc. **hybridize** (v), **hybridization** (n).

heterosis (n) the condition of a hybrid (↑) that is fitter than either of its parents. This is also called hybrid vigour (↓).

hybrid vigour = heterosis (↑).

generation (n) a set of individuals (p. 135) of roughly equal age or stage of development (p. 109). The parents are one generation, and the progeny (p. 59) are the next.

近交(名) 一個種(第134頁)的近親個體(第135頁)間進行許多世代(↓)的繁育(第59頁)。

遠交(名) 非近親個體(第135頁)間的繁育(第59頁)。

親和的(形) 指兩株植物彼此能繁育(第59頁)。(名詞為 compatibility)

自交親和的(形) 指一個個體(第135頁)植物的雌配子(第61頁)能被自身的雄配子受精(↑)。

不親和的(形) 指兩株植物彼此不能繁育(第59頁)。(名詞為 incompatibility)

自交不親和的(形) 指一個個體(第135頁)植物的雌配子(第61頁)不能被自身的雄配子受精(↑)。

雜種(名) 由兩個不同的種(第134頁)、亞種(第135頁)、變種(第135頁)和株(第135頁)等之間異體受精(↑)產生的植物。(動詞為 hybridize，名詞為 hybridization)

雜種優勢(名) 指雜種(↑)比其親代任一方更能適應，亦稱雜種精壯(↓)。

雜種精壯 同雜種優勢(↑)。

世代(名) 指年齡或發育(第109頁)階段約略相等的一組個體(第135頁)。親代是一個世代，後裔(第59頁)是次一個世代。

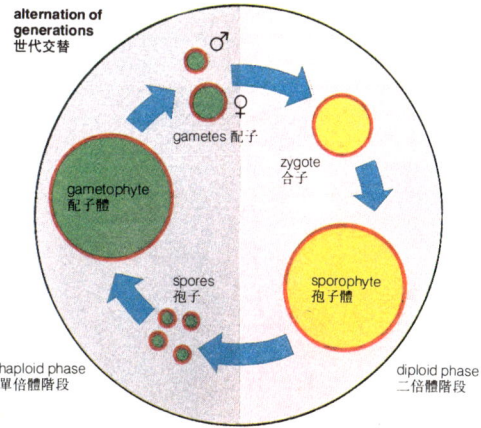

life cycle the complete set of changes that occur from any stage in the life of an organism to the same stage in the life of its offspring (p. 44). In bryophytes (p. 122), pteridophytes (p. 126) and spermatophytes (p. 128), the life cycle consists of an alternation of haploid (p. 50) and diploid (p. 50) generations (p. 63).

alternation of generations the life cycle (↑) of bryophytes (p. 122), pteridophytes (p. 126), and spermatophytes (p. 128), which consists of a haploid (p. 50) gametophyte (↓) producing gametes (p. 61) followed by a diploid (p. 50) sporophyte (↓) producing spores (p. 66).

haplont (*adj*) of the haploid (p. 50) stage in a life cycle (↑), ending with fertilization (p. 62), e.g. of the gametophyte (↓).

diplont (*adj*) of the diploid (p. 50) stage in a life cycle (↑), e.g. of the sporophyte (↓).

生活週期；生活史　從一個有機體生命的任何階段到其後代 (第 44 頁) 生命的相同階段所發生的一系列變化。苔蘚植物 (第 122 頁)、蕨類植物 (第 126 頁) 及種子植物 (第 128 頁) 的生活週期包括單倍體 (第 50 頁) 及二倍體 (第 50 頁) 的世代 (第 63 頁) 交替。

世代交替　苔蘚植物 (第 122 頁)、蕨類植物 (第 126 頁) 及種子植物 (第 128 頁) 的生活週期 (↑)，包括單倍體 (第 50 頁) 的配子體 (↓) 產生配子 (第 61 頁)，隨後二倍體 (第 50 頁) 的孢子體 (↓) 產生孢子 (第 66 頁)。

單倍體的 (形)　指生活週期 (↑) 的單倍體 (第 50 頁) 階段，以受精 (第 62 頁) 為終結。例如配子體 (↓) 階段。

二倍體的 (形)　指生活週期 (↑) 的二倍體 (第 50 頁) 階段。例如孢子體 (↓) 階段。

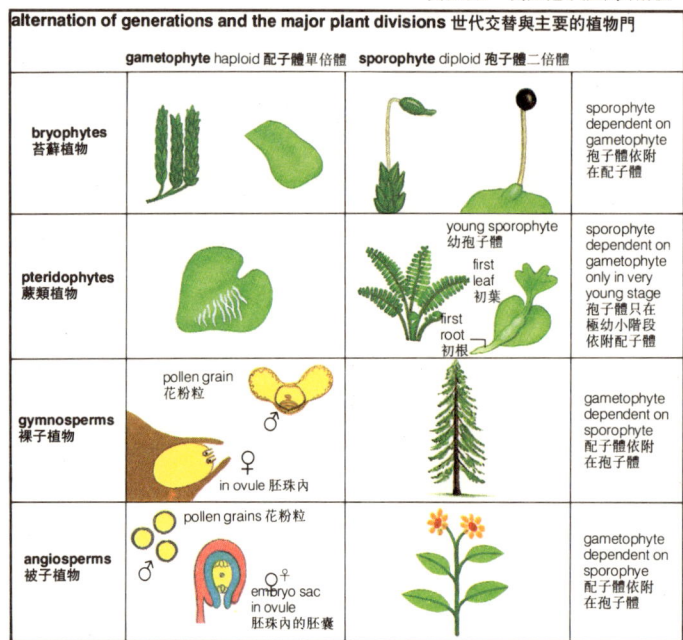

alternation of generations and the major plant divisions 世代交替與主要的植物門

REPRODUCTION/GAMETES AND GAMETANGIA 生殖/配子及配子囊

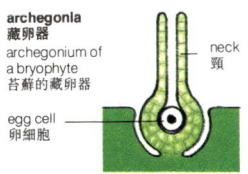

archegonia
藏卵器
archegonium of a bryophyte
苔蘚的藏卵器
neck
頸
egg cell
卵細胞

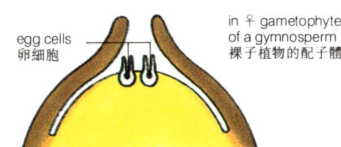

egg cells
卵細胞
in ♀ gametophyte of a gymnosperm
裸子植物的配子體

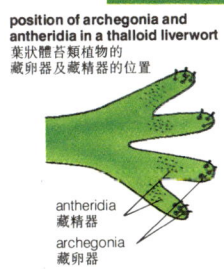

position of archegonia and antheridia in a thalloid liverwort
葉狀體苔類植物的藏卵器及藏精器的位置
antheridia
藏精器
archegonia
藏卵器

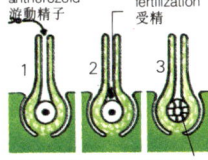

sexual reproduction in a bryophyte 苔蘚植物的有性生殖
antherozoid
游動精子
fertilization
受精
1 2 3
archegonium
藏卵器
young sporophyte
幼孢子體
developing sporophyte
發育中的孢子體
calyptra 帽狀體 (gametophyte tissue)
(配子體組織)
capsule
蒴
4
seta
蒴柄
foot
蒴座

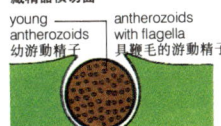

cross section of antheridium of a liverwort
苔類植物的藏精器橫切面
young antherozoids
幼游動精子
antherozoids with flagella
具鞭毛的游動精子

gametophyte (n) the haploid (p. 50) generation (p. 63) in an alternation of generations (↑). The gametophyte is the generation producing gametes (p. 61). In bryophytes (p. 122), the gametophyte is the main vegetative (p. 60) stage. In angiosperms (p. 130), the gametophyte is very small, contained in the ovules (p. 78) and pollen (p. 74) grains.

sporophyte (n) the diploid (p. 50) generation (p. 63) in an alternation of generations (↑). The sporophyte is the generation producing spores (p. 66). In angiosperms (p. 130), gymnosperms (p. 128) and pteridophytes (p. 126), the sporophyte is the main vegetative (p. 60) stage. In bryophytes (p. 122), the sporophyte grows directly from the archegonium (↓) of the gametophyte (↑), and depends on the gametophyte for its nutrition (p. 111).

gametangium (n) any organ (p. 88) which produces gametes (p. 61). **gametangia** (pl.).

archegonium (n) the flask-shaped female organ (p. 88) of bryophytes (p. 122), pteridophytes (p. 126) and gymnosperms (p. 128). The archegonium consists of a hollow neck, whose wall is one cell thick, and a swollen base containing the ovum (p. 61). The antherozoid (↓) swims down the neck to reach the ovum. **archegonia** (pl.), **archegoniate** (adj).

antheridium (n) the organ producing male gametes (p. 61) in bryophytes (p. 122) and ferns (p. 126). **antheridia** (pl.).

antherozoid (n) a flagellate (p. 121), motile (p. 121) male gamete (p. 61) of bryophytes (p. 122) and some ferns (p. 126). Antherozoids are produced in antheridia (↑).

spermatozoid (n) a motile (p. 121) male gamete (p. 61), or antherozoid (↑), in bryophytes (p. 122), ferns (p. 126) and many algae (p. 119).

配子體（名） 世代交替（↑）中的單倍體（第 50 頁）世代（第 63 頁）。配子體是產生配子（第 61 頁）的世代。在苔蘚植物（第 122 頁），配子體是主要的營養（第 60 頁）階段。被子植物（第 130 頁）的配子體非常細小，包含在胚珠（第 78 頁）及花粉（第 74 頁）粒之中。

孢子體（名） 世代交替（↑）中的二倍體（第 50 頁）世代（第 63 頁）。孢子體是產生孢子（第 66 頁）的世代。在被子植物（第 130 頁）、裸子植物（第 128 頁）及蕨類植物（第 126 頁），配子體是主要的營養（第 60 頁）階段。苔蘚植物（第 122 頁）的孢子體直接從配子體（↑）的藏卵器（↓）長出，並依賴配子體取得營養（第 111 頁）。

配子囊（名） 產生配子（第 61 頁）的任何器官（第 88 頁）。（複數為 gametangia）

藏卵器（名） 苔蘚植物（第 122 頁）、蕨類植物（第 126 頁）及裸子植物（第 128 頁）的瓶狀雌性器官（第 88 頁）。藏卵器由一個中空的頸構成，頸壁厚一個細胞，並有一個膨大而含卵（第 61 頁）的基部，游動精子（↓）向頸下游達卵。（複數為 archegonia，形容詞為 archegoniate）

藏精器（名） 苔蘚植物（第 122 頁）及蕨類植物（第 126 頁）產生雄配子（第 61 頁）的器官。（複數為 antheridia）

游動精子（名） 苔蘚植物（第 122 頁）及某些蕨類植物（第 126 頁）中具鞭毛（第 121 頁），能運動（第 121 頁）的雄配子（第 61 頁）。游動精子是在藏精器（↑）中產生的。

游走精子（名） 苔蘚植物（第 122 頁）、蕨類植物（第 126 頁）及多數藻類（第 119 頁）的能游動（第 121 頁）雄配子（第 61 頁）或游動精子（↑）。

REPRODUCTION/SPORES AND SPORANGIA 生殖／孢子及孢子囊

spore (n) a small round cell with a thick wall from which a whole new plant is produced. In bryophytes (p. 122), pteridophytes (p. 126) and spermatophytes (p. 128), spores are haploid (p. 50) and are produced by the sporophyte (p. 65). In bryophytes and pteridophytes, dispersal (p. 84) is achieved by spores. In angiosperms (p. 130), the spores develop (p. 109) into small gametophytes (p. 65) in the ovules (p. 78) and pollen (p. 74) grains. In all these plants, spores are produced as a result of meiosis (p. 49). Fungi (p. 163) also produce spores, but these are of many kinds and are different from those of green plants (see p. 163).

spore mother cell a cell which divides by meiosis (p. 49) to produce spores (↑).

tetrad[2] (n) a group of four haploid (p. 50) spores (↑), which are the product of meiosis (p. 49) of the spore mother cell (↑).

sporogenous (adj) of tissues (p. 88) in which spores (↑) are produced.

sporulation (n) the process of releasing spores (↑) for dispersal (p. 84). **sporulate** (v).

孢子（名）　厚壁的小圓球形細胞，孢子可長成全新的植株。苔蘚植物（第122頁）、蕨類植物（第126頁）及種子植物（第128頁）的孢子為單倍體（第50頁），是由孢子囊（第65頁）產生的。苔蘚植物及蕨類植物由孢子實現散播（第84頁）。被子植物（第130頁）的孢子在胚珠（第78頁）及花粉（第74頁）粒內發育（第109頁）成細小的配子體（第65頁），所有這些植物都是由減數分裂（第49頁）產生孢子。真菌（第163頁）也產生孢子，但其孢子有多種類型，且與綠色植物（第163頁）的孢子不同。

孢子母細胞　經減數分裂（第49頁）產生孢子（↑）的細胞。

四分體（名）　孢子母細胞（↑）經減數分裂（第49頁）所產生的一組四個單倍體（第50頁）孢子（↑）。

產孢子的（形）　描述產生孢子（↑）的組織（第88頁）。

孢子形成（名）　釋放孢子（↑）以供散播（第84頁）的過程。（動詞為 sporulate）

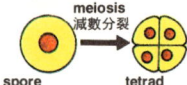

production of haploid spores in sporangia of vascular plants 在維管植物孢子囊形成單倍體的孢子

spore mother cell (diploid nucleus) 孢子母細胞（二倍體的細胞核）

meiosis 減數分裂

tetrad of spores (haploid nuclei) 四分體孢子單倍體的核

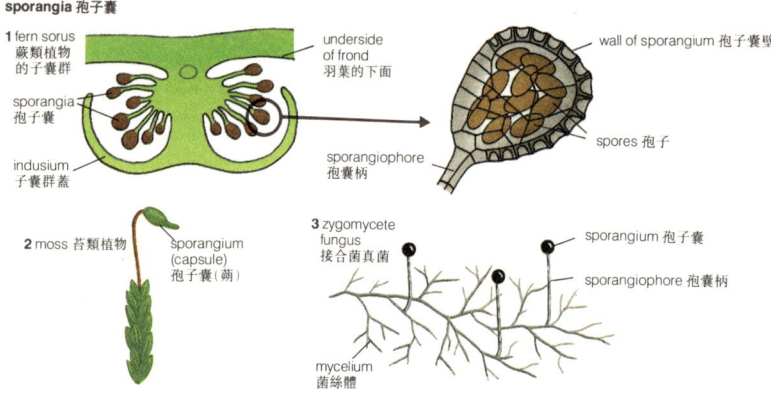

sporangia 孢子囊

1 fern sorus 蕨類植物的子囊群
sporangia 孢子囊
indusium 子囊群蓋
underside of frond 羽葉的下面
sporangiophore 孢囊柄
wall of sporangium 孢子囊壁
spores 孢子

2 moss 苔類植物
sporangium (capsule) 孢子囊（蒴）

3 zygomycete fungus 接合菌真菌
mycelium 菌絲體
sporangium 孢子囊
sporangiophore 孢囊柄

sporangium (n) a small round organ (p. 88) in which spores (↑) are produced, by meiosis (p. 49), from spore mother cells (↑). **sporangia** (pl.).

sporangiophore (n) the stalk of a sporangium (↑).

孢子囊（名）　由孢子母細胞（↑）經減數分裂（第49頁）產生孢子（↑）的圓形細小器官（第88頁）。（複數為 sporangia）

孢囊柄（名）　孢子囊（↑）的柄。

REPRODUCTION/SPORES AND SPORANGIA 生殖／孢子及孢子囊 · 67

sporophyll (n) a modified leaf, whose function is to produce sporangia (↑) and spores (↑). Sporophylls may be similar to vegetative (p. 60) leaves, as in many pteridophytes (p. 126), or organized into cones (p. 68), as in gymnosperms (p. 128). The sporophylls of angiosperms (p. 130) are the stamens (p. 73) and carpels (p. 75).

homosporous (adj) of plants whose spores (↑) are all the same, as in bryophytes (p. 122) and true ferns (p. 126). **homospory** (n).

heterosporous (adj) of plants which produce spores (↑) of two different sizes, as in some pteridophytes (p. 126) and all spermatophytes (p. 128). The large spore develops into a female gametophyte (p. 65), and the small spore develops into a male gametophyte. **heterospory** (n).

microspore (n) a small spore (↑), produced in a microsporangium (↓), in heterosporous (↑) plants. The microspore develops into the male gametophyte (p. 65). In angiosperms (p. 130), the microspore is the pollen (p. 74) grain.

孢子葉（名）　一種變態葉，其功能是產生孢子囊（↑）及孢子（↑）。孢子葉可以和營養葉（第60頁）相似，如在許多蕨類植物（第126頁）所見，或者可以形成果球（第68頁），如在裸子植物（第128頁）所見。被子植物（第130頁）的孢子葉是雄蕊（第73頁）和心皮（第75頁）。

具同形孢子的（形）　指所具孢子（↑）完全相同的植物，如在苔蘚植物（第122頁）及真蕨類（第126頁）所見。（名詞為 homospory）

具異形孢子的（形）　指形成兩種不同大小的孢子（↑）的植物，如在某些蕨類植物（第126頁）及一切種子植物（第128頁）所見。大孢子發育成雌配子體（第65頁），小孢子發育成雄配子體。（名詞為 heterospory）

小孢子（名）　具異形孢子（↑）的植物中，其小孢子囊（↓）產生細小的孢子（↑）。小孢子發育成雄配子體（第65頁）。被子植物（第130頁）的小孢子就是花粉（第74頁）粒。

homospory and heterospory in vascular plants
維管植物的孢子同形及孢子異形

homosporous 具同形孢子的
bryophytes 苔蘚相物
some pteridophytes (e.g. ferns) 某些蕨類植物（如羊齒植物）

heterosporous 具異形孢子的
some pteridophytes (e.g. clubmosses) 某些蕨類植物（如石松）
gymnosperms 裸子植物
angiosperms 被子植物

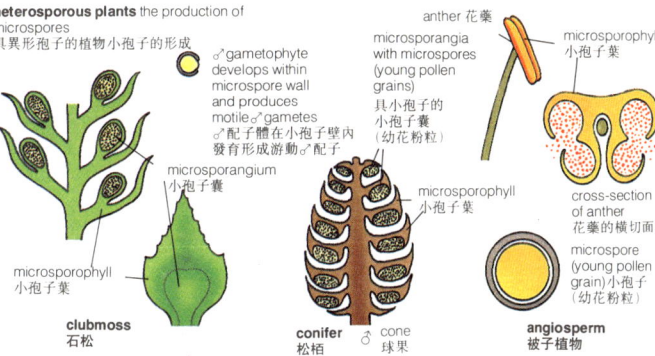

heterosporous plants the production of microspores 具異形孢子的植物小孢子的形成

clubmoss 石松 — ♂gametophyte develops within microspore wall and produces motile ♂gametes ♂配子體在小孢子壁內發育形成游動♂配子; microsporangium 小孢子囊; microsporophyll 小孢子葉

conifer 松栢 — ♂cone 球果; microsporangia with microspores (young pollen grains) 具小孢子的小孢子囊（幼花粉粒）; microsporophyll 小孢子葉

angiosperm 被子植物 — anther 花藥; cross-section of anther 花藥的橫切面; microspore (young pollen grain) 小孢子（幼花粉粒）; microsporophyll 小孢子葉

microsporangium (n) a sporangium (↑) producing microspores (↑) in a heterosporous (↑) plant. Microsporangia (pl.) usually produce many more spores than megasporangia (p. 68).

microsporophyll (n) a sporophyll (↑) bearing microsporangia (↑).

小孢子囊（名）　具異形孢子（↑）的植物中，產生小孢子（↑）的孢子囊（↑）。小孢子囊產生的孢子通常比大孢子囊（第68頁）多。

小孢子葉（名）　具有小孢子囊（↑）的孢子葉（↑）。

68 · REPRODUCTION/SPORES AND SPORANGIA 生殖／孢子及孢子囊

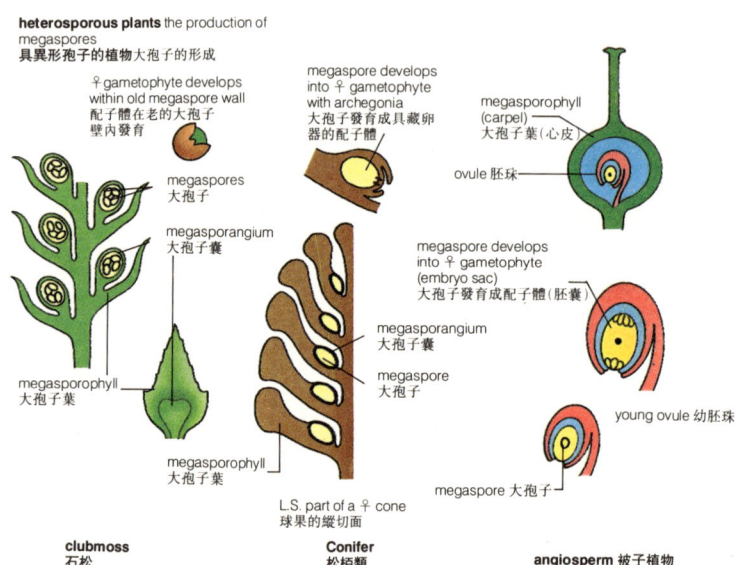

megaspore (*n*) a large spore (p. 66) produced in a megasporangium (↓), in heterosporous (p. 67) plants. The megaspore develops into the female gametophyte (p. 65). In angiosperms (p. 130), the megaspore is the embryo sac (p. 78).
megasporangium (*n*) a sporangium (p. 66) producing megaspores (↑), in heterosporous (p. 67) plants. **megasporangia** (*pl.*).
megasporophyll (*n*) a sporophyll (p. 67) bearing megasporangia (↑). In angiosperms (p. 130), the carpels (p. 75) are the megasporophylls.
cone (*n*) a group of sporophylls (p. 67) closely packed together around a central axis (p. 92). Cones are the reproductive (p. 59) structures of all gymnosperms (p. 128) and many pteridophytes (p. 126). In many cone-bearing plants, the male and female cones are separate.
strobilus (*n*) a reproductive (p. 59) organ (p. 88) consisting of overlapping scales (p. 100), as in some pteridophytes (p. 126) and the cones (↑) of gymnosperms (p. 128). **strobili** (*pl.*).

大孢子（名）異形孢子（第67頁）植物的大孢子囊（↓）所產生的大的孢子（第66頁）。大孢子發育成雌配子體（第65頁）。被子胚囊（第78頁）是被子植物（第130頁）的大孢子。

大孢子囊（名）異形孢子（第67頁）植物所產生大孢子（↑）的孢子囊（第66頁）。（複數為megasporangia）

大孢子葉（名）具有大孢子囊（↑）的孢子葉（第67頁）。大孢子葉是被子植物（第130頁）的心皮（第75頁）

球果（名）圍繞一個中心軸（第92頁）緊密堆疊在一起的一簇孢子葉（第67頁）。球果是一切裸子植物（第128頁）及許多蕨類植物（第126頁）的生殖（第59頁）結構。許多生有球果的植物，其雄球果與雌球果是分開的。

孢子葉球（名）由一些重疊鱗片（第100頁）組成的生殖（第59頁）器官（第88頁），如在某些蕨類植物（第126頁）所見及裸子植物（第128頁）的球果（↑）。（複數為strobili）

REPRODUCTION/PROPAGATION 生殖／繁殖・69

propagation (n) the process of reproduction (p. 59), either by natural or artificial means. **propagate** (v).

繁殖(名) 以自然方法或人工方法的生殖(第 59 頁)過程。(動詞為 propagate)

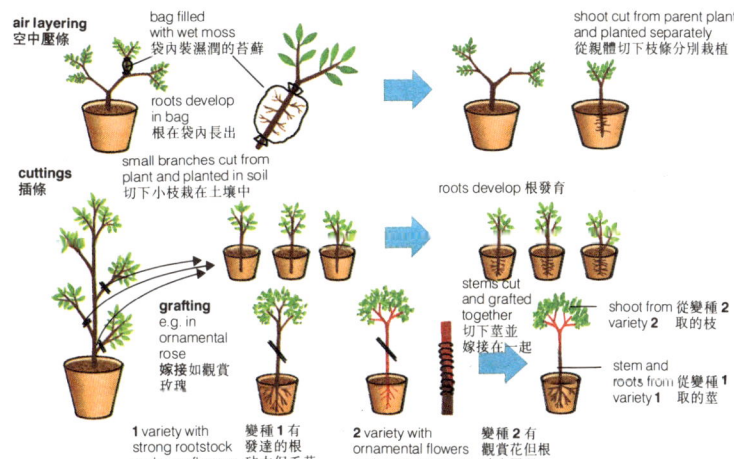

air layering a way of causing the formation of roots from nodes (p. 90) on a shoot. Wet moss (p. 124) is wrapped around the shoot; when roots have formed, the shoot can be cut from the plant and be grown as a new individual (p. 135).

cutting (n) a piece of shoot cut from a plant, which grows roots from its nodes (p. 90) when placed in soil.

graft (v) to join together artificially parts from two different plants, e.g. the shoot of one variety (p. 135) of a species (p. 134) onto the rootstock (↓) of another variety. **graft** (n).

rootstock (n) the roots of a plant.

tissue culture a process in which cells from an organism are grown on a medium (p. 171) in isolation from the organism from which they were taken. Plant tissue cultures, which usually consist of calluses (↓) of undifferentiated (p. 110) cells, are sometimes used for the production of drugs.

callus[1] (n) a lump of undifferentiated (p. 110) cells in a tissue culture (↑).

空中壓條 使枝條節(第 90 頁)長出根的一種方法。濕潤的苔蘚(第 124 頁)包著枝條。待長出新根後切下枝條，作為一株新個體(第 135 頁)培養。

插條(名) 從一株植物切下一段枝條，插在土壤中，枝條的節(第 90 頁)上即長出根。

嫁接(動) 以人工方法使取自兩株不同植物的部分連接起來。例如將某一個種(第 134 頁)的變種(第 135 頁)枝條接到另一變種的根砧木(↓)上。(名詞為 graft)

根砧木(名) 植物的根部。

組織培養 從有機體取出細胞，使之在一種培養基(第 171 頁)內生長的過程。植物組織培養(通常都含有未分化(第 110 頁)細胞的愈傷組織(↓)，此方法有時用於藥物生產。

愈傷組織(名) 組織培養(↑)中未分化(第 110 頁)的細胞塊。

flower¹ (n) the reproductive (p. 59) shoot of an angiosperm (p. 130), consisting usually of four sets of modified leaves arranged in whorls (p. 98). These are the sepals (↓), petals (↓), stamens (p. 73) and carpels (p. 75). The function of a flower is to produce male gametes (p. 61), in pollen (p. 74), and female gametes, in ovules (p. 78). After fertilization (p. 62), the ovules develop into seeds. The reproductive shoots of conifers (p. 128) are also sometimes called flowers. **floral** (adj).

花(名) 被子植物(第130頁)的生殖(第59頁)枝，通常由四組變態葉排列成幾個輪生體(第98頁)構成，這四種變態葉是萼片(↓)、花瓣(↓)、雄蕊(第73頁)和心皮(第75頁)。花的功能是在花粉(第74頁)產生雄配子(第61頁)，及在胚珠(第78頁)內產生雌配子。受精(第62頁)之後胚珠發育成種子。松柏類(第128頁)的生殖枝有時亦稱為花。(形容詞為floral)

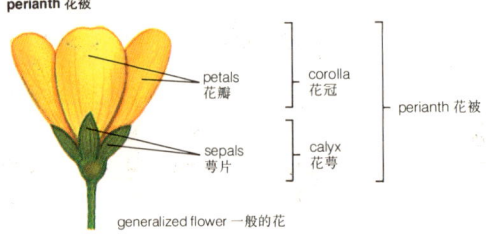
generalized flower 一般的花

perianth (n) the outer whorls (p. 98) of the flower (↑), i.e. the calyx (↓) and corolla (↓), which are the parts of the flower not concerned with the production of gametes (p. 61). The function of the perianth is to protect the reproductive (p. 59) organs (p. 88), and to attract pollinators (p. 74) to the flower.

calyx (n) the outer whorl (p. 98) of the perianth (↑), consisting of sepals (↓).

sepal (n) a usually green, leaf-like organ (p. 88). A whorl (p. 98) of sepals forms the calyx (↑) of a flower (↑). The sepals are the outer layer of the flower bud (p. 110) before it opens.

corolla (n) the inner whorl (p. 98) of the perianth (↑) of a flower (↑), composed of petals (↓).

petal (n) an often brightly coloured leaf-like organ (p. 88). A whorl (p. 98) of petals forms the corolla (↑) of a flower (↑). The function of coloured petals is often to attract pollinators (p. 74) to the flower.

tepal (n) an organ (p. 88) of a perianth (↑) in which there is no difference between the calyx (↑) and corolla (↑), e.g. in tulips.

花被(名) 花(↑)的外輪(第98頁)，即花萼(↓)和花冠(↓)，這是和產生配子(第61頁)無關的花部。花被的功能是保護生殖(第59頁)器官(第88頁)，並吸引傳粉者(第74頁)到花上。

花萼(名) 花被(↑)的外輪(第98頁)，由萼片(↓)組成。

萼片(名) 通常是一個綠色的葉狀器官(第88頁)。由一輪(第98頁)萼片組成花(↑)的花萼(↑)。萼片是花芽(第110頁)開放前的外層。

花冠(名) 花(↑)的花被(↑)的內輪(第98頁)，由花瓣(↓)組成。

花瓣(名) 常為色彩鮮艷的葉狀器官(第88頁)，由一輪(第98頁)花瓣組成花(↑)的花冠(↑)。彩色花瓣的功能是吸引傳粉者(第74頁)到花上。

花被片(名) 花萼(↑)及花冠(↑)間無差別的花被(↑)的器官(第88頁)。例如鬱金香。

FLOWER BIOLOGY/FLOWER TYPES 花的生物學/花的形式 · 71

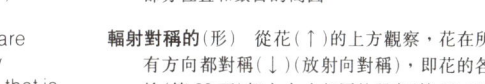

L.S. of flower 花的縱切面
stamen 雄蕊
filament 花絲
anther 花藥
stigma 柱頭
style 花柱
petal 花瓣
ovules 胚珠
sepal 萼片
ovary 子房
receptacle 花托

floral diagram 花圖式
a flower with 6 petals 具 6 花瓣、6 雄 6 stamens, 6 sepals 蕊、6 萼片的花
ovary 子房
stamens 雄蕊
petals 花瓣
sepals 萼片

floral diagram a diagram showing the position and number of all the parts of a flower (↑) in transverse section (p. 171).

actinomorphic (adj) of flowers (↑) which are symmetrical (↓) in all directions (radially symmetrical) when viewed from above, that is, with each whorl (p. 98) consisting of organs (p. 88) of the same size.

actinomorphic flower 輻射對稱的花 (radial symmetry) (放射對稱)

zygomorphic flower 兩側對稱的花 (bilateral symmetry) (兩邊對稱)

zygomorphic (adj) of flowers (↑) which are symmetrical (↓) in one direction only (bilaterally symmetrical), e.g. the flowers of orchids, often due to differences in sizes and shapes of petals (↑) and/or sepals (↑).

symmetrical (adj) of structures whose parts are arranged equally and regularly on either side of a line or plane (bilaterally symmetrical), e.g. in a zygomorphic (↑) flower, or around a central point (radially symmetrical), e.g. in an actinomorphic (↑) flower. **symmetry** (n).

asymmetrical (adj.) not symmetrical (↑).

apetalous (adj) of flowers (↑) without petals (↑). Apetalous flowers are often pollinated (p. 74) by wind.

gamopetalous (adj) of flowers (↑) in which the corolla (↑) is a tube.

polypetalous (adj) of flowers (↑) with petals (↑) that are not united.

sympetalous (adj) = gamopetalous (↑).

gamosepalous (adj) of flowers (↑) in which the sepals (↑) are united at their margins (p. 97).

polysepalous (adj) of flowers (↑) with sepals (↑) that are not united.

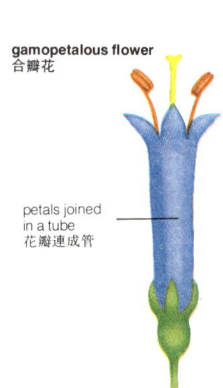

gamopetalous flower 合瓣花
petals joined in a tube 花瓣連成管

花圖式　以橫切面(第 171 頁)顯示花(↑)各組成部分位置和數目的簡圖。

輻射對稱的(形)　從花(↑)的上方觀察，花在所有方向都對稱(↓)(放射向對稱)，即花的各輪(第 98 頁)都由大小相同的器官(第 88 頁)組成。

兩側對稱的(形)　指只有一個方向對稱(↓)(即兩邊對稱)的花(↑)。例如蘭花。常因花瓣(↑)或萼片(↑)或花瓣和萼片兩者的大小和形狀差異而引起。

對稱的(形)　花的各組成部分在一條線或一個平面的每一邊都相等而規則地排列的結構(兩邊對稱)。例如兩側對稱(↑)的花，或者圍繞一個中心點相等而規則排列的結構(放射對稱)。例如輻射對稱(↑)的花。(形容詞為 symmetry)

不對稱的(形)　不是對稱的(↑)。

無瓣的(形)　指沒有花瓣(↑)的花(↑)。無花瓣的花常靠風傳粉(第 74 頁)。

花瓣相連的(形)　指花冠(↑)是管狀的花(↑)。

離瓣的(形)　指花瓣(↑)不連合的花(↑)。

合瓣的(形)　同花瓣相連的(↑)。

合萼的(形)　指萼片(↑)的邊緣(第 97 頁)互相連合的花(↑)。

離萼的(形)　指萼片(↑)不連合的花(↑)。

receptacle (n) the top of the stalk of a flower, bearing the perianth (p. 70), stamens (↓) and pistil (p. 75).
torus (n) the name sometimes given to the receptacle (↑) of a flower.
disk (n) a flat, circular receptacle (↑).
aestivation (n) the way in which parts of the flower, i.e. calyx (p. 70), corolla (p. 70), stamens (↓) and pistil (p. 75) are arranged with respect to each other.
hypogynous (adj) of flowers in which the stamens (↓), petals (p. 70) and sepals (p. 70) grow from below the gynoecium (p. 75) on the receptacle (↑). **hypogyny** (n).

花托(名) 花柄的頂端，生有花被(第70頁)、雄蕊(↓)及雌蕊(第75頁)。

英文有時亦稱花的花托(↑)為 **torus**。

花盤(名) 扁平的環狀花托(↑)。

花被捲疊式(名) 花的各部分，即花萼(第70頁)、花冠(第70頁)、雄蕊(↓)及雌蕊(第75頁)彼此相對的排列方式。

下位的(形) 指花的雄蕊(↓)、花瓣(第70頁)及萼片(第70頁)長在雌蕊群(第75頁)下的花托(↑)上。(名詞 hypogyny 意為下位花)

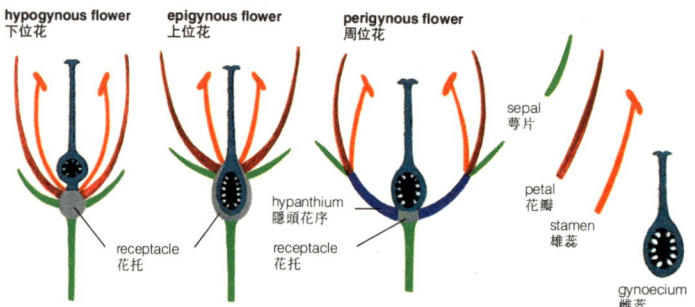

epigynous (adj) of flowers in which the ovary (p. 76) is within the receptacle (↑), and the other floral parts attached above it. **epigyny** (n).
hypanthium (n) a tube which results from growth at the edge of the receptacle (↑), in some plants. The perianth (p. 70) and the stamens (↓) grow from the top of the hypanthium.
perigynous (adj) of flowers with a hypanthium (↑). **perigyny** (n).
nectary (n) a gland (p. 112) which secretes (p. 112) nectar (↓). Many angiosperms (p. 130) have nectaries in their flowers; animals feed on the nectar and at the same time carry pollen (p. 74) from one flower to another. Some plants have extrafloral (↓) nectaries, providing food for ants which protect the plant against herbivores (p. 153).

上位的(形) 指花的子房(第76頁)長在花托(↑)之內，其他花部連在花托之上(名詞 epigyny 意為上位花)

隱頭花序(名) 在某些植物中，從花托(↑)邊緣長成的花管，花被(第70頁)及雄蕊(↓)長在其頂端。

周位的(形) 具有隱頭花序(↑)的花。(名詞 perigyny 意為周位花)

蜜腺(名) 分泌(第112頁)蜜汁(↓)的腺體(第112頁)。許多被子植物的花都生有蜜腺，動物採蜜的同時把花粉(第74頁)從一朵花傳到另一朵花。有些植物生有花外(↓)蜜腺，給那些可防止食草類動物(第153頁)啃嚼此植物的螞蟻提供食物。

FLOWER BIOLOGY/MALE PARTS 花的生物學／雄性部分 · 73

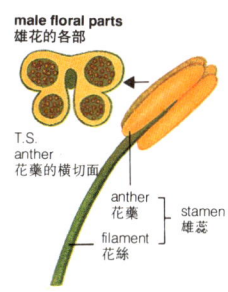

male floral parts
雄花的各部

T.S. anther
花藥的橫切面

anther 花藥
filament 花絲
stamen 雄蕊

nectar (n) a liquid containing sugars, amino acids (p. 56), and other organic (p. 11) compounds. Nectar is secreted (p. 112) by nectaries (↑).

extrafloral (adj) positioned away from the flower, e.g. an extrafloral nectary (↑).

anthesis (n) flower-opening.

androecium (n) the male part of a flower, consisting of stamens (↓). The function of the androecium is to produce the male gametes (p. 61) contained in pollen (p. 74).

stamen (n) the male reproductive (p. 59) organ (p. 88) of a flower, consisting of a filament (↓) bearing an anther (↓). The stamen is attached to the receptacle (↑) between the petals (p. 70) and the pistil (p. 75). The number, shape, and position of the stamens in a flower are important characters in the classification (p. 132) of angiosperms (p. 130). **staminal** (adj).

staminate (adj) of flowers which have stamens (↑) but no pistil (p. 75), i.e. male flowers.

staminode (n) a sterile (p. 62) stamen (↑), which does not produce pollen (p. 74).

anther (n) the part of a stamen (↑) in which pollen (p. 74) is produced. The anther is attached to the receptacle by the filament (↓). Anthers are hollow organs (p. 88) which dehisce (p. 84) along one side to release pollen.

filament (n) the stalk of a stamen (↑). The filament attaches the anther (↑) to the receptacle (↑) of the flower.

basifixed (adj) of an organ (p. 88) which is attached to another organ by its base. This is one way in which anthers (↑) are attached to filaments (↑).

monadelphous (adj) having all stamens (↑) joined together in a tube which surrounds the style (p. 76), e.g. lupin.

diadelphous (adj) having stamens (↑) in two groups, with the stamens in each group joined together by their filaments (↑), e.g. pea.

polyadelphous (adj) of stamens (↑) which are united by their filaments (↑) into three or more groups in a flower.

dimorphic (adj) having two shapes, e.g. two different kinds of stamen (↑) in one flower.

花蜜（名） 一種含有多種糖、氨基酸（第56頁）及其他有機（第11頁）化合物的液體。蜜腺（↑）分泌（第112頁）花蜜。

花外的（形） 位於花之外的，例如花外蜜腺（↑）。

開花期（名） 指花朵開放。

雄蕊群（名） 花的雄性部分，由雄蕊（↓）組成。雄蕊群的功能是產生花粉（第74頁）中所含的雄配子（第61頁）。

雄蕊（名） 花的雄性生殖（第59頁）器官（第88頁），由花絲（↓）構成，花絲支撐著花藥（↓）。雄蕊附於花托（↑）上，位於花瓣（第70頁）和雌蕊（第75頁）之間。花的雄蕊的數目、形狀及位置，在被子植物（第130頁）分類（第132頁）上是重要的特徵。（形容詞為staminal）

雄蕊的（形） 指只有雄蕊（↑）沒有雌蕊（第75頁）的花，例如雄花。

退化雄蕊（名） 一種不育（第62頁）的雄蕊（↑），它不能產生花粉（第74頁）。

花藥（名） 雄蕊（↑）產生花粉（第74頁）的部分，花藥以花絲（↓）附在花托上。花藥是個中空器官（第88頁），它沿一側開裂（第84頁），釋放出花粉。

花絲（名） 雄蕊（↑）的柄。花絲將花藥（↑）與花的花托（↑）相連。

基生的（形） 指一個器官（第88頁）以其基部附在另一個器官。這是花藥（↑）附於花絲（↑）的一種形式。

單體雄蕊的（形） 全部雄蕊（↑）圍繞花柱（第76頁）連成管狀。例如羽扇豆。

兩體雄蕊的（形） 雄蕊（↑）分成兩群，每群中的雄蕊（↑）以其花絲（↑）連在一起。例如豌豆。

多體雄蕊的（形） 雄蕊（↑）在花上以其花絲（↑）連結成三群或更多群。

兩形的（形） 具有兩種形狀的。例如在一朵花上有兩種不同的雄蕊（↑）。

pollen (*n*) small grains containing the male gametophyte (p. 65), in seed plants (p. 128). A pollen grain has a hard wall, or exine (↓). The gametophyte consists of just three cells in angiosperms (p. 130), and between four and forty cells in gymnosperms (p. 128); in both cases, only two cells are gametes (p. 61). The pollen grain protects the male gametophyte during its journey to the female reproductive (p. 59) organs (p. 88). In angiosperms, pollen is produced in the anthers (p. 73); in gymnosperms it is produced in male cones (p. 68).

exine (*n*) the hard outer coat of a pollen (↑) grain. The patterns of the surface of the exine are often used as characters in the classification (p. 132) of seed plants (p. 128).

sporopollenin (*n*) the material contained in the exine (↑) of pollen (↑) grains. Sporopollenin is resistant to decay and under the right conditions the exine may last for many thousands of years, although its contents die.

pollen sac the hollow space inside an anther (p. 73), where pollen (↑) grains are produced.

pollen tube a thread of cytoplasm (p. 18), covered by a membrane (p. 18), which grows from the pollen (↑) grain into the micropyle (p. 85) of the ovule (p. 78), through the tissues (p. 88) of the style (p. 76). In angiosperms (p. 130), the pollen tube carries two haploid (p. 50) nuclei (p. 19) into the ovule, one to fertilize (p. 62) the ovum (p. 61), and the other to fertilize the endosperm (p. 86) nucleus. The pollen tube will only grow if the pollen grain has landed on the stigma (p. 76).

pollination (*n*) the process in which pollen (↑) is carried from the anther (p. 73) to the surface of the stigma (p. 76), in angiosperms (p. 130), or from the male cone (p. 68) to the female cone in gymnosperms (p. 128). This can happen in several ways, e.g. by wind, water, insects, birds, bats, or even non-flying mammals, depending on the species (p. 134) of the plant. **pollinate** (*v*), **pollinator** (*n*).

cross-pollination (*n*) pollination (↑) of one plant by pollen (↑) from another individual (p. 135).

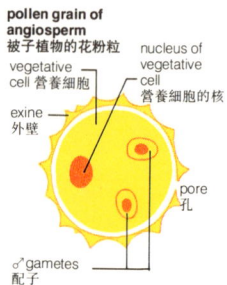

self-pollination (n) the pollination (↑) of an ovule (p. 78) by pollen (↑) from the same flower or the same individual (p. 135).

vector[1] (n) anything which carries pollen (↑) from one plant to another, e.g. insects, birds, wind, etc.

entomophily (n) pollination (↑) by insects. Flowers pollinated by insects are usually brightly coloured and scented. If they are pollinated by bees, they usually produce large amounts of pollen (↑) which the bees collect. If they are pollinated by butterflies or moths, they produce nectar (p. 73). **entomophilous** (adj).

honey guides coloured spots or lines on the petals (p. 70) of a flower, which may guide pollinating animals towards the sources of pollen (↑) and nectar (p. 73).

ornithophily (n) pollination (↑) by birds. Ornithophilous flowers are usually brightly coloured and secrete (p. 112) nectar (p. 73) on which the birds feed. **ornithophilous** (adj).

anemophily (n) pollination (↑) by wind. Plants which are pollinated by wind produce large amounts of pollen (↑). They are not usually scented, do not produce nectar (p. 73), and are sometimes apetalous (p. 71). **anemophilous** (adj).

pollinium (n) a large group of pollen (↑) grains, which are carried together during pollination (↑), as in the orchid family (p. 134) Orchidaceae. **pollinia** (pl.).

gynoecium (n) the female part of a flower, consisting of one or more pistils (↓).

pistil (n) the female reproductive (p. 59) organ (p. 88) of a flower, consisting of the ovary (p. 76), style (p. 76) and stigma (p. 76).

pistillate (adj) of flowers which have pistils (↑) but no stamens (p. 73), i.e. female flowers.

carpel (n) the female reproductive (p. 59) unit of a flower, consisting of the ovary (p. 76) with ovules (p. 78). The carpels are the sporophylls (p. 67) of angiosperms (p. 130), and are like highly modified leaves. Many angiosperms have several carpels, which are joined together at their margins (p. 97) to form the ovary.

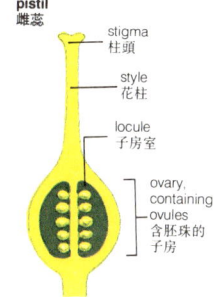

pistil 雌蕊
stigma 柱頭
style 花柱
locule 子房室
ovary, containing ovules 含胚珠的子房

style (*n*) a long upgrowth at the top of a carpel (p. 75), bearing the stigma (↓) at its tip. The style positions the stigma so that it is likely to receive pollen (p. 74). After pollen has reached the stigma, pollen tubes (p. 74) grow down through the style to the ovary (↓).

stigma (*n*) the tip of the style (↑) of a flower. Pollen (p. 74) must reach the stigma if successful pollination (p. 74) is to occur. **stigmatic** (*adj*).

homostylous (*adj*) of plant species (p. 134) having styles (↑) all of the same length on all individuals (p. 135). **homostyly** (*n*).

heterostylous (*adj*) of plant species (p. 134) having two or more different lengths of style (↑) on different individuals (p. 135). **heterostyly** (*n*).

ovary (*n*) the interior of the carpel (p. 75) of a flower, containing the ovules (p. 78). The ovary has a thick wall, which develops into the fruit after the ovules inside it have been fertilized (p. 62) by the male gametes (p. 61) carried in the pollen (p. 74) grains.

locule (*n*) the space inside an ovary (↑).

syncarpous (*adj*) of ovaries (↑) formed by two or more carpels (p. 75) joined together. This is an important character in the classification (p. 132) of angiosperms (p. 130).

apocarpous (*adj*) of ovaries (↑) in separate carpels (p. 75), not joined together at their margins (p. 97). This is characteristic of many primitive (p. 141) flowers.

花柱(名) 在心皮(第 75 頁)頂端生長的細長生長物，其尖端支撐著柱頭(↓)。花柱上柱頭的位置便於接受花粉(第 74 頁)。花粉到達柱頭之後，花粉管(第 74 頁)向下生長通過花柱到達子房(↓)。

柱頭(名) 花的花柱(↑)的尖端部位，花粉(第 74 頁)必須到達柱頭才能成功出現授粉(第 74 頁)。(形容詞為 stigmatic)

花柱同長的(形) 指植物種(第 134 頁)中所有個體(第 135 頁)的花柱(↑)長度都相同。(名詞為 homostyly)

花柱異長的(形) 指植物種(第 134 頁)的不同的個體(第 135 頁)有兩種或多種長度不同的花柱(↑)。(名詞為 heterostyly)

子房(名) 花的心皮(第 75 頁)內部，其中含有胚珠(第 78 頁)。子房有厚的壁，花粉(第 74 頁)粒帶來的雄配子(第 61 頁)使子房內的胚珠受精(第 62 頁)之後，子房發育成果實。

子房室(名) 子房(↑)內的空間。

合心皮的(形) 兩個或多個心皮(第 75 頁)連結一起形成的子房(↑)。在被子植物(第 130 頁)分類(第 132 頁)上這是一個重要的特徵。

離心皮的(形) 心皮(第 75 頁)分離，邊緣(第 97 頁)不連結的子房(↑)，這是許多原始(第 141 頁)花的特徵。

heterostyly
花柱異長

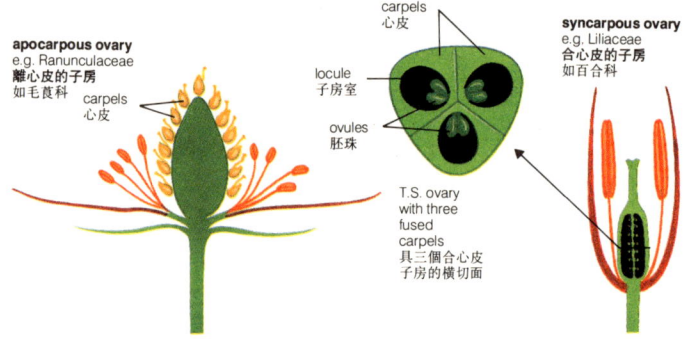

apocarpous ovary
e.g. Ranunculaceae
離心皮的子房
如毛茛科
carpels 心皮

carpels 心皮
locule 子房室
ovules 胚珠
T.S. ovary with three fused carpels
具三個合心皮子房的橫切面

syncarpous ovary
e.g. Liliaceae
合心皮的子房
如百合科

FLOWER BIOLOGY/OVARIES 花的生物學／子房

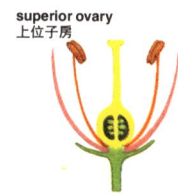

superior ovary
上位子房

perianth and stamens
attached to receptacle
below ovary
花被及雄蕊著生於
子房之下的花托

inferior ovary
下位子房

perianth and stamens
attached to receptacle
above ovary
花被及雄蕊著生
於子房之上的花托

inferior ovary one which is beneath the point of attachment of the calyx (p. 70), corolla (p. 70) and stamens (p. 73) of the flower.

superior ovary one which is attached to the receptacle (p. 72) above the stamens (p. 73) and the perianth (p. 70).

placentation (n) the arrangement of ovules (p. 78) in the ovary (↑). Because the ovules are attached to the margins (p. 97) of the carpels (p. 75), placentation depends on the way in which the carpels are joined together. Common types of placentation are axile (↓), parietal (↓) and free central (↓). This is an important character in the classification (p. 132) of angiosperms (p. 130).

下位子房　在花的花萼（第 70 頁）、花冠（第 70 頁）及雄蕊（第 73 頁）著生點之下的一種子房。

上位子房　在雄蕊（第 73 頁）及花被（第 70 頁）之上，著生於花托（第 72 頁）的一種子房。

胎座式（名）　胚珠（第 78 頁）在子房（↑）內的排列方式。由於胚珠著生於心皮（第 75 頁）的邊緣（第 97 頁），因此胎座式決定於心皮相連一起的方式。常見的胎座式類型有中軸（↓）胎座式、側膜（↓）胎座式及分離中央（↓）胎座式。在被子植物（第 130 頁）分類（第 132 頁）上這是一種重要特徵。

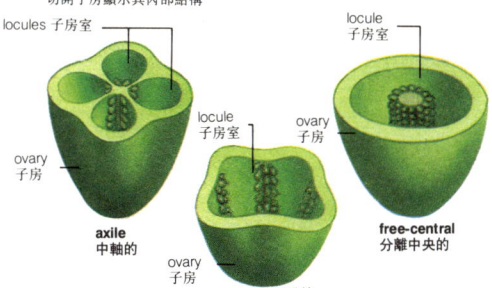

placentation types 胎座式的類型

ovaries cut through to show internal structure
切開子房顯示其內部結構

locules 子房室　　locule 子房室

ovary 子房　　ovary 子房

axile 中軸的　　**free-central** 分離中央的

ovary 子房

parietal 側膜的

placenta (n) the margin (p. 97) of a carpel (p. 75), where the ovules (p. 78) are attached.

axile (adj) of a kind of placentation (↑) in which the margins (p. 97) of the carpels (p. 75) grow inwards to the centre of the ovary (↑), forming several locules (↑), so that the ovules (p. 78) are arranged in a divided central column.

free central (adj) of a kind of placentation (↑) in which the ovules (p. 78) are borne on a central growth from the bottom of the ovary (↑).

parietal (adj) of a kind of placentation (↑) in which the ovules (p. 78) are arranged in rows down the wall of the ovary (↑). The rows mark the lines where the margins (p. 97) of the carpels (p. 75) are joined together.

胎座（名）　胚珠（第 78 頁）著生的心皮（第 75 頁）邊緣（第 97 頁）。

中軸的（形）　胎座式（↑）的一種類型，心皮（第 75 頁）的邊緣（第 97 頁）向內長到子房（↑）中央形成幾個子房室（↑）。胚珠（第 78 頁）排列在分隔子房室的中柱上。

分離中央（形）　胎座式（↑）的一種類型，胚珠（第 78 頁）長在子房（↑）底部生出的中央胎座上。

側膜的（形）　胎座式（↑）的一種類型，胚珠（第 78 頁）著生在子房（↑）側壁下行排列成行。這些行列顯示心皮（第 75 頁）邊緣（第 97 頁）連結在一起的接線。

78 · FLOWER BIOLOGY/OVULES 花的生物學／胚珠

ovule (n) a small body in the ovary (p. 76), which contains the female gamete (p. 61), in seed plants (p. 128). After the gamete is fertilized (p. 62) by one pollen (p. 74) nucleus, the ovule develops into a seed.

funicle (n) the stalk of the ovule (↑), attaching it to the wall of the ovary (p. 76). After the ovule is fertilized (p. 62), the funicle becomes the stalk of the seed.

chalaza (n) a tissue (p. 88) in the region where the funicle (↑) is attached to the ovule (↑).

integuments (n.pl.) the outermost layers of the ovule (↑), which become the coat of the seed after the ovule is fertilized (p. 62).

nucellus (n) tissue (p. 88) of the ovule (↑), between the integuments (↑) and the embryo sac (↓).

orthotropous (adj) of ovules (↑) which are borne erect on the funicle (↑), with the micropyle (p. 85) pointing away from the placenta (p. 77).

campylotropous (adj) of ovules (↑) with the funicle (↑) attached at one side, between the chalaza (↑) and the micropyle (p. 85).

anatropous (adj) of ovules (↑) with the funicle (↑) bent back on itself and the micropyle (p. 85) facing the placenta (p. 77).

embryo sac the female gametophyte (p. 65) in angiosperms (p. 130), consisting of 8 haploid (p. 50) cells including the ovum (p. 61), three antipodal cells (↓), two synergids (↓) and two endosperm (p. 86) nuclei (p. 19) which fuse (p. 61) before fertilization (p. 62). The embryo sac is contained inside the ovule (↑).

synergids (n.pl.) a group of two cells next to the ovum (p. 61) in the female gametophyte (p. 65) of an angiosperm (p. 130).

antipodal cells the three cells at the other end of the embryo sac (↑) from the ovum (p. 61) in the female gametophyte (p. 65) of an angiosperm (p. 130).

double fertilization fertilization (p. 62) of the ovum (p. 61) by one pollen (p. 74) nucleus (p. 19), and of the endosperm mother cell (↓) by another. This happens in all angiosperms (p. 130), but not in other plants.

胚珠（名） 種子植物（第 128 頁）子房（第 76 頁）內的一個小體，含有雌配子（第 61 頁）。配子被一個花粉（第 74 頁）核受精（第 62 頁）之後，胚珠發育成種子。

珠柄（名） 胚珠（↑）的柄，使子房（第 76 頁）和子房壁相連。胚珠受精（第 62 頁）之後，珠柄成為種子的柄。

合點（名） 珠柄（↑）和胚珠（↑）相連處的一種組織（第 88 頁）。

珠被（名、複） 胚珠（↑）的最外層，胚珠受精（第 62 頁）後，發育成種衣。

珠心（名） 位於珠被（↑）及胚囊（↓）之間的胚珠（↑）組織（第 88 頁）。

直生的（形） 指胚珠（↑）直立生長在珠柄（↑）上，珠孔（第 85 頁）背向胎座（第 77 頁）。

彎生的（形） 指胚珠（↑）和珠柄（↑）以一端著生在合點（↑）與珠孔（第 85 頁）之間。

倒生的（形） 指胚珠（↑）有珠柄（↑）向本身彎曲倒轉，而珠孔（第 85 頁）面對著胎座（第 77 頁）。

胚囊 被子植物（第 130 頁）的雌配子體（第 65 頁），具有八個單倍體（第 50 頁）細胞，包括卵（第 61 頁）、三個反足細胞（↓）、兩個助細胞（↓）以及兩個胚乳（第 86 頁）核（第 19 頁）。這兩個胚乳核在受精（第 62 頁）之前已融合（第 61 頁）。胚囊藏在胚珠（↑）內。

助細胞（名、複） 被子植物（第 130 頁）雌配子體（第 65 頁）內兩個或三個靠近卵細胞（第 61 頁）的細胞群。

反足細胞 被子植物（第 130 頁）雌配子體（第 65 頁）內在卵細胞（第 61 頁）胚囊（↑）另一端的三個細胞。

雙受精作用 卵（第 61 頁）被一個花粉（第 74 頁）核（第 19 頁）受精（第 62 頁），胚乳母細胞（↓）則被另一個花粉核受精。全部被子植物（第 130 頁）都有雙受精作用，其他類植物則無。

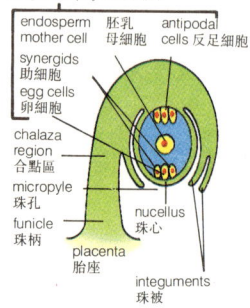

ovule structure 胚珠的構造

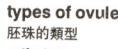

types of ovule 胚珠的類型
orthotropous 直生胚珠

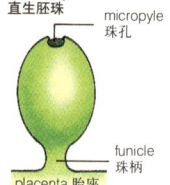

campylotropous 彎生胚珠

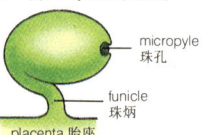

anatropous 倒生胚珠

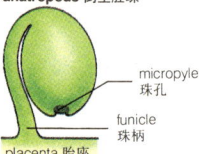

FLOWER BIOLOGY/FLOWER SEXES 花的生物學／花的性別

endosperm mother cell the cell formed in the embryo sac (↑) by the fusion (p. 61) of the two haploid (p. 50) endosperm (p. 86) nuclei (p. 19). The mother cell is diploid (p. 50), and it is fertilized (p. 62) by a pollen (p. 74) nucleus, in angiosperms (p. 130), to form the triploid (p. 50) endosperm.

hermaphrodite (adj) of flowers with male and female reproductive (p. 59) organs (p. 88).

perfect (adj) of flowers with male and female reproductive (p. 59) organs (p. 88), i.e. hermaphrodite (↑) flowers.

dioecious (adj) having male and female flowers on different individuals (p. 135) of the same plant species (p. 134). This is a way of avoiding self-fertilization (p. 62). **dioecy** (n).

monoecious (adj) having separate male and female flowers on the same individual (p. 135). **monoecy** (n).

gynodioecious (adj) having female and hermaphrodite (↑) flowers separately on different individuals (p. 135) of a plant species (p. 134). **gynodioecy** (n).

andromonoecious (adj) having male and hermaphrodite (↑) flowers on the same individual (p. 135). **andromonoecy** (n).

polygamous (adj) of plants which bear male, female and hermaphrodite (↑) flowers at the same time. **polygamy** (n).

homogamous (adj) having male and female floral parts functioning at the same time. **homogamy** (n).

dichogamous (adj) of flowers in which the male and female parts become functional at different times. This is a way of avoiding self-fertilization (p. 62). **dichogamy** (n).

protogynous (adj) of flowers in which the female parts become functional before the male parts. This is a way of avoiding self-fertilization (p. 62). **protogyny** (n).

protandrous (adj) of flowers whose anthers (p. 73) produce pollen (p. 74) before the ovules (p. 78) or stigma (p. 76) of the same flower are functional. This is a way of avoiding self-fertilization (p. 62). **protandry** (n).

胚乳母細胞　胚囊（↑）內兩個單倍體（第 50 頁）胚乳（第 86 頁）核（第 19 頁）融合（第 61 頁）形成的細胞。母細胞為二倍體（第 50 頁），在被子植物（第 130 頁），它由一個花粉（第 74 頁）核受精（第 62 頁）形成三倍體（第 50 頁）胚乳。

兩性花的（形）　指具有雄性和雌性的生殖（第 59 頁）器官（第 88 頁）的花。

完全花的（形）　指具有雄性和雌性的生殖（第 59 頁）器官（第 88 頁）的花，亦即兩性（↑）花。

雌雄異株的（形）　雄花與雌花著生於同一植物種（第 134 頁）的不同個體（第 135 頁）上，這是避免自花受精（第 62 頁）的一種方式。（名詞為 dioecy）

雌雄同株的（形）　雄花與雌花分開著生於同一個體（第 135 頁）上。（名詞為 monoecy）

雌花兩性花異株的（形）　雌花與兩性（↑）花分別著生在同一植物種（第 134 頁）的不同個體（第 135 頁）上。（名詞為 gynodioecy）

雄花兩性花同株的（形）　雄花和兩性（↑）花著生在同一個體（第 135 頁）上。（名詞為 andromonoecy）

雜性的（形）　指同時長有雄花、雌花及兩性（↑）花的植物。（名詞為 polygamy）

雌雄同熟的（形）　指同時具有雄性花部及雌性花部的功能。（名詞為 homogamy）

雌雄蕊異熟的（形）　指花的雄性部及雌性部在不同時間實現它的功能，這是避免自花受精（第 62 頁）的一種方式。（名詞為 dichogamy）

雌蕊先熟的（形）　花的雌蕊部分在雄蕊之前實現它的功能。這是避免自花受精（第 62 頁）的一種方式。（名詞為 protogyny）

雄蕊先熟的（形）　花的花藥（第 73 頁）在同一朵花的胚珠（第 78 頁）或柱頭（第 76 頁）產生花粉（第 74 頁）之前起作用。這是避免自花受精（第 62 頁）的一種方式。（名詞為 protandry）

inflorescence (n) a shoot bearing flowers and no leaves. An inflorescence can have one to many flowers.
peduncle (n) the stalk of a whole inflorescence (↑).
pedicel (n) the stalk of a single flower in an inflorescence (↑).
scape (n) a flower stalk growing from the level of the ground, as in herbaceous (p. 136) plants whose leaves form a rosette (p. 99).
raceme (n) a kind of inflorescence (↑) with a central axis (p. 92) bearing flowers along its length. **racemose** (adj).
panicle (n) a branched inflorescence (↑), consisting of a number of racemes (↑), as in many grasses (p. 130).
corymb (n) a raceme (↑) whose lower stalks are longer than the upper ones, so the inflorescence (↑) has a flat top. **corymbose** (adj).
cyme (n) a sympodial (p. 93) inflorescence (↑) growing by means of lateral (p. 92) branches, each with a flower at its apex (p. 90). **cymose** (adj).
umbel (n) an inflorescence (↑) in which all the pedicels (↑) are the same length and arise from the same point.
spike (n) an inflorescence (↑) with a long central axis (p. 92) and sessile (p. 100) flowers, as in many grasses (p. 130).
catkin (n) a spike (↑) of small, either male or female flowers, falling entire from the plant, e.g. in the willow family (p. 134), Salicaceae.

花序(名) 長有花而不長葉片的枝條。一個花序可以有一朵至多朵花。
總花梗(名) 花序(↑)的總柄。
花梗(名) 花序(↑)上一朵花的柄。
花葶(名) 從地面上長出的花柄。如在草本(第136頁)植物所見，其葉片形成一個蓮座葉叢(第99頁)。
總狀花序(名) 一種花序(↑)，花沿著中軸(第92頁)等長分佈。(形容詞為 racemose)
圓錐花序(名) 一種分枝式花序(↑)，由許多總狀花序(↑)組成，如在許多種禾草類(第130頁)所見。
繖房花序(名) 一種總狀花序(↑)，其下部花梗比上部的長，花序(↑)有一個平頂。(形容詞為 corymbose)
聚繖花序(名) 藉側(第92頁)枝生長而成的合軸(第93頁)花序(↑)，每一側枝尖端(第90頁)長一朵花。(形容詞為 cymose)
繖形花序(名) 全部花梗(↑)等長，且都從同一點伸出的花序(↑)。
穗狀花序(名) 有一枝長中軸(第92頁)及無柄(第100頁)花的一種花序(↑)。如在許多禾草類(第130頁)所見。
柔荑花序(名) 一種小穗狀花序(↑)，雄花或雌花整個地從植物體脫落。例如楊柳科(第134頁)的花序。

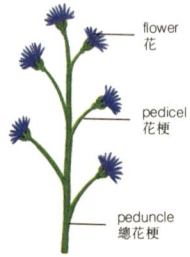

inflorescence 花序 — flower 花, pedicel 花梗, peduncle 總花梗

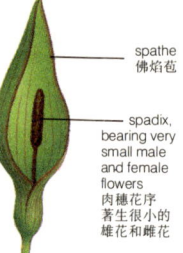

inflorescence in Araceae 天南星科的花序 — spathe 佛焰苞, spadix, bearing very small male and female flowers 肉穗花序 著生很小的雄花和雌花

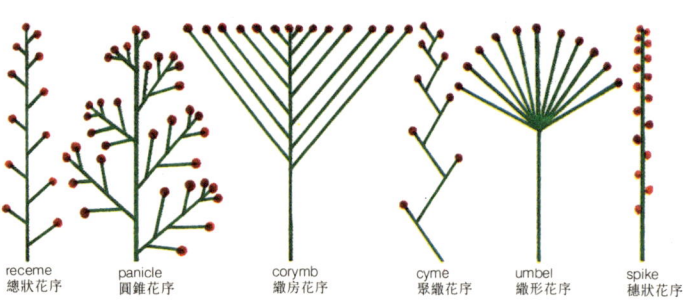

inflorescence types 花序的類型 — raceme 總狀花序, panicle 圓錐花序, corymb 繖房花序, cyme 聚繖花序, umbel 繖形花序, spike 穗狀花序

FLOWER BIOLOGY/INFLORESCENCES 花的生物學／花序 · 81

capitulum (*n*) an inflorescence (↑) like a head, consisting of many sessile (p. 100) flowers, e.g. in Compositae. **capitula** (*pl.*).
capitate (*adj*) like a head, e.g. when many flowers are clustered together in an inflorescence (↑).

頭狀花序(名) 似頭狀的一種花序(↑)，由許多無柄(第 100 頁)花組成。如菊科。(複數為 capitula)

頭狀的(形) 形狀似頭，例如許多花簇生在一個花序(↑)上。

composite, capitate inflorescence 菊花的頭狀花序
many flowers in a single capitulum
許多花在一個單獨的頭狀花序中

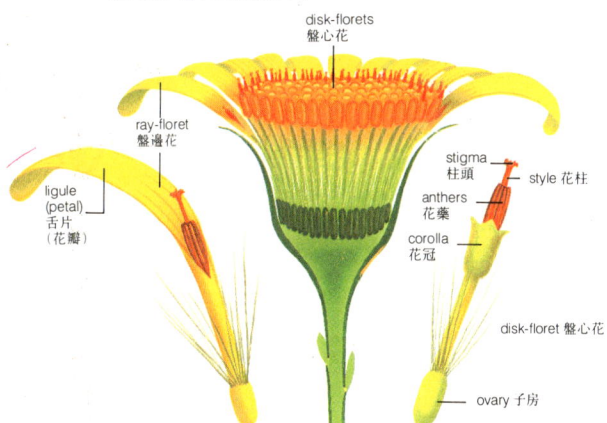

composite (*adj*) of inflorescences (↑) where many small flowers are grouped together in a head, looking like a large single flower, e.g. in the daisy family (p. 134), Compositae.
floret (*n*) a small flower, usually in a large or composite (↑) inflorescence (↑).
disk-floret a flower in the central part of a composite (↑) inflorescence (↑).
ray-floret a flower at the edge of a composite (↑) inflorescence (↑). Most ray-florets have a single petal (p. 70), called a ligule (↓).
ligule[1] (*n*) the corolla (p. 70) of a ray-floret (↑) in a composite (↑) inflorescence (↑).
involucre (*n*) a structure which protects or encloses another organ (p. 88), e.g. the bracts (p. 99) enclosing the developing inflorescence (↑) in Compositae, or the leaves joined together to protect the sex organs in leafy liverworts (p. 123).

菊科複合花序的(形) 許多小花集合在一個頭狀花序(↑)上，形似一個大型單花。如在菊科(第 134 頁)植物所見。

小花(名) 一種小的花，常見於一個大的花序(↑)或菊科(↑)植物花序。

盤心花 菊科(↑)花序(↑)中央部分的一朵花。

盤邊花 菊科(↑)花序(↑)邊緣的花。盤邊花大都有單獨的一片花瓣(第 70 頁)，稱為舌片(↓)。

舌片(名) 菊科(↑)花序(↑)盤邊花(↑)的花冠(第 70 頁)。

總苞(名) 保護或包圍另一器官(第 88 頁)的一種構造。例如菊科植物包著開放中花序(↑)的苞片(第 99 頁)，或者葉狀地衣類(第 123 頁)連結在一起的保護性器官的葉片。

spadix (n) an inflorescence (p. 80) consisting of a single fleshy (p. 99) axis (p. 92) with many small sessile (p. 100) flowers, as in the monocotyledon (p. 130) family (p. 134) Araceae. **spadices** (pl.).

spathe (n) the single large bract (p. 99) which encloses a young spadix (↑).

佛焰花序(名) 由一個單一的肉質(第99頁)軸(第92頁)和許多無柄(第100頁)花組成的一種花序(第80頁)。見於單子葉(第130頁)天南星科(第134頁)植物。(複數為 spadices)

佛焰苞(名) 包著一個幼嫩佛焰花序(↑)的獨生大型苞片(第99頁)。

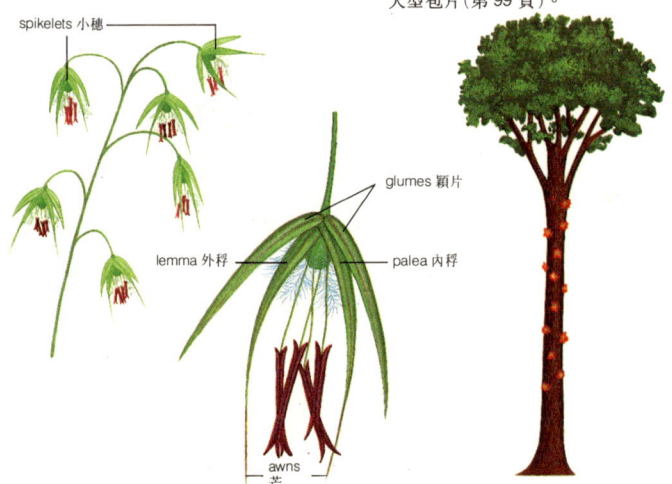

grass inflorescence 禾草類植物的花序

spikelet (n) a small branch of the spike (p. 80) in grasses (p. 130), bearing a few flowers.

lemma (n) one of the pair of inner bracts (p. 99) at the base of a grass (p. 130) spikelet (↑). **lemmas** (pl.)

glumes (n.pl.) the pair of outer bracts (p. 99) at the base of a grass (p. 130) spikelet (↑).

palea (n) one of the pair of inner bracts (p. 99) at the base of a grass (p. 130) spikelet (↑).

awn (n) a long, thin pointed tip, e.g. of the lemma (↑) of a grass (p. 130) flower.

cauliflorous (adj) of plants with flowers or inflorescences (p. 80) on the stem or trunk (p. 92). **cauliflory** (n).

solitary (adj) of organs (p. 88) which are borne singly, on their own, e.g. a flower in a one-flowered inflorescence (p. 80).

小穗(名) 禾草類(第130頁)植物穗狀花序(第80頁)的小枝條,長有一些花。

外稃(名) 禾草類(第130頁)植物小穗(↑)基部一對內苞片(第99頁)之一。(複數為lemmas)

穎片(名、複) 禾草類(第130頁)植物小穗(↑)基部外側的一對苞片(第99頁)。

內稃(名) 禾草類(第130頁)植物小穗(↑)基部內側的一對苞片(第99頁)之一。

芒(名) 一條纖細長尖的刺。例如禾草類(第130頁)植物花外稃(↑)的尖刺。

莖花的(形) 指花或花序(第80頁)長在莖上或樹幹(第92頁)上的植物。(名詞為 cauliflory)

單生的(形) 指單獨著生的植物的器官(第88頁)。例如一個開花的花序(第80頁)上的一朵花。

FRUITS AND SEEDS/FRUITS 果實及種子／果實

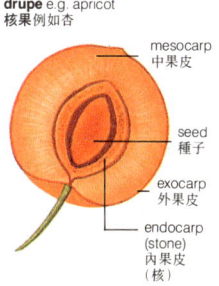

berry e.g. tomato
漿果例如番茄

- exocarp 外果皮
- seeds 種子
- mesocarp 中果皮
- endocarp 內果皮

drupe e.g. apricot
核果例如杏

- mesocarp 中果皮
- seed 種子
- exocarp 外果皮
- endocarp (stone) 內果皮 (核)

fruit (n) the organ (p. 88) of angiosperms (p. 130) containing the seeds. A true fruit is the product of the development of the ovary (p. 76) wall, and the seeds are fertilized (p. 62) ovules (p. 78). The function of the fruit is to protect the seeds as they develop and to help in their dispersal (p. 84). The term fruit or *fruiting body* can be used to describe any organ containing propagules (p. 59) in members of the plant kingdom (p. 134).

pome (n) = pseudocarp (↓).

pseudocarp (n) a false 'fruit' which has developed from the receptacle (p. 72), not from the ovary (p. 76), e.g. apple.

pericarp (n) the whole wall of the ripe (↓) ovary (p. 76) or fruit (↑), usually consisting of exocarp (↓), mesocarp (↓) and endocarp (↓).

exocarp (n) the outer layer of tissue (p. 88) of the fruit (↑). The exocarp is often hard or skin-like.

epicarp (n) = exocarp (↑).

mesocarp (n) the layer of tissue (p. 88) in a fruit (↑), between the exocarp (↑) and the endocarp (↑). The mesocarp is often fleshy (p. 99) or succulent (p. 99).

pulp (n) the succulent (p. 99) part of a fruit (↑).

endocarp (n) the innermost layer of tissue (p. 88) in a fruit (↑), surrounding the seeds.

ripe (adj) of fruits (↑) which are ready to release their seeds, or of a seed which has finished growing in the fruit. **ripen** (v).

monocarpic (adj) of plants which produce fruit (↑) only once in their life-cycle (p. 64), e.g. most annual (p. 117) plants. **monocarpy** (n).

parthenocarpic (adj) of plants whose fruits (↑) develop without seeds; this occurs naturally in some plants when fertilization (p. 62) has not taken place. **parthenocarpy** (n).

berry (n) a succulent (p. 99) or juicy fruit (↑), with many, usually small, seeds.

drupe (n) a fruit (↑) with seeds which are covered by a hard, stony endocarp (↑). Drupes usually have a fleshy (p. 99) mesocarp (↑).

kernel (n) the seed in a drupe (↑).

stone (n) the hard endocarp (↑) of a drupe (↑), containing the seed.

pyrene (n) a single stone (↑) in a small drupe (↑).

果實（名） 被子植物（第130頁）含種子的器官（第88頁）。真果是子房（第76頁）壁發育而成的果實，而種子是受精（第62頁）的胚珠（第78頁）。果實的功能是保護種子發育及幫助種子散播（第84頁）。果實或子實體這個術語可用以描述植物界（第134頁）中各種植物所具備繁殖體（第59頁）的任何器官。

梨果（名） 同假果（↓）。

假果（名） 指由花托（第72頁）而不是由子房（第76頁）發育成的假的"果實"。例如蘋果。

果皮（名） 成熟的（↓）子房（第76頁）或果實（↑）的整個子房壁，通常由外果皮（↓）、中果皮（↓）及內果皮（↓）組成。

外果皮（名） 果實（↑）組織（第88頁）的外層，外果皮通常堅硬或為硬殼狀。

外果皮（↑）的另一英文名稱為 epicarp。

中果皮（名） 果實（↑）內介於外果皮（↑）與內果皮（↑）之間的組織（第88頁）層。中果皮常為肉質（第99頁）或多汁（第99頁）。

果肉（名） 果實（↑）的多肉（第99頁）部分。

內果皮（名） 果實（↑）的最內層組織（第88頁），包住種子。

成熟的（形） 指將放出種子的果實（↑），或者已在果實內生長的種子。（動詞為 ripen）

結一次果的（形） 指在生活週期（第64頁）內僅生一次果實（↑）的植物。例如大多數一年生（第117頁）植物。（名詞為 monocarpy）

單性結實的（形） 指果實（↑）不經種子發育的植物，某些植物在沒有發生受精（第62頁）時自然地出現這種情形。（名詞為 parthenocarpy）

漿果（名） 富含肉質（第99頁）而多汁，通常有許多細小種子的果實（↑）。

核果（名） 種子被一層堅硬如石的內果皮（↑）包著的果實（↑）。核果常有肉質（第99頁）的中果皮（↑）。

果仁（名） 核果（↑）內的種子。

核（名） 核果（↑）堅硬的內果皮（↑），含種子。

小堅果（名） 小核果（↑）內一個單獨的核果（↑）。

legume (n) a dehiscent (↓) pod (↓) containing seeds (↓), developed (p. 109) from a single carpel (p. 75). The fruit of the family (p. 134) Leguminosae (beans, clovers, acacias, etc.).

pod (n) a long, thin, dry fruit, developed from a single carpel (p. 75), splitting down the side where the margins (p. 97) of the carpel (p. 75) were joined.

dehisce (v) to split open along a line. Many fruits, especially dry fruits, dehisce to release their seeds (↓). Anthers (p. 73) dehisce to release their pollen (p. 74). **dehiscent** (adj).

indehiscent (adj) = not dehiscent (↑).

capsule[1] (n) a dehiscent (↑) dry fruit, with at least one carpel (p. 75), often with many small seeds (↓), e.g. in the family (p. 134) Orchidaceae.

loculicidal (adj) of the dehiscence (↑) of a capsule (↑) with several carpels (p. 75), which split lengthways, exposing the seeds (↓) in each locule (p. 76).

nut (n) a dry, indehiscent (↑) fruit with a hard wall, containing one seed (↓).

follicle (n) a dry, dehiscent (↑) fruit formed from a single carpel (p. 75).

achene (n) a dry fruit with one seed (↓), the product of one carpel (p. 75).

samara (n) a small dry fruit or achene (↑), with wing-like outgrowths which assist in dispersal (↓) by wind.

schizocarp (n) a dry fruit developed from a syncarpous (p. 76) ovary (p. 76). A schizocarp splits into achene (↑)-like units when ripe (p. 83). Each unit is a single carpel (p. 75).

silicula (n) a long dry fruit, developed from an ovary (p. 76) consisting of two carpels (p. 75), as in the family (p. 134) Cruciferae.

siliqua (n) = silicula (↑).

pappus (n) a group of fine hairs on a small dry fruit, which helps in dispersal (↓) by wind, e.g. in the family (p. 134) Compositae.

dispersal (n) the movement of propagules (p. 59) away from the parent plant, e.g. by wind or birds. Dispersal is the way in which plants can spread. Fruits and seeds have many different adaptations (p. 141) for different kinds of dispersal.

莢果（名） 由單心皮（第75頁）發育（第109頁）而來的含種子（↓）的開裂（↓）豆莢果（↓）。豆科（第134頁）植物的果實（豆類、苜蓿、金合歡等）。

豆莢果（名） 一種細長的乾果，由單心皮（第75頁）發育而成，在心皮（第75頁）邊緣（第97頁）合縫處裂開。

開裂（動） 沿着一條線裂開。許多果實，特別是乾果，開裂並放出其種子（↓）。花藥（第73頁）則開裂放出其花粉（第74頁）。（形容詞為 dehiscent）

不裂的（形） 同不開裂（↑）。

蒴果（名） 一種開裂的（↑）乾果，至少有一個心皮（第75頁），通常有許多細小的種子（↓）。例如蘭科（第134頁）植物的種子。

室背開裂的（形） 指具有幾個心皮（第75頁）的蒴果（↑）的開裂方式，它縱向裂開，露出各個子房室（第76頁）的種子（↓）。

堅果（名） 果殼堅硬，內藏有一粒種子（↓），不裂開的（↑）乾果。

蓇葖果（名） 由一個單心皮（第75頁）組成的開裂的（↑）乾果。

瘦果（名） 由一個心皮（第75頁）形成含有一粒種子（↓）的乾果。

翅果（名） 一種具翅狀贅生物，能借風散播（↓）的乾果或瘦果（↑）。

離果（名） 從一個合心皮（第76頁）的子房（第76頁）發育而成的一種乾果，離果成熟（第83頁）後，裂成幾個瘦果（↑）狀單元。每一單元是一個單獨的心皮（第75頁）。

短角果（名） 由兩個心皮（第75頁）組成的子房（第76頁）發育而成的乾果。如十字花科（第134頁）植物。

長角果（名） 其構造與短角果（↑）相似。

冠毛（名） 細小乾果上長的一束細毛。冠毛有助於借風散播（↓）。例如菊科（第134頁）植物所見。

散播（名） 繁殖體（第59頁）離開親本植物的運動。如借助風力或鳥。散播是植物散佈的方式。果實和種子具有各種適應（第141頁）的散播方式。

legume e.g. pea
莢果例如豌豆
seeds 種子
pod 豆莢

capsule e.g. poppy
蒴果例如罌粟

achene e.g. strawberry
瘦果例如草莓
achene 瘦果
receptacle 花托
achenes 瘦果

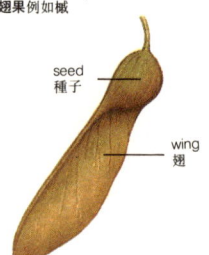

samara e.g. sycamore
翅果例如槭
seed 種子
wing 翅

seed (*n*) a fertilized (p. 62) ripe (p. 83) ovule (p. 78) of an angiosperm (p. 130) or gymnosperm (p. 128). The seed is the product of sexual (p. 59) reproduction (p. 59), and the means by which the progeny (p. 59) of a plant can be spread. The seed is covered by a testa (↓), and contains an embryo (↓) and endosperm (p. 86). The seeds of angiosperms are produced in fruits, and those of gymnosperms are produced in cones (p. 68) or strobili (p. 68).

testa (*n*) the hard, outer coat of a seed (↑), which protects the embryo (↓) and prevents water from entering the seed until it is ready to germinate (p. 87).

hilum (*n*) the place on the seed (↑) marking the point where the funicle (p. 78) was attached to the ovule (p. 78).

micropyle (*n*) a hollow tube or pore (p. 19) at the tip of an ovule (p. 78), through which the pollen tube (p. 74) enters. The micropyle can be seen in the testa (↑) of the mature seed (↑). Water enters the micropyle at the beginning of germination (p. 87).

raphe (*n*) a long ridge on the coat of a seed (↑) which has developed from an anatropous (p. 78) ovule (p. 78). The raphe marks the position where the funicle (p. 78) of the ovule used to be.

embryo (*n*) the young plant contained in the seed (↑). The embryo is the product of repeated mitotic (p. 45) divisions of the zygote (p. 61). It consists of cotyledons (p. 86), a plumule (p. 86), a hypocotyl (p. 86) and a radicle (p. 86).

embryonic (*adj*).

種子（名） 是被子植物（第 130 頁）或裸子植物（第 128 頁）受精後（第 62 頁）發育成熟（第 83 頁）的胚珠（第 78 頁）。種子是有性（第 59 頁）生殖（第 59 頁）的產物，也是植物延續後代（第 59 頁）的方式。種子外包一層種皮（↓），內含胚（↓）及胚乳（第 86 頁）。被子植物的種子生於果實內，而裸子植物的種子則生於球果（第 68 頁）或孢子葉球（第 68 頁）內。

外種皮（名） 種子（↑）中保護胚（↓）的堅韌包被，萌發（第 87 頁）前防止水分進入種子。

種臍（名） 種子上（↑）遺留下的一個瘢痕，是珠柄（第 78 頁）和胚珠（第 78 頁）相連處的標誌。

珠孔（名） 胚珠（第 78 頁）頂端的一個中空管或孔（第 19 頁），是花粉管（第 74 頁）進入的通道。在成熟種子（↑）的種皮（↑）上可見到珠孔。在發芽（第 87 頁）開始時，水分從珠孔進入。

種脊（名） 種子（↑）的種衣上的長脊，係由倒生（第 78 頁）胚珠（第 78 頁）發育而成。種脊標明胚珠的珠柄（第 78 頁）所在的位置。

胚（名） 藏在種子（↑）內的幼小植物。胚是合子（第 61 頁）反復有絲分裂（第 45 頁）的產物。胚由子葉（第 86 頁）、胚芽（第 86 頁）、下胚軸（第 86 頁）及胚根（第 86 頁）組成。（形容詞為 embryonic）

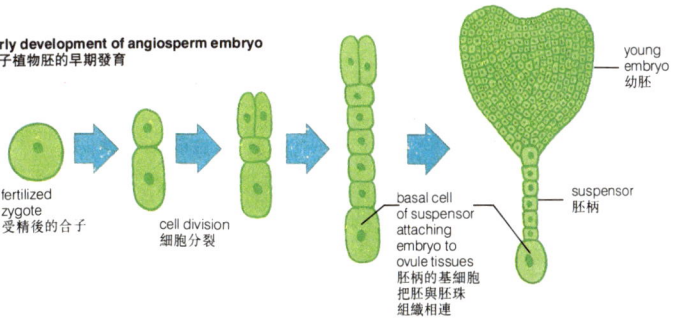

early development of angiosperm embryo
被子植物胚的早期發育

fertilized zygote 受精後的合子

cell division 細胞分裂

basal cell of suspensor attaching embryo to ovule tissues 胚柄的基細胞把胚與胚珠組織相連

young embryo 幼胚

suspensor 胚柄

86 · FRUITS AND SEEDS/SEEDS 果實及種子／種子

aril (n) an extra seed envelope, often coloured and fleshy (p. 99), found in some angiosperms (p. 130). The aril is produced from the tissues (p. 88) of the funicle (p. 78) or base of the ovule (p. 78). **arillate** (adj)

suspensor[1] (n) a group or chain of cells developed from the fertilized (p. 62) ovum (p. 61) in seed plants (p. 128), which attaches the embryo (p. 85) to the wall of the embryo sac (p. 78).

cotyledon (n) part of the embryo (p. 85) of a seed plant (p. 128). The cotyledon sometimes becomes the first photosynthetic (p. 32) organ (p. 88) of the young seedling (↓). Some plants, e.g. Leguminosae, have large cotyledons which store food. Angiosperms (p. 130) have either one or two cotyledons; gymnosperms (p. 128) have more. Angiosperms are classified (p. 132) into two classes (p. 134), the monocotyledons (p. 130) which have one cotyledon and the dicotyledons (p. 131) which have two.

seed leaf = a cotyledon (↑).

epicotyl (n) part of the embryo (p. 85) and seedling (↓) above the cotyledons (↑). The first true leaves are produced on the epicotyl after germination (↓).

plumule (n) the apical (p. 90) part of the epicotyl (↑) of an embryo (p. 85), from which the first true leaves of the seedling (↓) develop.

hypocotyl (n) the part of the embryo (p. 85) and seedling (↓) below the cotyledons (↑), bearing the radicle (↓) at its end.

radicle (n) the part of the embryo (p. 85) that develops into the root of the seedling (↓).

endosperm (n) triploid (p. 50) tissue (p. 88) in the seed, resulting from double fertilization (p. 78). The function of the endosperm is to store food for the seedling (↓).

albumen (n) the endosperm (↑) of a seed. **albuminous** (adj)

exalbuminous (adj) of seeds without albumen (↑).

aleurone layer the outer layer of thick-walled cells in the endosperm (↑) of the seeds of many grasses (p. 130). The aleurone layer is rich in protein (p. 56).

假種皮（名）　在某些被子植物（第130頁）所見的種子的附加被膜，常為有色肉質（第99頁），係由珠柄（第78頁）組織或胚珠（第78頁）基部組織（第88頁）生成的。（形容詞為arillate）

胚柄（名）　種子植物（第128頁）的卵（第61頁）受精（第62頁）後發育成的一團細胞或細胞鏈，胚柄使胚（第85頁）與胚囊（第78頁）壁相連。

子葉（名）　種子植物（第128頁）胚（第85頁）的一部分。子葉有時成為幼苗（↓）最初的光合作用（第32頁）器官（第88頁）。某些植物，如豆科植物具有貯存食物的大子葉。被子植物（第130頁）具有一片或兩片子葉。裸子植物（第128頁）子葉較多。被子植物分類（第132頁）為兩綱（第134頁）；即有一片子葉的單子葉植物（第130頁）及有兩片子葉的雙子葉植物（第131頁）。

種子葉　同子葉（↑）。

上胚軸（名）　胚（第85頁）及幼苗（↓）位於子葉（↑）以上的軸部分。第一片真葉是在發芽（↓）之後由上胚軸長出。

胚芽（名）　胚（第85頁）的上胚軸（↑）頂端（第90頁）部分，幼苗（↓）的第一片真葉就從此處長出。

下胚軸（名）　胚（第85頁）及幼苗（↓）位於子葉（↑）以下的部分，其末端長胚根（↓）。

胚根（名）　胚（第85頁）發育成幼苗（↓）的根部分。

內胚乳（名）　種子內的三倍體（第50頁）組織（第88頁），是雙受精（第78頁）產生的。內胚乳的功能是為幼苗（↓）貯存食物。

胚乳（名）　即種子的內胚乳（↑）。（形容詞為albuminous）

無胚乳的（形）　指缺胚乳（↑）的種子。

糊粉層　許多禾草類植物（第130頁）種子胚乳（↑）內厚壁細胞的外層。糊粉層富含蛋白質（第56頁）。

aril 假種皮

seed 種子
aril 假種皮 (fleshy outer covering)（肉質外覆層）

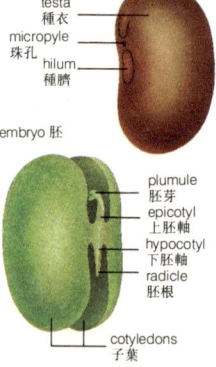

exalbuminous seed 無胚乳的種子
food stored in cotyledons e.g. bean
食物貯存在子葉如豆類

testa 種衣
micropyle 珠孔
hilum 種臍
embryo 胚
plumule 胚芽
epicotyl 上胚軸
hypocotyl 下胚軸
radicle 胚根
cotyledons 子葉

albuminous seed 有胚乳的種子
most food stored in endosperm e.g. maize
大部分食物貯存在胚乳內如玉米

endosperm 胚乳
aleurone layer 糊粉層
testa 種衣
embryo 胚
coleoptile 胚芽鞘
epicotyl 上胚軸
hypocotyl 下胚軸
radicle 胚根
cotyledon 子葉

germination (n) the first stage in the growth of a seed into a seedling (↓), or a spore (p. 66) into a young plant. In seed plants (p. 128) germination begins with the imbibition (↓) of water and ends with the production of the first true leaves.
germinate (v).
imbibition (n) the process in which water is taken up by a seed at the beginning of germination (↑).
epigeal (adj) of the kind of germination (↑) in which the cotyledons (↑) are borne above ground level, becoming the first photosynthetic (p. 32) organs (p. 88) of the seedling (↓).

萌發；發芽（名） 種子生長成幼苗（↓）的最早階段，或孢子（第66頁）長成幼小植物的最早階段。在種子植物（第128頁），發芽始於吸脹作用（↓），止於第一片真葉生出。（動詞為 germinate）

吸脹作用（名） 萌發（↑）開始時種子吸入水分的過程。

出土的（形） 萌發（↑）的一種類型，子葉（↑）長出地面，成為幼苗（↓）最早的光合作用（第32頁）器官（第88頁）。

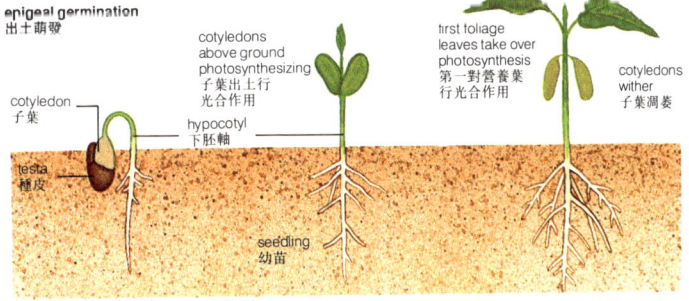

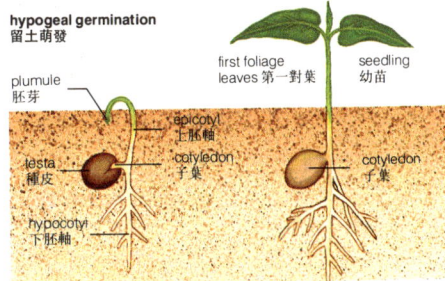

hypogeal (adj) of the kind of germination (↑) in which the cotyledons (↑) remain below ground. Their stored food is used up in the early growth of the epicotyl (↑) and the hypocotyl (↑).
seedling (n) a young plant growing from its seed. It is usually called a seedling until it loses its cotyledons (↑).

留土的（形） 萌發（↑）的一種類型，子葉（↑）留在土裏。其貯存的養份在上胚軸（↑）及下胚軸（↑）早期生長中被用盡。

幼苗（名） 由種子長成的幼小植物，在子葉（↑）消失前常稱為幼苗。

ANATOMY AND MORPHOLOGY/GENERAL 解剖學與形態學/概述

morphology (n) the study of the shape and arrangement of organs (↓) and tissues (↓).

anatomy (n) the study of the way in which tissues (↓) and organs (↓) are arranged in organisms.

anatomical (adj)

tissue (n) a group of cells, of similar shape and size, which all have the same function. Plant organs (↓) usually have several different kinds of tissue, e.g. leaves have epidermal (p. 90), mesophyll (p. 95) and vascular (p. 122) tissue.

organ (n) a group of cells or tissues (↑), forming part of an organism, with a special function, e.g. a leaf, a stamen (p. 73).

形態學(名) 研究器官(↓)及組織(↓)的形狀和排列的學科。

解剖學(名) 研究生物體內組織(↓)及器官(↓)排列方式的學科。(形容詞為 anatomical)

組織(名) 形狀及大小都相似且都具有相同功能的一群細胞。植物器官(↓)通常都具幾種不同類型的組織。例如葉有表皮(第90頁)組織、葉肉(第95頁)，及維管束(第122頁)組織。

器官(名) 組成生物體的一部分並具有某種特定功能的一群細胞或組織(↑)。例如葉、雄蕊(第73頁)。

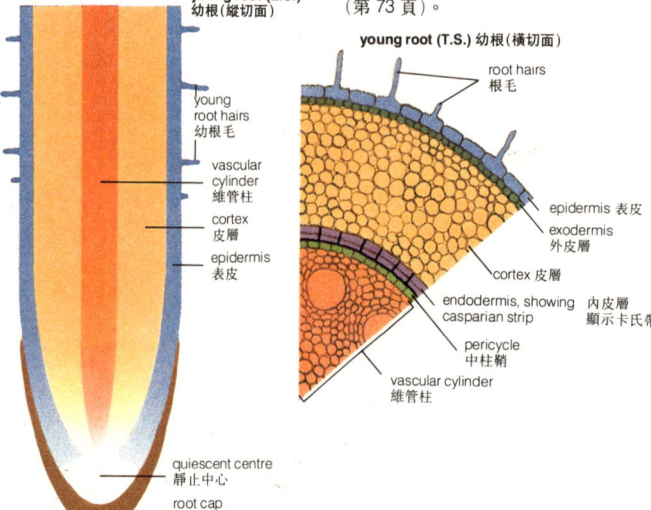

root (n) the organ (↑) of a plant, that grows down into the soil. Roots anchor the plant in the ground and take up water and nutrients (p. 111) from the soil. In some plants the roots also store food. They differ from stems in not having nodes (p. 90) and leaves.

radical (adj) of roots.

taproot (n) the main, primary root of a plant which shows apical dominance (p. 114).

根(名) 植物向下生長入土的器官(↑)。根使植物固定在土中，並從土壤吸取水和養料(第111頁)。某些植物根也貯藏食物。根和莖的差別在於：根無節(第90頁)亦無葉片。

根的(形) 指根。

直根(名) 植物主要的初生根，它顯現頂端優勢(第114頁)。

ANATOMY AND MORPHOLOGY/ROOTS 解剖學與形態學／根 · 89

stilt roots
支柱根

aerial roots
氣生根
e.g. orchid
例如蘭花

e.g. fig 例如：榕樹

aerial roots growing down from branches
氣生根從枝向下生長

adventitious root any root which grows from a tissue (↑) other than the pericycle (↓) or endodermis (↓) of an older root.

stilt root a root which grows out from near the bottom of the trunk (p. 92), in some trees, into the ground. Its function is support. Many palms (p. 130) have stilt roots. They are sometimes known as prop roots.

prop root = stilt root (↑).

aerial root a root growing from a part of a plant which is above the ground.

velamen (n) the tissue (↑) of dead cells underneath the epidermis (p. 90) of the aerial roots (↑) of some plants, e.g. the orchid family (p. 134), Orchidaceae. The velamen absorbs water.

root cap a layer of cells on the surface of the root tip, which protects the root as it grows and lubricates its passage through the soil.

quiescent centre a region of cells in the root tip, at the end of the stele (p. 105), where no cell division (p. 45) takes place.

piliferous layer the layer of cells, in the epidermis (p. 90) of the root, which bears root hairs (↓).

root hair a thread-like outgrowth of a cell in the epidermis (p. 90) of a root. Root hairs increase the surface area of a root and help in the uptake (p. 101) of water and nutrients (p. 111).

endodermis (n) the innermost layer of the cortex (↓) of a root, surrounding the vascular cylinder (p. 105) in all vascular (p. 122) plants. **endodermal** (adj).

casparian strip a band of suberin (p. 94) around the cells of the endodermis (↑) of the root, which stops the movement of substances from the cortex (↓) to the vascular cylinder (p. 105) other than through the cytoplasm (p. 18) of the endodermal cells.

pericycle (n) a layer of cells lying inside the endodermis (↑), on the surface of the vascular cylinder (p. 105) of a root.

cortex (n) the tissue (↑) between the vascular cylinder (p. 105) and the epidermis (p. 90) of a root or stem. The cortex usually has many layers of cells. **cortical** (adj).

不定根　一切不從老根的中柱鞘(↓)或內皮層(↓)的組織(↑)長出的根。

支柱根　某些植物接近幹(第 92 頁)基處長出插入土中的根。支柱根起支持作用。許多棕櫚(第 130 頁)都有支柱根。有時也稱為支持根。

支持根　同支柱根(↑)。

氣生根　在地面以上從植物某一部分長出的根。

根被(名)　某些植物氣生根(↑)表皮(第 90 頁)下面的死細胞組織(↑)，如蘭科植物(第 134 頁)所見。根被的功能是吸水。

根冠　根尖表面的一層細胞，它有保護根生長的作用並潤滑根穿過土壤的通道。

靜止中心　根尖中柱(第 105 頁)端的細胞區。此處不出現細胞分裂(第 45 頁)。

根毛層　在根的表皮(第 90 頁)內的細胞層。此層長有根毛(↓)。

根毛　在根的表皮(第 90 頁)內細胞的線狀生長物。根毛增加根的表面積，有助於吸收(第 101 頁)水分和養料(第 111 頁)。

內皮層(名)　根的皮層(↓)的最內層。在一切維管束(第 122 頁)植物都有內皮層包著維管柱(第 105 頁)。(形容詞為 endodermal)

卡氏帶　圍繞根的內皮層(↑)細胞的一條木栓質(第 94 頁)帶，它阻止物質從皮層(↓)移向維管柱(第 105 頁)，不讓通過內皮層細胞的細胞質(第 18 頁)。

中柱鞘(名)　內皮層(↑)內側的一層細胞，在根的維管柱(第 105 頁)的表面上。

皮層(名)　在根或莖的維管柱(第 105 頁)與表皮(第 90 頁)之間的組織(↑)。皮層常有多層細胞。(形容詞為 cortical)

ANATOMY AND MORPHOLOGY/TISSUES, SHOOTS 解剖學與形態學／組織、枝條

epidermis (n) the outer layer of cells of leaves, green stems, young roots, etc. **epidermal** (adj).

exodermis (n) layer of cortical (p. 89) cells, with suberin (p. 94) in their cell walls (p. 17). The exodermis is on the outer surface of the cortex, underneath the epidermis (↑). **exodermal** (adj).

parenchyma (n) the general name for tissues (p. 88) of cells with thin cell walls (p. 17), often with intercellular spaces (p. 95), e.g. the spongy mesophyll (p. 95) of leaves, or the cortex (p. 89) of stems and roots.

medulla (n) (1) the parenchyma (↑) or sclerenchyma (↓) inside the vascular cylinder (p. 105) of a stem or root. Its function is the storage of food. (2) the name given to the central part of the thallus (p. 122) of some algae (p. 119) and lichens (p. 147).

ray (n) a band of parenchyma (↑) and/or sclerenchyma (↓) cells, running from the cortex (p. 89) towards the centre of a stem.

shoot (n) the general name for any stem above the surface of the ground.

apex (n) the tip of a root or shoot. **apical** (adj).

表皮(名)　葉片、綠色莖、幼根等的最外層細胞。(形容詞為 epidermal)

外皮層(名)　皮層(第 89 頁)的細胞層，其細胞壁(第 17 頁)含有木栓質(第 94 頁)。外皮層在皮層的外面，近靠表皮(↑)之下。(形容詞為 exodermal)

薄壁組織(名)　具薄細胞壁(第 17 頁)的細胞組織(第 88 頁)的通稱，通常有細胞間隙(第 95 頁)。例如葉片的海綿狀葉肉(第 95 頁)或莖和根的皮層(第 89 頁)。

髓部(名)　(1)莖或根維管柱(第 105 頁)內側的薄壁組織(↑)或厚壁組織(↓)，其功能是貯藏養分；(2)給某些藻類(第 119 頁)或地衣(第 147 頁)葉狀體(第 122 頁)中央部分所取的名稱。

射線(名)　從皮層(第 89 頁)向莖中心延伸的薄壁組織(↑)帶或厚壁組織(↓)帶或兩者的組織帶。

枝條(名)　地面以上的一切莖的總稱。

尖端、頂端(名)　莖或枝條的尖端。(形容詞為

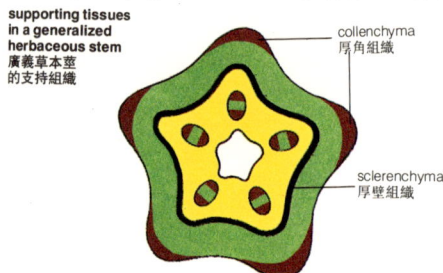

supporting tissues in a generalized herbaceous stem
廣義草本莖的支持組織

collenchyma 厚角組織
sclerenchyma 厚壁組織

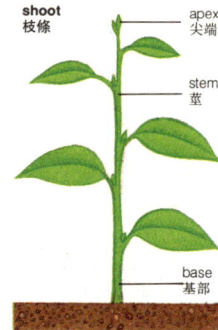

shoot 枝條
apex 尖端
stem 莖
base 基部

stem (n) the part of a plant with nodes (↓), buds (p. 110) and leaves. Most stems are above the ground, but some, e.g. rhizomes (p. 60), are underground.

node (n) the point on a stem from which a leaf grows. Nodes are spaced along stems, with internodes (↓) between them. **nodal** (adj).

internode (n) the space on a stem between two nodes (↑).

莖(名)　植物上具有節(↓)、芽(第 110 頁)及葉的部分。大多數莖長在地面以上，但有些莖如根狀莖(第 60 頁)長在地面下。

節(名)　葉片在莖上生長的部位。節沿着莖分隔，節與節之間稱為節間(↓)。(形容詞為 nodal)

節間(名)　兩個節(↑)之間的間隔。

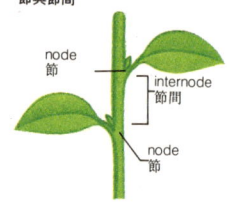

nodes and internodes 節與節間
node 節
internode 節間
node 節

ANATOMY AND MORPHOLOGY/TISSUES 解剖學與形態學／組織・91

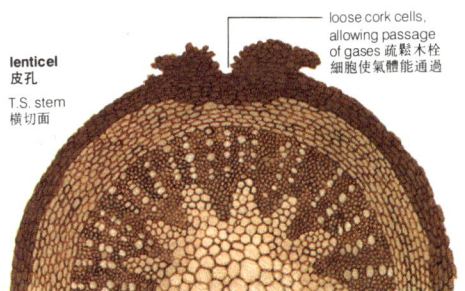

lenticel 皮孔
T.S. stem 橫切面

loose cork cells, allowing passage of gases 疏鬆木栓細胞使氣體能通過

lenticel (*n*) a pore (p. 19) on the surface of the stems of some plants, allowing gas exchange between the stem and the atmosphere.
lenticellate (*adj*).
culm (*n*) the stem of a grass.
sclerenchyma (*n*) a hard, lignified (p. 93) tissue (p. 88) consisting of fibres (↓) and sclereids (↓). It is found in the stems, roots, leaves or fruits of many plants, and its function is support.

皮孔(名)　某些植物莖表面上的孔(第19頁)，皮孔使莖與大氣之間能進行氣體交換。(形容詞為 lenticellate)
空心稈(名)　禾草類的莖。
厚壁組織(名)　由纖維(↓)及短石細胞(↓)組成的堅硬的木質化(第93頁)組織(第88頁)。許多植物的根、莖、葉或果實都有此種組織，它起支持作用。

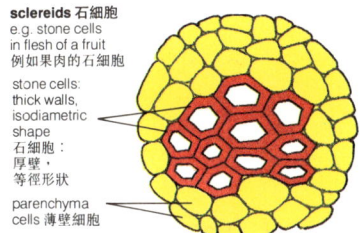

sclereids 石細胞
e.g. stone cells in flesh of a fruit
例如果肉的石細胞
stone cells: thick walls, isodiametric shape
石細胞：厚壁，等徑形狀
parenchyma cells 薄壁細胞

sclereid (*n*) kind of cell found in the sclerenchyma (↑) of some plants, with heavily lignified (p. 93) walls. Sclereids are usually found in groups.
stone cell an isodiametric (↓) sclereid (↑).
isodiametric (*adj*) of cells or structures with sides of equal length.
fibre (*n*) a long, thick-walled cell in the sclerenchyma (↑).
sclerophyllous (*adj*) of plants whose leaves contain sclerenchyma (↑). Such leaves are usually thick and leathery.

石細胞(名)　某些植物厚壁組織(↑)中所見的細胞類型，具有嚴重木質化(第93頁)的胞壁。石細胞常成群存在。
短石細胞　一種等徑的(↓)石細胞(↑)。
等徑的(形)　指各邊長度相等的細胞或構造。
纖維(名)　厚壁組織(↑)中的一種長形厚壁細胞。
硬葉的(形)　指葉片含厚壁組織(↑)的植物。通常是厚而革質的葉片。

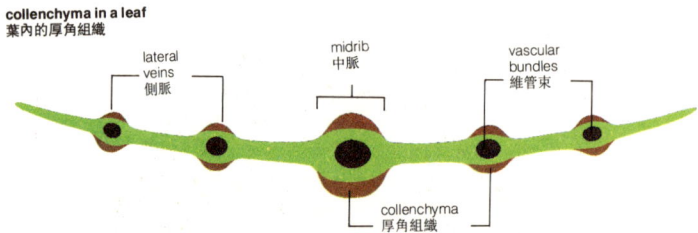

collenchyma in a leaf 葉內的厚角組織
lateral veins 側脈
midrib 中脈
vascular bundles 維管束
collenchyma 厚角組織

collenchyma (n) tissue (p. 88) of cells with thick cellulose (p. 17) cell walls (p. 17), especially at the angles of the cells, found in the stems of many herbs (p. 136), and in leaves. Its function is support.

aerenchyma (n) tissue (p. 88) with air-filled spaces between its cells, common in aquatic (p. 161) plants.

pith (n) tissue (p. 88), sometimes soft, in the centre of the stem of a non-woody (↓) dicotyledon (p. 131). Its function is to store food.

axis (n) general term for any stalk, stem, or long central organ (p. 88), from which other organs grow, e.g. the trunk (↓) of a tree.

trunk (n) the main woody stem of a tree, consisting of heartwood (↓), sapwood (↓) and bark (p. 94).

buttress (n) a large, flattened woody structure, growing outwards and downwards from near the base of the trunk (↑) of a tree. Buttresses are found especially in very large trees in tropical (p. 162) rain forests (p. 158).

bole (n) the trunk (↑) of a tree.

branch (n) a lateral (↓) shoot on a main axis (↑), e.g. the trunk (↑) of a tree.

lateral (adj) at, on, or of the side.

architecture (n) the way in which the branches of a tree are arranged on the trunk (↑), and the way in which the vegetative (p. 60) and reproductive (p. 59) axes (↑) are arranged on the branches.

crown (n) the top of a tree, including the branches and leaves.

厚角組織（名） 具有增厚纖維素（第17頁）細胞壁（第17頁），特別是在細胞的角隅處的細胞所組成的組織（第88頁），見於許多草本（第136頁）莖及葉，它起支持作用。

通氣組織（名） 細胞之間有充氣間隙的組織（第88頁），常見於水生（第161頁）植物。

髓部（名） 在非木質（↓）雙子葉植物（第131頁）莖中心部位，有時是鬆軟的組織（第88頁），其功能是貯存養分。

軸（名） 任何柄、莖或長形的中心器官（第88頁）的通稱。其他一些器官都是從軸長出，例如喬木的主幹（↓）就是軸。

主幹（名） 喬木的木質主莖，由心材（↓）、邊材（↓）和樹皮（第94頁）構成。

板狀根（名） 從樹幹（↑）基部往外及向下生長的大而平的木質構造，板狀根特別常見於熱帶雨（第162頁）林（第158頁）的巨樹。

樹幹（名） 喬木的主幹（↑）。

枝條 主軸（↑）（例如喬木主幹（↑））的側（↓）枝。

側面的（形） 在側面；位於側面；或者屬於側面的。

結構（名） 樹枝在樹幹（↑）上的排列方式，以及營養（第60頁）軸和生殖（第59頁）軸（↑）在枝條上的排列方式。

樹冠（名） 喬木的頂部，包括枝和葉。

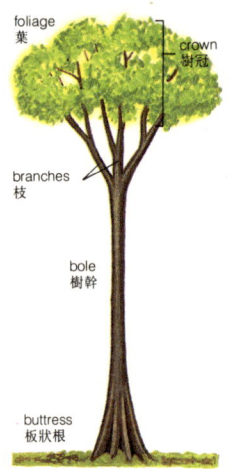

parts of a tree 喬木的各部分
foliage 葉
crown 樹冠
branches 枝
bole 樹幹
buttress 板狀根

ANATOMY AND MORPHOLOGY/GROWTH, WOOD 解剖學與形態學／生長、木材・93

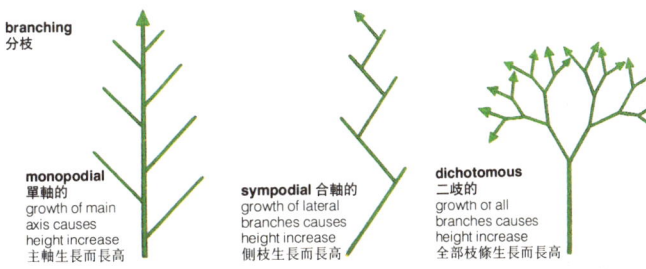

sympodial (adj) of a kind of growth in which the main axis (↑) of the plant is formed by the growth of lateral (↑) buds (p. 110) near the apex (p. 90) of the shoot, instead of by continuous growth from the apex.
monopodial (adj) of a kind of growth in which the main axis (↑) of the plant is formed by continuous growth of the same shoot apex (p. 90), with lateral (↑) branches arising from it.
dichotomous (adj) dividing equally into two, especially of branches.
orthotropic (adj) of axes (↑) growing upwards.
plagiotropic (adj) of branches which grow more or less parallel (p. 110) to the ground.
wood (n) hard tissue (p. 88) made of the remains of dead xylem (p. 106) cells in the stems of perennial (p. 117) plants. Wood contains lignin (↓), and its function is to support the plant and to conduct water. **woody** (adj).
lignin (n) a complex aromatic (p. 31) compound which is deposited in the cellulose (p. 17) cell walls (p. 17) of the xylem (p. 106) and sclerenchyma (p. 91) during the process of secondary thickening (p. 94). Wood is made mostly of lignin. **lignify** (v), **lignified** (adj).
sapwood (n) the outer part of the xylem (p. 106) of a stem, containing some living cells. The sapwood lies outside the heartwood (↓), and its main function is translocation (p. 101).
heartwood (n) the wood in the centre of a trunk (↑) or branch. Heartwood is usually dense and compact, and it helps to support the tree. It is often darker than the outer wood, and is unable to conduct sap (p. 102).

合軸的（形）指植物主軸（↑）由靠近枝條頂端（第 90 頁）的側（↑）芽（第 110 頁）生長而成，而不是頂端不斷生長所形成的一種生長形式。

單軸的（形）指植物主軸（↑）由同一枝條頂端（第 90 頁）不斷生長而成的一種生長形式，側（↑）枝由主軸頂端長出。

二歧的（形）指分為相等的兩枝，尤其是枝條。

直生的（形）指主軸（↑）向上生長。

斜生的（形）指枝的生長或多或少與地面平行（第 110 頁）。

木質（名）多年生（第 117 頁）植物莖內死亡木質部（第 106 頁）細胞遺留下來的堅硬組織（第 88 頁）。木質含有木質素（↓），功能是支持植物及傳導水分。（形容詞為 woody）

木質素（名）次生增厚（第 94 頁）過程中，木質部（第 106 頁）及厚壁組織（第 91 頁）纖維素（第 17 頁）細胞壁（第 17 頁）上沉積的一種複雜的芳香族（第 31 頁）化合物。木質主要是由木質素組成。（動詞為 lignify，形容詞為 lignified）

邊材（名）莖的木質部（第 106 頁）外側部分，含有一些生活細胞。邊材位於心材（↓）外側，其主要功能是運輸（第 101 頁）。

心材（名）主幹（↑）或枝條的中心部分的木材。心材通常是緻密而壓實的，心材有助於支持喬木，其色澤通常比外側木材深，不能輸導汁液（第 102 頁）。

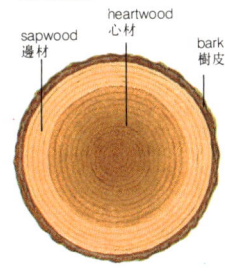

heartwood and sapwood 心材與邊材

primary thickening the thickening of a stem or root which occurs near the growing apex (p. 90).

secondary thickening the thickening of a stem or root because of the activity of the cambium (p. 108) to give xylem (p. 106) and phloem (p. 108). It provides the plant with extra support and vascular (p. 122) tissue.

pachycaul (*adj*) having fat stems due to massive primary thickening (↑), e.g. in many palms (p. 130). **pachycauly** (*n*).

leptocaul (*adj*) having thin stems, without heavy primary thickening (↑), as in most trees. **leptocauly** (*n*).

初生增厚　靠近生長頂端(第 90 頁)發生的莖或根的增厚。

次生增厚　莖或根由於形成層(第 108 頁)使木質部(第 106 頁)及韌皮部(第 108 頁)活動而致的增厚。它為植物提供額外的支持及維管(第 122 頁)組織。

粗短莖的(形)　指由於大量初生增厚(↑)而出現的肥厚莖。例如在許多種棕櫚(第 130 頁)所見。(名詞為 pachycauly)

細瘦莖的(形)　指沒有大量初生增厚(↑)的瘦弱莖，如在多數喬木所見。(名詞為 leptocauly)

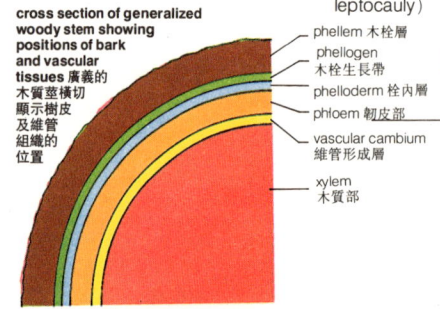

cross section of generalized woody stem showing positions of bark and vascular tissues 廣義的木質莖橫切顯示樹皮及維管組織的位置

phellem 木栓層
phellogen 木栓生長帶
phelloderm 栓內層 } periderm 周皮 } bark cambium 樹皮形成層
phloem 韌皮部
vascular cambium 維管形成層
xylem 木質部

bark (*n*) tissue (p. 88) usually made of dead cork (↓) cells and phloem (p. 108), on the outside of woody stems. Its function is to protect the stem.

cork (*n*) tissue (p. 88) of dead cells with suberin (↓) cell walls (p. 17), forming part of the bark (p. 94).

periderm (*n*) tissue (p. 88) forming part of the bark (p. 94), consisting of phelloderm (↓), phellogen (↓) and phellem (↓).

phellem (*n*) = cork (↑).

phelloderm (*n*) the inner layer of the periderm (↑), inside the cork (↑).

phellogen (*n*) the cambium (p. 108) which produces cork (↑) and phelloderm (↑). It is sometimes called the cork cambium.

suberin (*n*) a mixture of substances formed from fatty acids (p. 31), which is found in cork (↑) cell walls (p. 17). Suberin prevents water from passing through the cork.

樹皮(名)　木質莖外側通常由死木栓(↓)細胞及韌皮部(第 108 頁)組成的組織(第 88 頁)，有保護莖的功能。

木栓(名)　具木栓質(↓)細胞壁(第 17 頁)的死細胞組織(第 88 頁)，為樹皮(第 94 頁)部分。

周皮(名)　組成樹皮(第 94 頁)部分的組織(第 88 頁)，由栓內層(↓)、木栓形成層(↓)及木栓層(↓)組成。

木栓層(名)　同木栓(↑)。

栓內層(名)　周皮(↑)的內層，在木栓(↑)內側。

木栓生長帶(名)　形成木栓(↑)及栓內層(↑)的形成層(第 108 頁)，亦稱木栓形成層。

木栓質(名)　由脂肪酸(第 31 頁)形成的混合物質，見於木栓(↑)細胞的細胞壁(第 17 頁)。木栓質能阻止水分通過木栓層。

ANATOMY AND MORPHOLOGY/LEAF TISSUES 解剖學與形態學／葉的組織 · 95

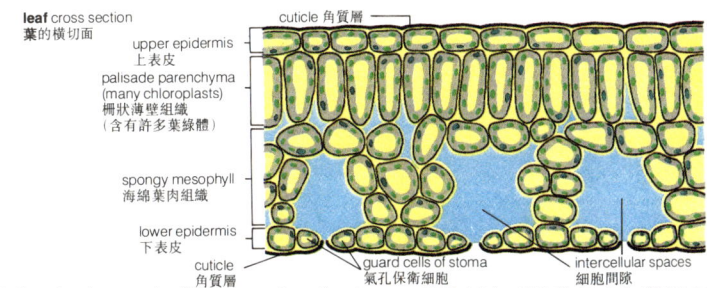

leaf (n) the plant organ (p. 88) whose function is photosynthesis (p. 32) and transpiration (p. 101). Leaves are produced from buds (p. 110) on the stem. Leaves have a wide range of form, but they nearly all share the inability to produce new growth from their apices (p. 90). In perennial (p. 117) plants, old leaves are replaced by new ones as the plant grows.

mesophyll (n) the tissue (p. 88) between the upper and lower epidermis (p. 90) of a leaf. In dicotyledons (p. 131) it is differentiated (p. 110) into palisade parenchyma (↓) and spongy mesophyll (↓), but in most monocotyledons (p. 130) it is undifferentiated.

spongy mesophyll a tissue (p. 88) in the leaves of many plants, e.g. dicotyledons (p. 131), lying underneath the palisade parenchyma (↓). It is composed of large cells with many intercellular spaces (↓) between them.

intercellular space the spaces between cells. In some tissues (p. 88), e.g. the spongy mesophyll (↑) of leaves, the intercellular spaces are large and filled with air.

palisade parenchyma layer of upright cells below the upper epidermis (p. 90) of leaves, especially in dicotyledons (p. 131). The cells are rich in chloroplasts (p. 32), and their main function is photosynthesis (p. 32).

cuticle (n) layer of cutin (p. 96) on the surface of leaves and green stems, which prevents evaporation (p. 12) and protects the plant against attack from herbivores (p. 153) and pathogens (p. 144).

葉片（名）　植物進行光合（第 32 頁）和蒸散（第 101 頁）功能的器官（第 88 頁）。葉由莖上的芽（第 110 頁）生出。葉片雖有多種不同的形狀，但都不能由葉端（第 90 頁）生出新的附生物。在多年生（第 117 頁）植物的生長過程中新葉代替老葉。

葉肉（名）　葉上、下表皮（第 90 頁）之間的組織（第 88 頁）。雙子葉植物（第 131 頁）的葉片分化（第 110 頁）為柵狀薄壁組織（↓）和海綿葉肉（↓），而多數單子葉植物（第 130 頁）的葉片不分化。

海綿葉肉　許多種植物如雙子葉植物（第 131 頁）葉片內位於柵狀薄壁組織（↓）下的一種組織（第 88 頁）。它由細胞之間具有許多細胞間隙（↓）的大細胞組成。

細胞間隙　細胞之間的空隙。在某些組織（第 88 頁），例如葉片的海綿葉肉（↑），細胞間隙寬大並充滿空氣。

柵狀薄壁組織　葉片的上表皮（第 90 頁）之下的直立細胞層，尤以雙子葉植物（第 131 頁）為然。其細胞富含葉綠體（第 32 頁），主要功能是進行光合作用（第 32 頁）。

角質層（名）　葉面及綠色莖表面的角質（第 96 頁）層，它能阻止水分蒸發（第 12 頁），並保護植物避免食草動物（第 153 頁）及病原體（第 144 頁）侵襲。

cutin (n) substance made of fatty acid (p. 31) products, which is impermeable (p. 102) to water.

wax (n) substance covering the surfaces of many plants. It is composed of a variety of organic (p. 11) compounds and polymers (p. 10), many of which are derived from lipids (p. 31). Wax coverings help to reduce the evaporation (p. 12) of water from leaves, and can also reflect light.

stoma (n) a pore (p. 19) in the surface of a leaf, usually consisting of two guard cells (↓) with a space between them. Stomata (pl.) can be opened and closed, controlling the evaporation (p. 12) of water from the leaf, and the entry of CO_2 into the leaf.

guard cells the pair of cells forming a stoma (↑).

foliage (n) the leaves of a plant collectively.

hypodermis (n) an extra layer of protective cells beneath the epidermis (p. 90) in the leaves, stems and roots of some plants.

角質(名) 為各種脂肪酸(第31頁)產物構成的物質，角質不透(第102頁)水。

蠟(名) 覆蓋在許多種植物表面的物質。蠟是由多種有機(第11頁)化合物與聚合體(第10頁)構成的。其中有許多是衍生於脂類(第31頁)。蠟覆層有助減少葉面的水分蒸發(第12頁)及反射陽光。

氣孔(名) 葉面上的孔(第19頁)，通常包含兩個保衛細胞(↓)，兩細胞之間有一定間隙。氣孔可開可閉，控制葉的水分蒸發(第12頁)，以及讓 CO_2 進入葉內。

保衛細胞 組成一個氣孔(↑)的一對細胞。

葉(名) 植物葉片的總稱。

下皮(名) 某些植物葉片、莖和根表皮(第90頁)下的附加保護細胞層。

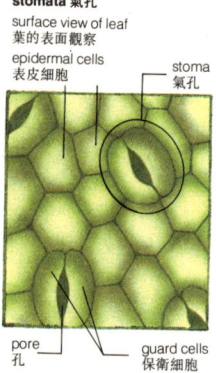

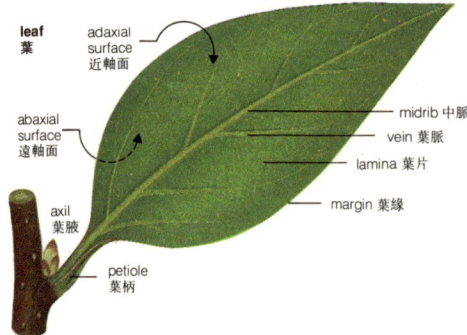

petiole (n) the stalk of a leaf, which joins it to a node (p. 90) on the stem.

axil (n) the point where the upper side of the petiole (↑) of a leaf joins the stem. **axillary** (adj).

lamina (n) the parts of the leaf on either side of the midrib (↓).

blade (n) either the whole leaf, excluding the petiole (↑), or all the parts of the leaf except the midrib (↓), i.e. the lamina (↑).

葉柄(名) 一片葉的柄，它將葉片與莖上的節(第90頁)相連。

葉腋(名) 葉柄(↑)上方和莖相連的點。(形容詞 axillary 意為腋生的)

葉面(名) 葉中脈(↓)任一側的葉部分。

葉片(名) 指葉柄(↑)以外的整片葉子，或者指除了中脈(↓)以外的葉的全部，即葉面(↑)。

midrib (*n*) the central vein (↓) of a leaf.
vein (*n*) one of the many lines which can be seen on the surface of a leaf, marking the position of the vascular bundle (p. 105).

中脈(名)　一片葉的中央葉脈(↓)。
葉脈(名)　葉面上能見的許多線條之一，葉脈標明維管束(第105頁)的位置。

venation 脈序
parallel 平行脈
reticulate 網狀脈

venation (*n*) the pattern of veins (↑) on the surface of a leaf. In most dicotyledons (p. 131) the venation is reticulate (↓), and in most monocotyledons (p. 130) it is parallel (p. 110).
reticulate (*adj*) of the veins (↑) of leaves, when their pattern is like a network.
margin (*n*) the edge, e.g. of a leaf.
adaxial (*adj*) on the top side of a leaf, that is, pointing towards the stem.
abaxial (*adj*) on the underside of a leaf, that is, pointing away from the stem.
simple (*adj*) of leaves which are not divided into leaflets (p. 98).
entire (*adj*) of leaves without lobes (↓).
digitate (*adj*) of leaves in which the lamina (↑) is divided like the fingers of a hand.
dissected (*adj*) of leaves which have many lobes (↓).
lobe (*n*) a flat, roundish piece of tissue (p. 88), as at the margin (↑) of a digitate (↑) or dissected (↑) leaf. Also, petals (p. 70) are sometimes called corolla (p. 70) lobes.

脈序(名)　葉片表面上葉脈(↑)的排列方式。大多數雙子葉植物(第131頁)的脈序是網狀(↓)的，而在大多數單子葉植物(第130頁)，則是平行(第110頁)的。
網狀的(形)　指葉的葉脈(↑)排列方式像一張網。
葉緣(名)　即葉片的邊緣。
近軸的(形)　在葉片的上表面，即面向莖的。
遠軸的(形)　在葉片的下表面，即遠離莖的。
單葉的(形)　指不分成小葉(第98頁)的葉片。
全緣的(形)　指葉片沒有裂片的(↓)。
掌狀的(形)　葉片(↑)分裂成手掌般的。
分裂的(形)　指葉片有許多裂片的(↓)。
裂片(名)　一種扁平略圓的組織(第88頁)，如在掌狀葉(↑)或分裂葉(↑)邊緣(↑)有時也將花瓣(第70頁)稱為花冠(第70頁)裂片。

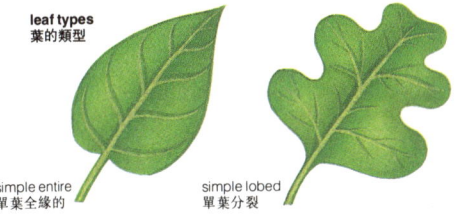

leaf types 葉的類型
simple entire 單葉全緣的
simple lobed 單葉分裂

compound[2] *(adj)* of leaves which are divided into several or many leaflets (↓) without axillary (p. 96) buds (p. 110).

leaflet *(n)* one of the small leaf-like divisions of a compound (↑) leaf.

rachis *(n)* the main axis (p. 92) of a pinnately (↓) compound (↑) leaf. The rachis is a continuation of the petiole (p. 96).

rhachis *(n)* = rachis (↑).

palmate *(adj)* of compound (↑) leaves with leaflets (↑) arising from a central point at the end of the petiole (p. 96), or, of simple (p. 97) leaves with lobes (p. 97), in which the main veins (p. 97) arise in the same way.

pinnate *(adj)* of a compound (↑) leaf with a central axis (p. 92) and leaflets (↑) - pinnae (↓) - on either side of it.

pinna *(n)* the leaflet (↑) of a pinnately (↑) compound (↑) leaf. **pinnae** *(pl.)*.

pinnule *(n)* the leaflet (↑) on the pinna (↑) of a bipinnate (↓) leaf, as in many members of the order (p. 134) Filicales, the ferns (p. 126).

bipinnate *(adj)* of pinnate (↑) leaves with their pinnae (↑) divided into pinnules (↑), as in many ferns (p. 126).

phyllotaxy *(n)* the arrangement of leaves on a stem, e.g. opposite (↓) phyllotaxy, alternate (↓) phyllotaxy, spiral (↓) phyllotaxy, whorled (↓) phyllotaxy. This is an important character in classification (p. 132).

spiral *(adj)* of nodes (p. 90) and leaves which are arranged on the stem in a helical (p. 51) way, or, of the thickening of the walls of xylem (p. 106) cells.

whorl *(n)* a group of three or more organs (p. 88) of the same kind, arising at the same level on a stem and arranged in a circle, e.g. the petals (p. 70) of a flower or the branches of a horsetail (p. 127). **whorled** *(adj)*.

alternate *(adj)* of an arrangement of leaves which arise singly on a stem, each one on the other side of the stem from the leaf below or above it.

opposite *(adj)* of an arrangement of two leaves which arise at the same node (p. 90), on either side of the stem.

複葉的(形) 指分裂為幾片或許多小葉(↓)，而沒有腋(第96頁)芽(第110頁)的葉片。

小葉(名) 指複(↑)葉的每片細小葉狀裂片。

葉軸(名) 羽狀(↓)複葉(↑)的主軸(第92頁)，葉軸是葉柄(第96頁)的延續。

葉軸(↑)英文亦拼為 **rhachis**。

掌狀的(形) 指一種複葉(↑)所具的小葉都從葉柄(第96頁)末端中心點生出；也指具裂片的(第97頁)單葉(第97頁)，其主脈(第97頁)以同樣的方式生出的。

羽狀的(形) 指具有中軸(第92頁)及其兩側有小葉(↑)和羽片(↓)的複葉(↑)。

羽片(名) 羽狀(↑)複葉(↑)的小葉(↑)。(複數為 pinnae)

小羽片(名) 二回羽狀複葉(↓)羽片(↑)的小葉(↑)。如在蕨綱的許多目(第134頁)、真蕨(第126頁)所見。

二回羽狀的(形) 指羽片(↑)分裂為小羽片(↑)的羽狀小葉(↑)。如在許多蕨類(第126頁)所見。

葉序(名) 葉在莖上的排列方式，如對生(↓)葉序、互生(↓)葉序、旋生(↓)葉序、輪生(↓)葉序。葉序在分類學上(第132頁)是重要的特徵。

旋生(形) 指節(第90頁)及葉片在莖上呈螺旋狀(第51頁)排列方式著生，或指木質部(第106頁)細胞壁以螺旋狀增厚。

輪生(名) 指在莖的相同高度上有三個或三個以上的同類器官(第88頁)，為一組排列成一圈長出。例如花的花瓣(第70頁)，或木賊屬植物(第127頁)的枝條。(形容詞為 whorled)

互生的(形) 指各片葉上方及下方莖的另一側只長單獨一片葉的排列方式。

對生的(形) 指兩片葉長在莖上同一節(第90頁)兩側的排列方式。

leaf types 葉的類型

compound palmate 掌狀複葉 — leaflets 小葉

compound pinnate 羽狀複葉 — leaflets 小葉, no terminal bud 無頂芽, no axillary buds 無腋芽

bipinnate 二回羽狀複葉 — rachis 葉軸, pinnae 羽片, pinnules 小羽片

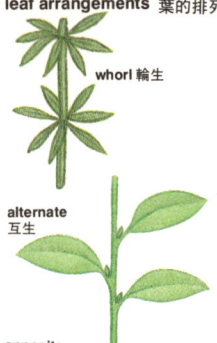

leaf arrangements 葉的排列

whorl 輪生

alternate 互生

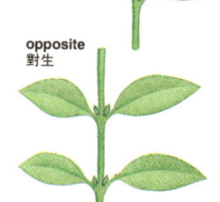

opposite 對生

ANATOMY AND MORPHOLOGY/LEAVES 解剖學與形態學／葉片・99

rosette 蓮座葉叢

variegated leaves 彩斑葉

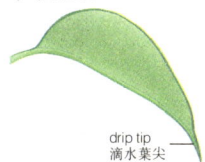
drip tip 滴水葉尖

needles 針狀葉

bract 苞片
e.g. on an inflorescence 例如在花序上
bracts 苞葉

rosette (n) a structure in which leaves are arranged in a tight spiral (↑) on a short stem with very short internodes (p. 90).

succulent (adj) of plants or parts of plants that are thick and fleshy (↓), owing to the presence of water-storing tissues (p. 88), e.g. in the cactus family (p. 134), Cactaceae.

fleshy (adj) of organs (p. 88) which are thick and often juicy.

coriaceous (adj) of leaves which are thick and stiff, like leather.

chartaceous (adj) of leaves which are like thick paper.

membranaceous (adj) of leaves which are very thin.

variegated (adj) of leaves with patches of different colour. **variegation** (n).

heterophyllous (adj) of plants which have two different kinds of leaves, e.g. when the leaves of the young plant are different from the leaves of the old plant, as in many species (p. 134) of the ivy family (p. 134), Araliaceae. **heterophylly** (n).

phyllode (n) a flat petiole (p. 96), which has the appearance of a leaf.

drip tip a long pointed tip to the leaf, which helps water to run off the leaf surface. Drip tips are common in wet tropical (p. 162) forests (p. 158).

needle (n) the long thin leaf of some conifers (p. 128)

bract (n) a small leaf, with a flower or part of an inflorescence (p. 80) growing from its axil (p96).

bracteole (n) a small bract (↑).

stipule (n) a small, leaf-like organ (p. 88), found in many plants, which grows at the base of a petiole (p. 96), sometimes protecting an axillary (p. 96) bud (p. 110).

exstipulate (adj) without stipules (↑).

蓮座葉叢（名） 葉在節間很短（第 90 頁）的短莖上緊密旋生（↑）排列的一種結構。

肉質的（形） 指植物或植物的部分由於存在貯水組織（第 88 頁）而呈肥厚且為肉質（↓）。例如仙人掌科植物（第 134 頁）。

多肉的（形） 指肥厚而多汁的器官（第 88 頁）。

革質的（形） 指葉厚而韌如皮革。

堅紙質的（形） 指葉如同厚紙。

膜質的（形） 指很薄的葉。

彩斑的（形） 形容葉有不同顏色的斑紋。（名詞為 variegation）

具異型葉的（形） 指具有兩種不同葉形的植物。例如幼嫩植物的葉與老植物的葉不同，常見於常春藤科（第 134 頁）、五加科植物的許多種（第 134 頁）。（名詞為 heterophylly）

葉狀柄（名） 呈葉片狀的一種扁平葉柄（第 96 頁）。

滴水葉尖 葉片長而尖的端部，便於水分從葉面滴落。滴水葉尖常見於潮濕的熱帶（第 162 頁）森林（第 158 頁）。

針狀葉（名） 某些針葉樹（第 128 頁）的細長葉。

苞葉（名） 有花或花序（第 80 頁）部從腋（第 96 頁）上長出的小葉片。

小苞片（名） 小的苞葉（↑）。

托葉（名） 許多種植物上從葉柄（第 96 頁）基部長出的細小葉片狀器官（第 88 頁），有時能保護腋生的（第 96 頁）芽（第 110 頁）。

無托葉的（形） 不具托葉（↑）的。

stipule 托葉
stipules 托葉

sheath (n) a protective covering, e.g. the lower part of the leaf of a grass, which is rolled around the stem.

coleoptile (n) a sheath (↑) which protects the young shoot tip in grasses (p. 130).

auricle (n) a small outgrowth at the side of the base of the leaf in some grasses (p. 130).

ligule[2] (n) a thin flap of tissue (p. 88) at the top of the leaf-sheath (↑) in many grasses (p. 130).

spine (n) a long, thin, sharp, stiff organ (p. 88) on the surface of the stems, and also sometimes of the leaves, of some plants, which is a defence against attack by herbivores (p. 153).

thorn (n) a sharp, pointed outgrowth on the surface of a plant, especially on the stem. Thorns can be simple outgrowths of the epidermis (p. 90), or can be modifications of other organs (p. 88), for example, the stipules (p. 99).

armed (adj) having thorns (↑) or spines (↑).

scale (n) a small outgrowth, e.g. on the petiole (p. 96) of a fern (p. 126) frond (p. 126).

trichome (n) a hair on the epidermis (p. 90) of a plant.

pubescent (adj) = hairy.

tomentose (adj) having a thick covering of very short hairs.

indumentum (n) the hairy part of a plant.

sessile (adj) of organs (p. 88) without a stalk, e.g. a leaf which has no petiole (p. 96) and is attached directly to the stem.

葉鞘(名) 一種保護性包覆層，如禾草類葉的下部包捲着莖的部分。

胚芽鞘(名) 禾草類植物(第 130 頁)中保護幼嫩枝條尖端的葉鞘(↑)。

葉耳(名) 在某些禾草類(第 130 頁)植物葉片基部側面的小突出物。

葉舌(名) 在許多禾草類(第 130 頁)植物葉鞘(↑)頂端的薄瓣組織(第 88 頁)。

針刺(名) 莖表面的一種細長尖銳而堅硬的器官(第 88 頁)，有時亦見於某些植物的葉片，針刺可防止食草動物(第 153 頁)侵襲。

棘刺(名) 植物表面上的尖銳針突，尤常見於莖上。棘刺可以是表皮(第 90 頁)的簡單突起，或者其他器官(第 88 頁)的變態物，如托葉(第 99 頁)。

具刺的(形) 指有棘刺(↑)或針刺(↑)的。

鱗片(名) 一種小突起物。如蕨類(第 126 頁)植物蕨葉(第 126 頁)葉柄(第 96 頁)的小突起物。

毛狀體(名) 植物表皮(第 90 頁)上的毛。

被短柔毛的(形) 即具有毛的。

被綿毛的(形) 被絨毛的，具有極短毛的厚被覆。

毛狀外被(名) 植物具毛部分。

無柄的(形) 指沒有柄的器官(第 88 頁)。例如沒有葉柄(第 96 頁)的葉片直接附於莖上。

leaf sheath 葉鞘
e.g. a grass 如禾草類
sheath of next leaf 下一葉的鞘
ligule 葉舌
leaf 葉
sheath covering stem 葉鞘包着莖

coleoptile 胚芽鞘
e.g. shoot apex of a grass seedling L.S. 如禾草幼苗枝頂縱切
shoot 枝條
coleoptile 胚芽鞘

spines 針刺
e.g. on stem of a palm 如在棕櫚莖上

thorns 棘刺
e.g. rose 如玫瑰

scale 鱗片
e.g. at base of a fern frond 如在蕨葉基部

trichomes 毛狀體
hairs on a leaf 葉上的毛

sessile 無柄的

leaves without petioles 無葉柄的葉

flowers without pedicels 無花梗的花

translocation (*n*) the movement of substances in the vascular system (p. 105) from one part of a plant to another. **translocate** (*v*).

轉移作用(名) 物質在維管系統(第 105 頁)內從植物體的某一部分移動到另一部分。(動詞為 translocate)

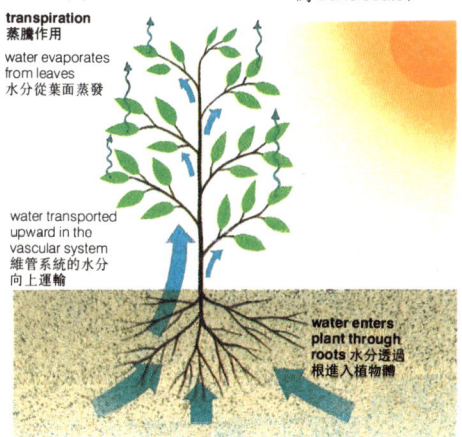

transpiration 蒸騰作用
water evaporates from leaves 水分從葉面蒸發
water transported upward in the vascular system 維管系統的水分向上運輸
water enters plant through roots 水分透過根進入植物體

transpiration (*n*) the process of upward movement of sap (p. 102) in the xylem (p. 106), due to the water potential (p. 103) gradient (p. 24) caused by the evaporation (p. 12) of water from the leaves. **transpire** (*v*).

potometer (*n*) an instrument for measuring the rate of transpiration (↑).

transpiration stream the upward-moving sap (p. 102) in the xylem (p. 106).

evapotranspiration (*n*) the process of losing water from vegetation (p. 150), caused by the evaporation (p. 12) of water at the surfaces of leaves.

uptake (*n*) the process of taking water and nutrients (p. 111) from the soil into the roots of a plant, or of taking substances into a cell or organelle (p. 16).

active transport the movement of substances across membranes (p. 18), using energy. This is necessary when a substance is being transported from the side of the membrane where it is less concentrated (p. 12) to the side where it is more concentrated.

蒸騰作用(名) 水分在葉面蒸發(第 12 頁)散失形成水勢(第 103 頁)梯度(第 24 頁),使木質部(第 106 頁)汁液(第 102 頁)向上移動的過程。(動詞為 transpire)

蒸騰計(名) 量度蒸騰(↑)速度的儀器。

蒸騰流動 汁液(第 102 頁)在木質部(第 106 頁)內向上移動。

蒸發蒸騰作用(名) 水分在葉面蒸發(第 12 頁)引致植被(第 150 頁)散失水分的過程。

吸收(名) 植物從土壤中接收水分及養分(第 111 頁)入根部的過程,或者接收物質入細胞或細胞器(第 16 頁)的過程。

主動運輸 細胞利用能量使物質橫過膜(第 18 頁)的運動。這對物質在濃度(第 12 頁)較低的膜一側向濃度較高的另一側運輸是必要的。

diffusion (n) the natural movement of molecules (p. 9) of a solute (p. 12) from regions of higher concentration (p. 12) to regions of lower concentration. **diffuse** (v).

apoplast (n) the non-living parts of a plant, i.e. the xylem (p. 106), the cellulose (p. 17) cell walls (p. 17) and the intercellular spaces (p. 95).

擴散作用（名） 溶質（第 12 頁）分子（第 9 頁）從高濃度（第 12 頁）區域向低濃度區域的自然運動。（動詞為 diffuse）

非原質體（名） 植物體的無生命部分，即木質部（第 106 頁）、纖維素（第 17 頁）、細胞壁（第 17 頁）及細胞間隙（第 95 頁）。

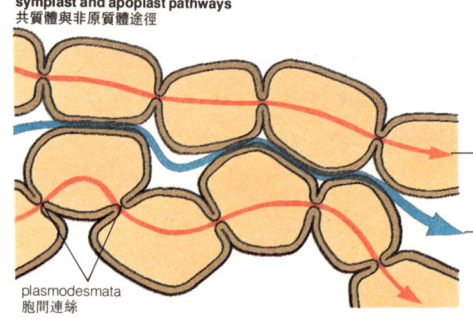

symplast and apoplast pathways
共質體與非原質體途徑

symplast substances translocated through living cells and plasmodesmata
共質體物質運輸通過活細胞及胞間聯絲

apoplast substances translocated through cell walls and intercellular spaces
非原質體物質運輸通過細胞壁和細胞間隙

plasmodesmata
胞間連絲

symplast (n) the living parts of a plant, i.e. the cells containing cytoplasm (p. 18).

root pressure the pressure which partly causes upward movement of xylem (p. 106) sap (↓), resulting from the active transport (p. 101) of solutes (p. 12) into the xylem which itself causes osmotic (↓) flow of water into the xylem.

sap (n) the water and nutrients (p. 111) contained and transported in the xylem (p. 106) or phloem (p. 108). Sap is also a general name for any liquid exuded (p. 112) from the plant when it is cut.

latex (n) a coloured, sticky or milky liquid, produced by specialized cells, which some plants exude (p. 112) when cut, e.g. rubber.

osmosis (n) the process by which water moves across semi-permeable (↓) membranes (p. 18) from a hypotonic (↓) solution (p. 12) to a hypertonic (↓) solution. **osmotic** (adj).

osmotic pressure the pressure needed to prevent osmotic (↑) movement of pure water across a semi-permeable (↓) membrane (p. 18) into a solution (p. 12). See also **osmotic potential** (↓).

共質體（名） 植物體的生命部分，即含細胞質（第 18 頁）的細胞部分。

根壓 由於溶質（第 12 頁）主動轉移（第 101 頁）入木質部，這本身又使水分滲透（↓）流入木質部。促使木質部（第 106 頁）汁液（↓）向上運動而造成的壓力。

汁液（名） 木質部（第 106 頁）或韌皮部（第 108 頁）所含有和運輸的水分及營養素（第 111 頁）。汁液亦是植物被切割時所滲出（第 112 頁）任何液體的通稱。

乳汁（名） 有色、黏稠或乳狀的液體，由特化細胞產生，有些植物被切割時有乳汁滲出（第 112 頁），如橡膠。

滲透作用（名） 水從低滲的（↓）溶液（第 12 頁）移動通過半透性（↓）膜（第 18 頁）向高滲的（↓）溶液的過程。（形容詞為 osmotic）

滲透壓 阻止純水滲透（↑）運動通過半透性（↓）膜（第 18 頁）進入某種溶液（第 12 頁）所需的壓力。參見滲透勢（↓）。

osmosis 滲透作用
semi-permeable membrane
半透性膜

pure water 純水 | concentrated solution 濃溶液

water diffuses across membrane until pressure from solution prevents further movement
水通過膜擴散直至溶液的濃度阻止再運動

permeable (adj) of membranes (p. 18) which allow the movement of substances from one side to the other. **permeability** (n).

impermeable (adj) of membranes (p. 18) which do not allow the movement of substances from one side to the other.

semipermeable (adj) of membranes (p. 18) which let some substances pass through them, but not others. The membranes in plant cells are mostly permeable (↑) to small molecules (p. 9), e.g. water (H_2O), monosaccharides (p. 28) and amino acids (p. 56), but not to large molecules, e.g. polypeptides (p. 56).

hypotonic (adj) less concentrated (p. 12).

hypertonic (adj) more concentrated (p. 12).

isotonic (adj) of two solutions (p. 12) with the same concentration (p. 12) of solute (p. 12) and the same osmotic pressure (↑).

water potential a measure, expressed in units of pressure, of the chemical difference between pure water and a solution (p. 12) in which water is the solvent (p. 12). Water diffuses (↑) from a solution of high water potential to a solution of low water potential if they are separated by a semi-permeable (↑) membrane (p. 18). In turgid (p. 104) plant cells, the water potential is equal to the sum of the osmotic pressure (↑), and matric potentials (↓).

osmotic potential a measure of the pressure that needs to be applied to a solution (p. 12) in order to make its water potential (↑) equal to that of pure water. When this pressure is applied to a solution, pure water will not pass into it across a semi-permeable (↑) membrane (p. 18). When there is no matric potential (↓) or pressure potential (↓), i.e. under experimental conditions, osmotic potential is equal to water potential (↑).

pressure potential a measure, in units of pressure, of the squeezing effect that the cell wall (p. 17) has on the contents of the cell, in turgid (p. 104) plant cells.

matric potential a measure, in units of pressure, of the attraction between water molecules (p. 9) and the organic (p. 11) compounds of the cell wall (p. 17) and cell.

渗透性的（形） 指容許物質從一側移動到另一側的膜（第18頁）。（名詞為 permeability）

不滲透性的（形） 指不容許物質從一側移動到另一側的膜（第18頁）。

半透性的（形） 指能讓某些物質通過而另一些物質不能通過的膜（第18頁）。小分子（第9頁）如水（H_2O）、單醣（第28頁）及氨基酸（第56頁）基本上可滲透（↑）植物的細胞膜，大分子如多肽（第56頁）則不能滲透。

低滲的（形） 指滲透濃度（第12頁）低的。

高滲的（形） 指滲透濃度（第12頁）高的。

等滲的（形） 指兩種溶液（第12頁）具有相同的溶質（第12頁）濃度（第12頁）及相等的滲透壓（↑）。

水勢 以水為溶劑（第12頁）的溶液（第12頁）和純水間化學勢差的量度，以壓力單位表示。如果用半透性（↑）膜（第18頁）將高水勢溶液和低水勢溶液分隔，則水從高水勢溶液擴散到低水勢溶液。緊脹（第104頁）的植物細胞內，水勢等於滲透壓（↑）與襯質勢（↓）的總和。

滲透勢 為使溶液的水勢（↑）與純水的水勢相等而需施予溶液（第12頁）的壓力的量度。當此壓力施予溶液時，純水不透過半透性（↑）膜（第18頁）進入此溶液。在實驗條件下，不存在襯質勢（↓）或壓力勢（↓）時，滲透勢等於水勢（↑）。

壓力勢 緊脹（第104頁）的植物細胞內，細胞壁（第17頁）對細胞內含物壓迫作用的量度，以壓力單位表示。

襯質勢 水分子（第9頁）與細胞壁（第17頁）及細胞的有機（第11頁）化合物間吸引力的量度，以壓力為單位表示。

104 · VASCULAR SYSTEMS/OSMOTIC PROCESSES 維管系統／滲透過程

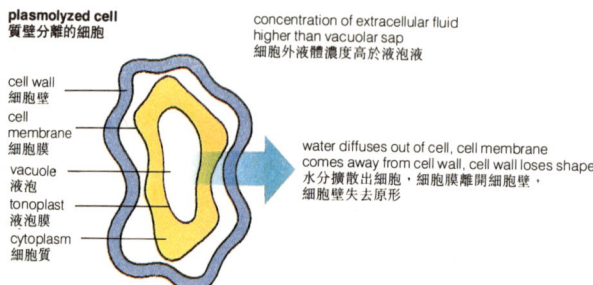

plasmolysis (n) the separation of the plasmalemma (p. 18) from the cell wall (p. 17) and the shrinking of the protoplast (p. 18) which occurs when water passes out of the cell due to the presence of a hypertonic (p. 103) solution (p. 12) outside it.

turgor (n) the tension on a cell wall (p. 17) due to the pressure of water inside the cell.

質壁分離(名) 由於細胞外存在高滲(第103頁)溶液(第12頁)使水分流出細胞外，原生質(第18頁)發生收縮，而使質膜(第18頁)從細胞壁(第17頁)分離。

膨壓(名) 細胞內的水壓對細胞壁(第17頁)的張力。

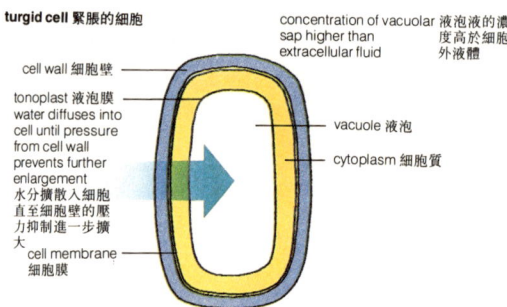

turgid (adj) of the state of a cell which can absorb no more water by osmosis (p. 102), because the cell wall (p. 17) prevents an increase in size.

wilt (v) to droop (of leaves and green stems of a plant) either because the rate of evaporation (p. 12) of water from the leaves is greater than the rate of uptake of water by the roots, or because of disease.

緊脹的(形) 由於細胞壁(第17頁)阻止細胞脹大，使細胞不能藉滲透作用(第102頁)而吸收更多水分的狀態。

萎蔫(動) 由於葉的水分蒸發(第12頁)速度大於根部吸水的速度，或者由於疾病，使植物的葉片及綠色莖下垂。

VASCULAR SYSTEMS/TISSUES 維管系統／組織 • 105

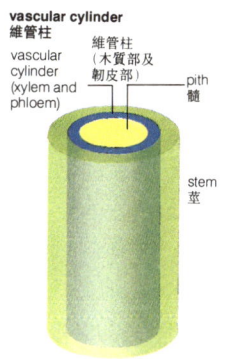

vascular cylinder 維管柱
vascular cylinder (xylem and phloem) 維管柱（木質部及韌皮部）
pith 髓
stem 莖

vascular system the tissues (p. 88) consisting of xylem (p. 106) and phloem (p. 108) cells, which translocate (p. 101) substances from one part of a plant to another. The development of vascular systems has made it possible for plants to evolve (p. 139) on land.

vascular cylinder a tube of vascular (p. 122) tissue (p. 88) consisting of xylem (p. 106) and phloem (p. 108) in a root or stem.

stele (n) the vascular cylinder (↑) of a stem or root, including the pith (p. 92) if this is present.

vascular bundle a thread of vascular (p. 122) tissue (p. 88) in the vein (p. 97) of a leaf, or in a stem.

維管系統　含木質部（第 106 頁）及韌皮部（第 108 頁）細胞的組織（第 88 頁）。它將物質從植物體某一個部分輸到（第 101 頁）其他部分，其發展使植物能在陸地上演變（第 139 頁）。

維管柱　根或莖內由木質部（第 106 頁）及韌皮部（第 108 頁）構成的維管（第 122 頁）組織（第 88 頁）柱狀體。

中柱（名）　莖或根的維管柱（↑），如有髓也包括髓部（第 92 頁）。

維管束　葉內的葉脈（第 97 頁）或莖內的一束維管（第 122 頁）組織（第 88 頁）

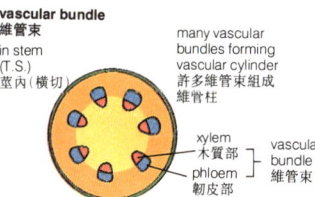

vascular bundle 維管束
in stem (T.S.) 莖內（橫切）
many vascular bundles forming vascular cylinder 許多維管束組成維管柱
xylem 木質部
phloem 韌皮部
vascular bundle 維管束

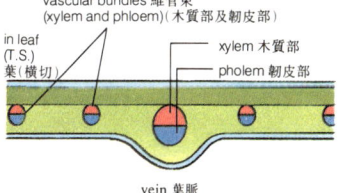

vascular bundles 維管束 (xylem and phloem)（木質部及韌皮部）
in leaf (T.S.) 葉（橫切）
xylem 木質部
pholem 韌皮部
vein 葉脈

葉跡　從莖分出枝並在節上（第 90 頁）進入葉的維管（第 122 頁）組織（第 88 頁）。

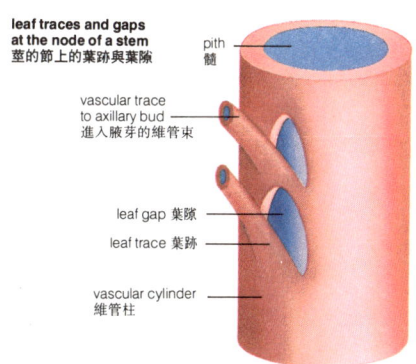

leaf traces and gaps at the node of a stem 莖的節上的葉跡與葉隙
pith 髓
vascular trace to axillary bud 進入腋芽的維管束
leaf gap 葉隙
leaf trace 葉跡
vascular cylinder 維管柱

leaf gap a gap in the vascular cylinder (↑) of a stem, just above a node (p. 90).

葉隙　莖的維管柱（↑）中的一個間隙，正好在一個節（第 90 頁）之上。

megaphyll (n) a leaf whose leaf trace (p. 105) makes a gap in the vascular system (p. 105) of the stem.

microphyll (n) a leaf whose leaf trace (p. 105) does not make a gap in the vascular system (p. 105) of the stem.

bundle sheath a layer of cells around the vascular bundle (p. 105) in a leaf.

xylem (n) tissue (p. 88) in the vascular system (p. 105) of a plant, consisting of tracheids (↓), vessels (↓), parenchyma (p. 90) and sclerenchyma (p. 91). The vessels, tracheids and sclerenchyma have lignified (p. 93) cell walls (p. 17). Most xylem cells are dead, and contain no cytoplasm (p. 18). The function of the xylem is to translocate (p. 101) water and nutrients (p. 111) from the roots to the stems and leaves.

巨型葉（名） 其葉跡（第105頁）使莖的維管系統（第105頁）出現一個間隙的葉。

小型葉（名） 其葉跡（第105頁）不使莖的維管系統（第105頁）出現間隙的葉。

維束鞘 葉內維管束（第105頁）周圍的一層細胞。

木質部（名） 植物維管系統（第105頁）的組織（第88頁），由管胞（↓）、導管（↓）、薄壁組織（第90頁）及厚壁組織（第91頁）構成。導管、管胞及厚壁組織的細胞壁（第17頁）已木化（第93頁）。木質部細胞多數已死亡，且不含細胞質（第18頁）。木質部的功能是從根部輸導（第101頁）水分及養分（第111頁）到莖和葉。

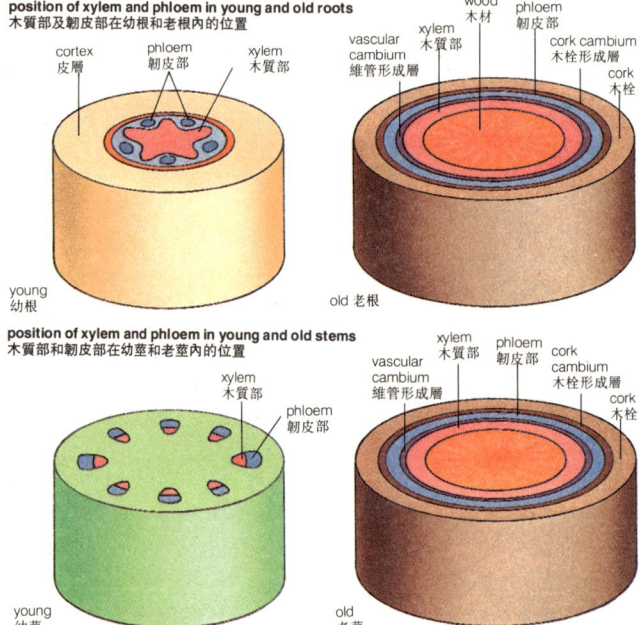

VASCULAR SYSTEMS/XYLEM 維管系統／木質部・107

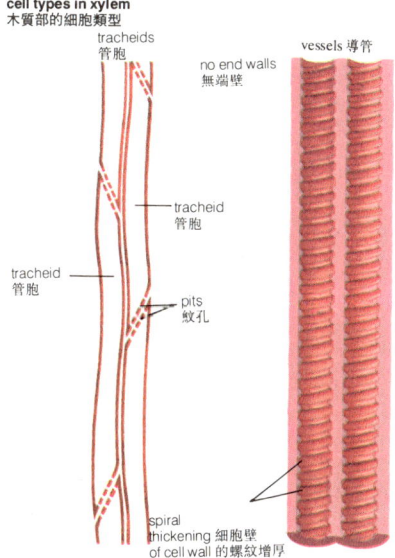

cell types in xylem
木質部的細胞類型
tracheids 管胞
no end walls 無端壁
vessels 導管
tracheid 管胞
tracheid 管胞
pits 紋孔
spiral thickening of cell wall 細胞壁的螺紋增厚

vessel (n) conducting tissue (p. 88) in the xylem (↑), consisting of vessel elements (↓), found mainly in angiosperms (p. 130).

vessel element a long, often thin, dead cell in the vessels (↑) of the xylem (↑). Vessel elements are arranged end to end, with large holes in the end walls, through which the xylem sap (p. 102) can pass. The cell walls (p. 17) of vessel elements are thickened with lignin (p. 93).

tracheid (n) a long, thin, dead cell in the xylem (↑), with closed ends and lignified (p. 93) walls. Xylem sap (p. 102) passes from one tracheid to the next through pits (↓) in the cell walls (p. 17).

scalariform (adj) of tracheids (↑) and vessels (↑) whose call walls (p. 17) have ladder-like ridges of thickening.

pit (n) an unthickened point in the cell wall (p. 17), which is usually next to a pit in the cell wall of the next cell. Pits enable the easy passage of substances from one cell to the next. They are common in the tracheids (↑) of the xylem (↑).

導管（名）木質部(↑)內的輸導組織（第88頁），由導管分子(↓)構成，主要存在於被子植物（第130頁）。

導管分子 木質部(↑)導管(↑)中的每一個長形、通常為薄壁的死細胞，導管分子端對端相接排列，端壁上有大孔洞，木質部的汁液（第102頁）可通過這些孔洞流通。導管分子的細胞壁（第17頁）以木質素（第93頁）增厚。

管胞（名）木質部(↑)中長形的薄壁死細胞，其兩端封閉，細胞壁已木化（第93頁）。木質部的汁液（第102頁）可經由細胞壁（第17頁）的紋孔(↓)從一個管胞流入另一個管胞。

梯紋的（形）指細胞壁（第17頁）具有梯狀增厚脊的管胞(↑)和導管(↑)。

紋孔（名）細胞壁（第17頁）上不增厚的點，它常鄰接於相鄰細胞細胞壁的紋孔。紋孔可成為物質從一個細胞流入鄰近細胞的通道。紋孔常見於木質部(↑)的管胞(↑)。

108 · VASCULAR SYSTEMS/PHLOEM 維管系統／韌皮部

phloem (n) one of the conducting tissues (p. 88) in the vascular system (p. 105). Phloem, unlike xylem (p. 106), is mainly a living tissue, whose cells contain cytoplasm (p. 18). It is made of sieve elements (↓) and companion cells (↓). The phloem can translocate (p. 101) substances in both directions, and its main function is to translocate the products of photosynthesis (p. 32) from leaves to other parts of the plant.

sieve-tube (n) the tissue (p. 88) in the phloem (↑) through which substances are translocated (p. 101). It consists of sieve elements (↓) with sieve plates (↓) between them.

sieve element a cell in the sieve-tube (↑) of the phloem (↑). Sieve elements are long, thin, living cells with thin cell walls (p. 17) and sieve plates (↓) at their ends. Translocation (p. 101) of substances takes place in the sieve elements.

sieve plate the wall at the end of a sieve element (↑), which has large pores (p. 19) through which substances can pass. Sieve plates contain callose (↓).

callose (n) a carbohydrate (p. 28) polymer (p. 10) which is found in sieve plates (↑), pollen tubes (p. 74) and on injured surfaces.

callus[2] (n) a tissue (p. 88) produced on injured plant surfaces. Callus tissue contains callose (↑).

companion cell a small, living cell next to the sieve elements (↑) in the phloem (↑).

韌皮部（名） 維管系統（第 105 頁）內的輸導組織（第 88 頁）。韌皮部與木質部（第 106 頁）不同，它主要是活細胞，細胞內含有細胞質（第 18 頁）。韌皮部由篩管分子（↓）及伴細胞（↓）組成。韌皮部可向兩個方向輸導（第 101 頁）物質，其主要功能是從葉輸導光合作用（第 32 頁）產物到植物的其他部分。

篩管（名） 韌皮部（↑）的組織（第 88 頁），物質可通過它輸導（第 101 頁）。篩管由篩管分子構成，篩管分子（↓）之間有篩板（↓）。

篩管分子 韌皮部（↑）篩管（↑）中的細胞。篩管分子是細長形具薄細胞壁（第 17 頁）的活細胞，其末端有篩板（↓）。在篩管分子內進行物質轉移（第 101 頁）。

篩板 篩管分子（↑）的端壁，壁上有大孔（第 19 頁），可讓物質通過。篩板含有胼胝質（↓）。

胼胝質（名） 一種碳水化合物（第 28 頁）聚合體（第 10 頁），它形成於篩板（↑）、花粉管（第 74 頁）及創傷面。

癒傷組織（名） 在植物受傷表面形成的組織（第 88 頁），癒傷組織含有胼胝質（↑）

伴細胞 韌皮部（↑）中與篩管分子（↑）毗隣的細小活細胞。

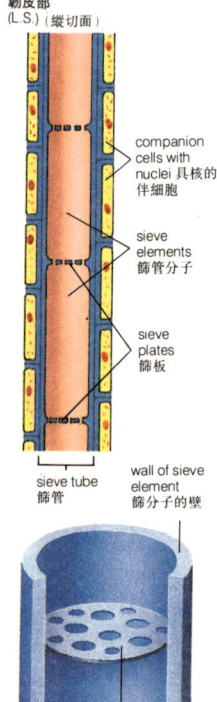

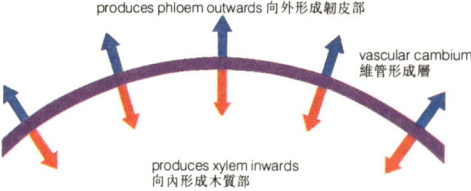

vascular cambium 維管形成層

cambium (n) a meristem (↓) in the vascular system (p. 105). In perennial (p. 117) plants, it produces a new layer of vascular (p. 122) tissue (p. 88) each year, producing xylem (p. 106) on the inside and phloem (↑) on the outside.

形成層（名） 維管系統（第 105 頁）中的分生組織（↓）。多年生（第 117 頁）植物的形成層，每年長一層新的維管（第 122 頁）組織（第 88 頁），在內側形成生木質部（第 106 頁），外側形成韌皮部（↑）。

growth (*n*) all the processes in an organism which result in an increase in size. The ability to grow is one of the important characteristics of living organisms.

vegetative growth growth of the tissues (p. 88) and organs (p. 88) not involved in sexual (p. 59) reproduction (p. 59). Vegetative growth occurs by mitosis (p. 45) and the lengthening and enlargement of cells.

development (*n*) the changes of structure and appearance of new organs (p. 88) and tissues (p. 88) as an organism grows. **develop** (*v*).

ontogeny (*n*) the process of development of an individual (p. 135) from zygote (p. 61) to adult.

morphogenesis (*n*) the development of shape and structure of organs (p. 88) and tissues (p. 88).

生長(名) 生物體增大體積和重量的全部過程。活生物體的重要特徵之一是具有生長能力。

營養生長 組織(第88頁)及器官(第88頁)與有性(第59頁)生殖(第59頁)無關的生長。營養生長由有絲分裂(第45頁)以及細胞的增長和增大所引起。

發育(名) 生物體的新器官(第88頁)及組織(第88頁)的構造及形貌隨生物體生長而出現的變化。(動詞為 develop)

個體發育(名) 一個個體(第135頁)從合子(第61頁)到成熟的整個發育過程。

形態發生(名) 器官(第88頁)及組織(第88頁)的形狀和構造的發育。

meristems 分生組織

apical meristems 頂生分生組織 (dark shading)（深色部分） zones of dividing cells in shoot and root tips resulting in growth 在枝尖及根尖的分裂細胞帶促成生長

intercalary meristems 居間分生組織 zones of dividing cells in monocotyledon leaf bases resulting in leaf growth 單子葉植物葉片基部的分裂細胞帶促使葉片生長

meristem (*n*) any tissue (p. 88) of actively dividing cells, which produces the cells of other plant tissues. Meristems in the apex (p. 90) of the root or shoot are called apical meristems. The meristem between the xylem (p. 106) and phloem (↑) is called the cambium (↑).

corpus (*n*) the inner layers of cells in the apical meristem (↑) of angiosperm (p. 130) shoots. Corpus cells divide anticlinally (p. 110), producing the inner tissues (p. 88) of the shoot.

tunica (*n*) the outer layer or layers of cells in the apical meristem (↑) of angiosperm (p. 130) shoots. Tunica cells divide periclinally (p. 110), producing the surface tissues (p. 88) of the shoot.

intercalary (*adj*) of the meristems (↑) at the base of leaves and stems in monocotyledons (p. 130).

分生組織(名) 活躍地分裂的細胞的任何組織(第88頁)，分生組織形成植物體其他組織的細胞。根尖端(第90頁)與枝頂的分生組織稱為頂生分生組織。在木質部(第106頁)與韌皮部(↑)之間的分生組織稱為形成層(↑)。

原體(名) 被子植物(第130頁) 枝條頂生分生組織(↑)中的內幾層細胞。原體細胞垂周(第110頁)分裂，形成枝條的內部組織(第88頁)。

原套(名) 被子植物(第130頁)枝條頂生分生組織(↑)中的外一層或幾層細胞。原套細胞平周(第110頁)分生。形成枝條的表面組織(第88頁)。

居間的(形) 指單子葉植物(第130頁)葉及莖基部的分生組織(↑)。

periclinal (*adj*) of cell divisions (p. 45) which occur parallel (↓) to the surface of the plant.
anticlinal (*adj*) of cell divisions (p. 45) which occur at right angles to the surface of the plant.
parallel (*adj*) of lines or planes running in the same direction and never meeting.
primordium (*n*) an undeveloped organ (p. 88), e.g. a leaf bud (↓) contains leaf primordia (*pl.*), a young flower bud contains the primordia of reproductive (p. 59) organs.
plastochrone (*n*) the time between the formation of one leaf primordium (↑) and the next.
bud (*n*) an undeveloped shoot covered with protecting scales (p. 100), consisting of a very short axis (p. 92) bearing primordia (↑) of leaves or flower parts.

平周的（形） 指細胞分裂（第 45 頁）出現與植物表面平行（↓）。
垂周的（形） 指細胞分裂（第 45 頁）出現與植物表面成直角。
平行的（形） 指線或平面向同一方向延伸而永不相交。
原基（名） 未發育的器官（第 88 頁）。例如一個葉芽（↓）含有葉原基，一個幼嫩花芽含有生殖（第 59 頁）器官的原基。
間隔期（名） 生成一個葉原基（↑）與下一個葉原基之間的時間。
芽（名） 保護性鱗片（第 100 頁）包覆著的未發育枝條，由一個極短的軸（第 92 頁）長著葉原基（↑）或花的各部原基組成。

primordium 原基
L.S. through dicotyledon shoot tip 通過雙子葉植物枝頂縱切面
apical meristem 頂端分生組織
young leaf 幼葉
leaf primordia will develop into leaves as plant grows 葉原基，當植物生長它將發展為葉
stem 莖

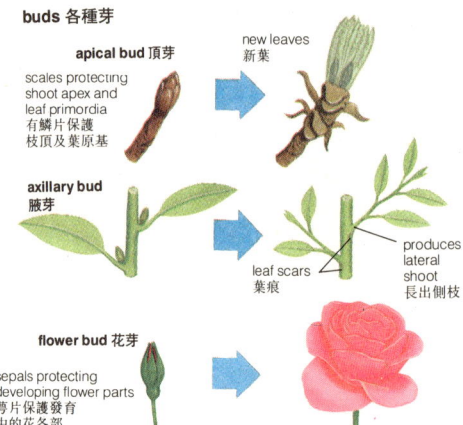

buds 各種芽
apical bud 頂芽
scales protecting shoot apex and leaf primordia 有鱗片保護枝頂及葉原基
new leaves 新葉
axillary bud 腋芽
leaf scars 葉痕
produces lateral shoot 長出側枝
flower bud 花芽
sepals protecting developing flower parts 萼片保護發育中的花各部

differentiated (*adj*) of cells which have developed a particular structure in relation to their function in a tissue (p. 88) or organ (p. 88). **differentiate** (*v*), **differentiation** (*n*).
undifferentiated (*adj*) of cells in an embryo (p. 85) or young part of a plant, e.g. a meristem (p. 109), which are all the same and have not developed into different tissues (p. 88). In many simple plants, e.g. the prothalli (p. 122) of ferns (p. 126), most of the cells are undifferentiated.

分化的（形） 指植物組織（第 88 頁）或器官（第 88 頁）的細胞已發育成一種與其功能有關的獨有結構。（動詞為 differentiate，名詞為 differentiation）
未分化的（形） 指植物胚（第 85 頁）細胞或植物幼嫩部分的細胞如分生組織（第 109 頁）的全部細胞都相同，未發育成不同的組織（第 88 頁）。許多簡單植物如蕨類（第 126 頁），其原葉體（第 122 頁）的絕大部分細胞未分化。

physiology (*n*) the study of the internal processes of organisms.
regeneration[1] (*n*) (1) the growth of new tissue (p. 88) on a part of a plant that has been damaged; (2) the growth of new plants from perennating (p. 117) organs (p. 88), e.g. rhizomes (p. 60). **regenerate** (*v*).

生理學（名） 研究生物體內各種過程的學科。
再生（名） （1）植物體受損傷的部分長出新組織（第 88 頁）；（2）從多年生（第 117 頁）器官（第 88 頁）如根狀莖（第 60 頁）長出新植物。（動詞為 regenerate）

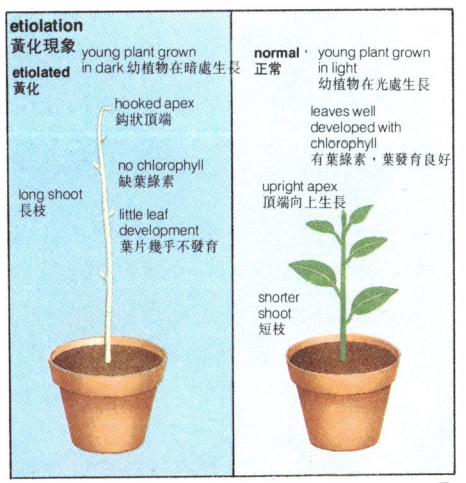

etiolation (*n*) the process of rapid growth, without the production of chlorophyll (p. 36), that occurs in shoots kept in the dark. Etiolated shoots are long, thin and pale, and their leaves are very small. **etiolate** (*v*).
nutrition (*n*) the process of taking up nutrients (↓) and using them in metabolism (p. 14).
nutrient (*n*) an inorganic (p. 11) substance which plants require for growth. Nutrients are taken up from the soil by the roots, e.g. nitrate (p. 13), phosphate (p. 13).
trace element an element (p. 8) required by a plant in very small amounts, e.g. boron, molybdenum.
deficiency (*n*) the lack of a nutrient (↑) required for growth and development. Deficiency can lead to poor growth and disease.

黃化現象（名） 植物的枝條在暗處不產生葉綠素（第 36 頁）而迅速生長的過程。黃化的枝條細長發白，葉片很小。（動詞為 etiolate）

營養（名） 吸收營養素（↓）用於新陳代謝（第 14 頁）的過程。

營養素（名） 植物生長所需的無機（第 11 頁）物質。植物根部從土壤吸收營養素，例如硝酸鹽（第 13 頁）、磷酸鹽（第 13 頁）。

微量元素 植物生長所需的極少量元素（第 8 頁），如硼、鉬。

養分缺乏（名） 缺少生長及發育所需的營養素（↑）。養分缺乏會引致生長不良和疾病。

secretion (n) the transport of a dissolved (p. 12) substance produced by a cell or organ (p. 88) out of that cell or organ. **secrete** (v).

excretion (n) the process of removing waste products of metabolism (p. 14) from a cell or organism. **excrete** (v).

gland (n) a group of cells on the surface of a plant, whose function is to secrete (↑) or excrete (↑) substances. **glandular** (adj).

exude (v) to secrete (↑) liquid from pores (p. 19), e.g. in guttation (↓), or from a cut surface.

exudate (n) the liquid exuded (↑) from pores (p. 19) and glands (↑) such as hydathodes (↓).

hydathode (n) a gland (↑) on the leaves of some plants, which exudes (↑) water.

guttation (n) the process of exuding (↑) sap (p. 102) or water through hydathodes (↑).

hormone (n) a substance which, in very small amounts, controls growth and development. Hormones are chemical messengers which are usually produced in one organ (p. 88) and transported to another part of the plant, where they have their effects. The five main groups of plant hormones are auxins (↓), gibberellins (↓), cytokinins (↓), ethene (p. 114) and abscisic acid (↓).

分泌作用(名) 細胞或器官(第88頁)所產生可溶解(第12頁)物質輸出該細胞或器官外。(動詞為 secrete)

排泄作用(名) 從細胞或有機體排除新陳代謝(第14頁)的廢物的過程。(動詞為 excrete)

腺體(名) 植物體表面具分泌(↑)或排泄(↑)物質功能的一群細胞。(形容詞為 glandular)

滲出(動) 從小孔(第19頁)分泌(↑)液體。例如在吐水(↓)中或從切割面分泌液體。

滲出液(名) 從小孔(第19頁)和腺體(↑)如水孔(↓)滲出(↑)的液體。

水孔(名) 某些植物葉上滲出(↑)水分的腺體(↑)。

吐水作用(名) 通過水孔(↑)滲出(↑)液汁(第102頁)或水分的過程。

激素(名) 植物體內控制生長和發育的一種物質,其量極少。激素是一種化學信使,通常在植物體的某一器官(第88頁)內產生後被輸送到其他部分發揮其作用。植物激素有植物生長素(↓)、赤霉素(↓)、細胞分裂素(↓)、乙烯(第114頁)及脫落酸(↓)五大類。

guttation 吐水作用

droplets of water exuded from hydathodes (ends of veins at leaf margin) 水滴從水孔流出(葉緣的葉脈端)

high humidity 濕度高

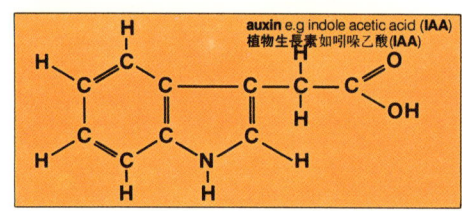

auxin e.g indole acetic acid (**IAA**)
植物生長素如吲哚乙酸(**IAA**)

auxin (n) general name for an important group of plant hormones (↑). Indole acetic acid, IAA (↓) is the most common auxin. Auxins affect many processes, e.g. tropisms (p. 115), fruit growth, apical dominance (p. 114), stem growth. They can also inhibit (p. 14) root growth.

IAA = indole acetic acid (↓).

indole acetic acid the most common auxin (↑). Indole acetic acid is produced in the shoot apex (p. 90).

植物生長素(名) 一組重要植物激素(↑)的通稱。吲哚乙酸(IAA)(↓)是最普通的植物生長素。植物生長素影響許多過程。例如影響向性(第115頁)、果實生長、頂端優勢(第114頁)和莖生長,亦可抑制(第14頁)根生長。

吲哚乙酸(↓)英文縮寫為 **IAA**。

吲哚乙酸 為最普通的植物生長素(↑)。吲哚乙酸在苗端(第90頁)生成。

GROWTH AND PHYSIOLOGY/HORMONES 生長與生理學／激素

gibberellin 赤黴素
e.g. gibberellic acid 1
(GA_1) 例如：赤黴酸 $1(GA_1)$

gibberellins (*n.pl.*) group of chemically complex plant hormones (↑), important in the control of tropisms (p. 115), the lengthening of cells during growth, germination (p. 87) and other processes.

赤黴素（名、複） 化學結構複雜的一類植物激素（↑），對控制向性（第 115 頁），影響生長時的細胞伸長，萌發（第 87 頁）及其他過程都有重要意義。

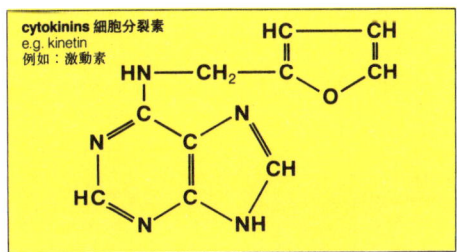

cytokinins 細胞分裂素
e.g. kinetin
例如：激動素

cytokinins (*n.pl.*) group of plant hormones (↑) which control cell division (p. 45).

abscisic acid a plant hormone (↑) which inhibits (p. 14) root growth and germination (p. 87), and which is important in the control of leaf abscission (p. 114).

細胞分裂素（名、複） 控制細胞分裂（第 45 頁）的一類植物激素（↑）。

脫落酸 抑制（第 14 頁）根生長及萌發（第 87 頁）的一種植物激素（↑），對控制葉的脫落（第 114 頁）有重要意義。

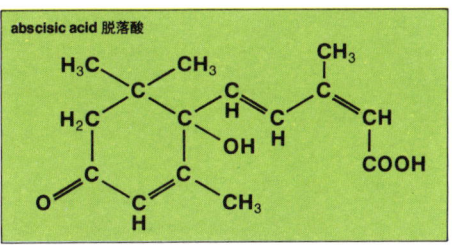

114 · GROWTH AND PHYSIOLOGY/HORMONES 生長與生理學／激素

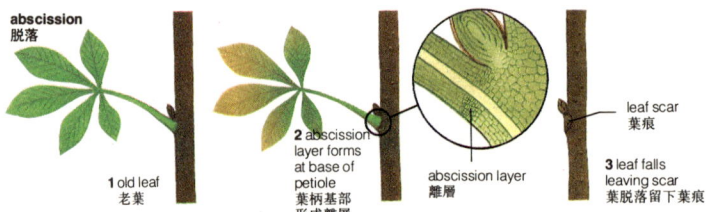

1 old leaf 老葉
2 abscission layer forms at base of petiole 葉柄基部形成離層
abscission layer 離層
leaf scar 葉痕
3 leaf falls leaving scar 葉脫落留下葉痕
abscission 脫落

ethene 乙烯

abscission (n) the process of separation of cells in the stalk of an organ (p. 88), e.g. the petiole (p. 96) of a leaf, leading to the dropping of the organ.

ethene (n) C_2H_4. A simple plant hormone (p. 112) which affects tropisms (↓), the inhibition (p. 14) of root growth, abscission (↑), the ripening (p. 83) of fruit, and other growth processes. Also known as **ethylene**.

florigen (n) a possible hormone (p. 112) involved in the production of flowers.

synergistic (adj) of processes in which one substance reinforces the action of another substance or substances. This is usually applied to plant hormones (p. 112), which often affect each other and control similar growth processes. **synergism** (n).

climacteric (n) a period of high CO_2 output, controlled by the hormone (p. 112) ethene (↑), at the beginning of fruit ripening (p. 83).

apical dominance the inhibition (p. 14) of the development of lateral (p. 92) buds (p. 110) by hormones (p. 112) produced in the shoot apex (p. 90). If the apex is cut off, the buds develop.

脫落（名） 細胞離開某個器官（第 88 頁）的柄。如葉的葉柄（第 96 頁）使該器官掉落的過程。

乙烯（名） C_2H_4。為一種簡單的植物激素（第 112 頁），它影響向性（↓）、抑制（第 14 頁）根生長、脫落（↑）、果實成熟（第 83 頁）及其他生長過程。英文亦拼寫為 **ethylene**。

成花素（名） 可能與開花有關的一種激素（第 112 頁）。

增效的（形） 指一種物質能加強另一種物質或幾種物質的作用的過程。這些過程常用於植物激素（第 112 頁），它們常相互作用，控制相似的生長過程。（名詞為 synergism）

果實成熟期（名） 二氧化碳高產的時期，在果實成熟（第 83 頁）開始時受激素（第 112 頁）乙烯（↑）控制。

頂端優勢 枝條頂端（第 90 頁）產生的激素（第 112 頁）抑制（第 14 頁）側（第 92 頁）芽（第 110 頁）的發育，如果切除頂端，芽體即發育。

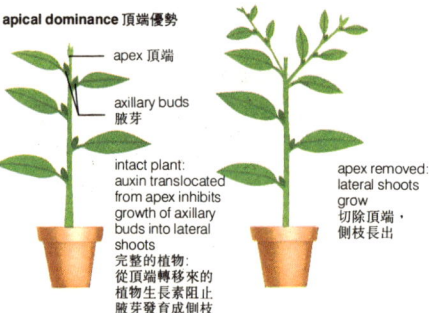

apical dominance 頂端優勢
apex 頂端
axillary buds 腋芽
intact plant: auxin translocated from apex inhibits growth of axillary buds into lateral shoots
完整的植物：從頂端轉移來的植物生長素阻止腋芽發育成側枝
apex removed: lateral shoots grow
切除頂端，側枝長出

GROWTH AND PHYSIOLOGY/TROPISMS 生長與生理學／向性 · 115

tropism (n) curving growth of a plant organ (p. 88) due to a stimulus (p. 170) coming from a particular direction, e.g. light or gravity.

向性（名） 植物器官（第 88 頁）由於受到來自獨有方向如光的方向及地心吸力的刺激（第 170 頁）而彎曲生長。

phototropism 向光性

geotropism 向地性
stem grows upwards 莖向上生長
gravity 重力
roots grow downwards 根向下生

auxin in phototropism 向光性中的生長素

1 shoot tip in dark, auxin evenly concentrated
1 枝尖在黑暗時，植物生長素濃度均勻

2 exposed to light from one side, auxin concentration increases on dark side and decreases on light side
2 光線從一側照射時，生長素在黑暗側濃度增加，受光一側減少

3 increased relative auxin concentration on dark side causes cells on dark side to elongate, and the shoot bends towards the light
3 黑暗一側生長素濃度相對地增加，使黑暗一側的細胞伸長，枝條彎向有光一側

phototropism (n) curving growth of a plant organ (p. 88), e.g. a shoot, towards light coming from a particular direction. phototropic (adj).

geotropism (n) curving growth of a plant organ (p. 88) due to gravity. Geotropism can be downwards (positive), e.g. in a taproot (p. 88), or upwards (negative), e.g. in the shoot of a seedling (p. 87). geotropic (adj).

statolith (n) a very small grain of starch (p. 30), surrounded by a membrane (p. 18), often found in cells of growing tissues (p. 88). Statoliths are thought to be important in the control of geotropism (↑).

plagiogeotropism (n) growth at an angle, in response to gravity, e.g. in lateral (p. 92) branches. plagiogeotropic (adj).

thigmotropism (n) curving growth due to contact with an object, e.g. the coiling of tendrils (p. 136) of a climbing plant around a stake. This is sometimes known as haptotropism.

chemotropism (n) curving growth of a plant organ (p. 88) in response to a chemical stimulus (p. 170) or gradient (p. 24). chemotropic (adj.).

hydrotropism (n) curved growth of a plant organ (p. 88) in response to the stimulus (p. 170) of moisture. hydrotropic (adj.).

nastic movement any plant movement caused by a diffuse stimulus (p. 170), e.g. the changes in leaf position which occur in some plants at night.

向光性（名） 植物器官（第 88 頁）如枝莖向着獨有方向照射的光彎曲生長。（形容詞為 phototropic）

向地性（名） 植物器官（第 88 頁）由於地心吸力而彎曲生長。向地性可以是向下（正向性），例如直根（第 88 頁）向下生長，也可以是向上（負向性），例如幼苗（第 87 頁）的枝條向上生長。（形容詞為 geotropic）

平衡石（名） 一種極細的澱粉（第 30 頁）粒，有一層膜（第 18 頁）包著，常見於生長組織（第 88 頁）的細胞內。平衡石對控制向地性（↑）有重要意義。

斜向地性（名） 對地心吸力的反應以某一個角度生長。例如在側（第 92 頁）枝所見。（形容詞為 plagiogeotropic）

向觸性（名） 由於觸及其他物體而彎曲生長。例如攀緣植物的捲鬚（第 136 頁）捲繞著一根柱。有時也稱為向實體性。

向化性（名） 植物器官（第 88 頁）對化學刺激（第 170 頁）或化學梯度（第 24 頁）刺激的反應而彎曲生長。（形容詞為 chemotropic）

向水性（名） 植物器官（第 88 頁）對濕氣刺激（第 170 頁）的反應而彎曲生長。（形容詞為 hydrotropic）

感性運動 擴散刺激（第 170 頁）引起的任何植物運動。例如某些植物在夜間出現的葉子位置變動。

plagiogeotropism 斜向地性
gravity 重力
branches grow at an angle 枝條以某角度生長

endogenous rhythm the repeated, regular, rhythmic changes of internal activity in an organism that are not due to external environmental (p. 149) factors.

photoperiod (n) the number of hours of daylight needed by a plant before it will begin to flower. See also **photoperiodism** (↓).

short-day plant a plant which will begin to flower when the length of the day, or photoperiod (↑), is shorter than a certain period.

long-day plant a plant which will begin to flower only when the length of the day, or photoperiod (↑), is longer than a certain period.

photoperiodism (n) the physiological (p. 111) responses of an organism to changes in the lengths of day and night. **photoperiodic** (adj).

內源節律 生物體內活動非因種種外環境(第149頁)因子而致的重複有規律節律變化。

光週期(名) 植物開始開花前所需的白天光照小時數。參見光週期性(↓)。

短日照植物 日照長度或光週期(↑)短時才開始開花的植物。

長日照植物 日照長度或光週期(↑)長時始開花的植物。

光週期性(名) 有機體對畫夜長度變化的生理(第111頁)反應。(形容詞為 photoperiodic)

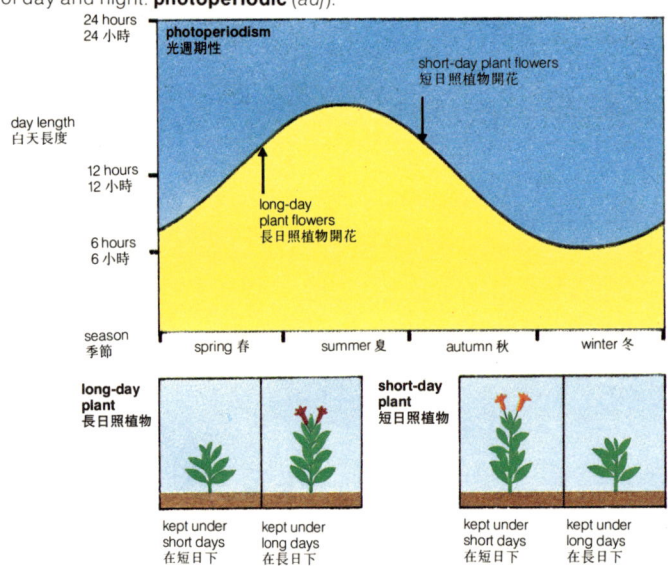

phytochrome (n) a pigment (p. 36) which controls many of the physiological (p. 111) responses of plants to light, e.g. photoperiodism (↑). Phytochrome absorbs the red and far-red wavelengths (p. 38) of light.

光敏色素(名) 控制植物對光的許多生理(第111頁)反應。如光週期性(↑)的一種色素(第36頁)。光敏色素吸收光線中的紅光及遠紅光波長(第38頁)。

GROWTH AND PHYSIOLOGY/GROWTH PERIOD 生長與生理學/生長週期・117

ephemeral (adj) of plants which germinate, grow, reproduce (p. 59) and die in a very short time.

短生的(形) 指在很短時間內發芽、生長、生殖(第59頁)及死亡的植物。

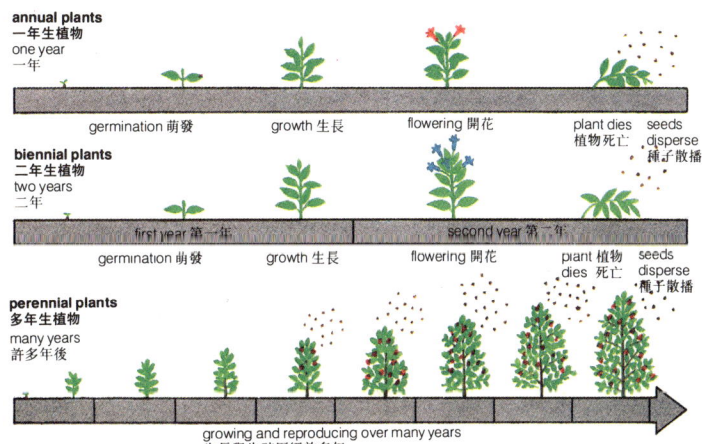

annual (adj) of plants which complete their entire life cycle (p. 64), from seed to reproduction (p. 59) to death, in one year.
biennial (adj) of plants which complete their life cycle (p. 64) in two years, growing in the first, reproducing (p. 59) and dying in the second.
perennial (adj) of plants which grow and reproduce (p. 59) for many years. Perennial plants are usually woody.
perennation (n) the survival of an individual (p. 135) over successive years, or of a dormant (↓) organ (p. 88) during unfavourable seasons. **perennate** (v).
dormant (adj) of cells, buds (p. 110), seeds, etc. during the period before growth begins. **dormancy** (n).
hibernation (n) the slowing-down of metabolism (p. 14) which occurs in many organisms during winter. **hibernate** (v).
vernalization (n) flowering due to treatment with low temperatures.
senescence (n) the process of growing old before death. **senescent** (adj).

一年生的(形) 指在一年內完成從種子到生殖(第59頁)直至死亡的整個生活週期(第64頁)的植物。

二年生的(形) 指在兩年內完成其生活週期(第64頁)的植物。第一年生長，生殖(第59頁)，第二年死亡。

多年生的(形) 指生長及生殖(第59頁)歷經許多年的植物。多年生植物多為木本植物。

多年生性(名) 指在不利的季節期間個體(第135頁)或休眠(↓)器官(第88頁)持續生存許多年。(動詞為 perennate)

休眠的(形) 指細胞、芽體(第110頁)、種子等經一段時間後才開始生長。(名詞為 dormancy)

冬眠(名) 在冬季期間許多生物體出現新陳代謝(第14頁)減慢。(動詞為 hibernate)

春化作用(名) 經過低溫處理促使開花。

衰老(名) 生物體進入老年直至死亡的過程。(形容詞為 senescent)

organism (n) any living thing. Organisms differ from non-living things in that they can grow and reproduce (p. 59).

microorganism (n) a very small organism e.g. a virus (↓), bacterium (↓) or yeast (p. 164).

plant (n) an organism which has most or all of the following characteristics: ability to synthesize (p. 13) carbohydrate (p. 28) by photosynthesis (p. 32), possession of cell walls (p. 17) containing cellulose (p. 17), a life cycle (p. 64) consisting of an alternation of generations (p. 64), and an inability to move.

生物體；有機體(名) 任何有生命的東西。生物體與無生命物體不同之處，在於生物體能夠生長和生殖(第 59 頁)。

微生物(名) 一種極微小的生物體，例如：病毒(↓)、細菌(↓)或酵母(第 164 頁)。

植物(名) 指具有大部分或全部下列特徵的生物：能行光合作用(第 32 頁)合成(第 13 頁)碳水化合物(第 28 頁)、具備含纖維素(第 17 頁)的細胞壁(第 17 頁)、生活週期(第 64 頁)由世代交替(第 64 頁)組成以及無行動能力。

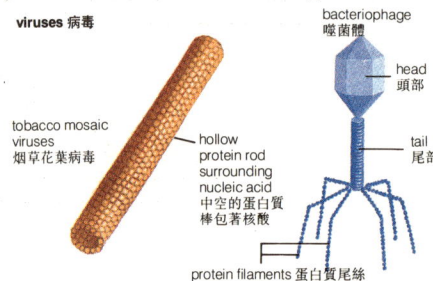

viruses 病毒
tobacco mosaic viruses 烟草花葉病毒
hollow protein rod surrounding nucleic acid 中空的蛋白質棒包著核酸
bacteriophage 噬菌體
head 頭部
tail 尾部
protein filaments 蛋白質尾絲

virus (n) a group of very simple organisms, which consist of a strand of nucleic acid (p. 51) surrounded by a coat of protein (p. 56). Viruses do not metabolize (p. 14), and can only reproduce (p. 59) inside the cells of other organisms, where the nucleic acid of the virus directs the synthesis (p. 13) of more viruses using the protein synthesis (p. 57) machinery already in the cell. Viruses can destroy cells in this way, and cause many diseases in other organisms. They are sometimes classified (p. 132) in a kingdom (p. 134) of their own. Viruses are very small, usually about 100 nm wide.

bacteriophage (n) a kind of virus (↑) which attacks the cells of bacteria (↓). Many bacteriophages have a 'head', consisting of a protein (p. 56) coat containing nucleic acid (p. 51), and a 'tail' of protein through which the nucleic acid is injected into a bacterial cell.

phage (n) = a bacteriophage (↑).

病毒(名) 一類極簡單的生物，由一股核酸(第 51 頁)分子組成，外有蛋白質(第 56 頁)外殼包圍著。病毒沒有新陳代謝(第 14 頁)，只能在其他生物體的細胞內部生殖(第 59 頁)，病毒的核酸利用已在細胞內的蛋白質合成(第 57 頁)機制引導更多的病毒進行合成(第 13 頁)。病毒能以這種方式破壞其他生物體的細胞，引起多種疾病。病毒往往自成一界(第 134 頁)分類(第 132 頁)。病毒極微細，通常約 100 毫微米寬。

噬菌體(名) 能侵入細菌(↓)細胞的一種病毒(↑)，許多噬菌體都有一個"頭部"和一個"尾部"；頸部由含核酸(第 51 頁)的蛋白質(第 56 頁)外殼組成，核酸通過蛋白質尾巴注入細菌細胞內。

噬體(名) 同噬菌體(↑)。

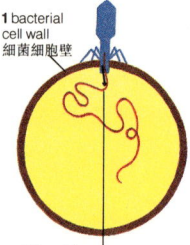

bacteriophage attacking a bacterium 噬菌體進攻細菌
1 bacterial cell wall 細菌細胞壁
nucleic acid injected 核酸注入

2 parts of new bacteriophages synthesized in bacterial cell 部分新噬菌體在細菌細胞內合成

3 bacterium destroyed, new bacteriophages released 細菌被破壞，新噬菌體釋放出來

PLANT KINGDOM/BACTERIA, ALGAE 植物界／細菌、藻類・119

bacteria 細菌
bacilli
桿菌

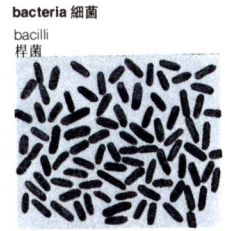

spirochaetes 螺旋體

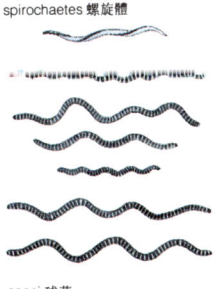

cocci 球菌

bacteria (n.pl.) a division (p. 134) of unicellular (↓) prokaryotic (p. 16) organisms, most of which are heterotrophic (p. 32). Bacterial cells are usually 0.5-2 μm across. Bacteria are important in the decay (p. 157) of organic (p. 11) matter in the soil. Many are parasitic (p. 144) on other organisms, often causing disease. **bacterium** (sing.), **bacterial** (adj)
bacillus (n) a rod-shaped bacterium (↑).
spirochaete (n) a spirally-shaped bacterium (↑), belonging to the order (p. 134) Spirochaetales.
cocci (n) a group of bacteria (↑) with spherically-shaped cells.
unicellular (adj) of organisms which consist of only one cell, e.g. euglenoids (p. 120), yeast (p. 164), bacteria (↑).
multicellular (adj) of organisms which consist of many cells, as in most plants.
algae (n.pl.) a large group of mainly aquatic (p. 161) plants, which differ from other plants by their lack of complex multicellular (↑) reproductive (p. 59) organs (p. 88). **alga** (sing.), **algal** (adj).
pyrenoid (n) a small grain of protein (p. 56) in the chloroplast (p. 32) of an algal (↑) cell, around which starch (p. 30) is deposited.
colony (n) a group of cells of the same kind, forming a single organism, as in many algae (↑). **colonial** (adj).
coenobium (n) a colony (↑) with a regular shape, consisting of cells which do not divide vegetatively (p. 60), e.g. the alga (↑) Volvox.
aggregation (n) a group of similar cells without a regular arrangement, as in many algae (↑).

細菌（名、複）　屬單細胞（↓）原核（第16頁）生物的一個門（第134頁），其中大多數是異養的（第32頁）。細菌的細胞通常為0.5-2微米寬。細菌顯著引起土壤中的有機（第11頁）物腐敗（第157頁）。許多細菌寄生（第144頁）在其他生物體上，常引起疾病。（單數為bacterium，形容詞為bacterial）
桿菌（名）　棒狀的細菌（↑）。
螺旋體（名）　螺旋狀的細菌（↑），屬於螺旋菌目（第134頁）。
球菌（名）　一群具球形細胞的細菌（↑）。
單細胞的（形）　指只由一個細胞組成的生物體。例如裸藻（眼蟲藻）（第120頁）、酵母（第164頁）、細菌（↑）。
多細胞的（形）　指由多個細胞組成的生物體。大多數植物是多細胞生物。
藻類植物（名、複）　一大群主要是水生的（第161頁）植物，和其他植物的區別在於沒有複雜的多細胞（↑）生殖（第59頁）器官（第88頁）。（單數為alga，形容詞為algal）
澱粉核（名）　藻類植物（↑）細胞葉綠體（第32頁）中的細小蛋白質（第56頁）粒體。澱粉（第30頁）沉積在粒體周圍。
群體（名）　一群同類細胞形成一個單個的機體。如在許多藻類（↑）所見。（形容詞為colonial）
定型群體（名）　具規則形狀的群體（↑），由不行營養性（第60頁）分裂的細胞組成。如藻類（↑）的團藻。
族聚（名）　一群相似的細胞，缺乏有規則的排列。如在許多藻類（↑）所見。

colonial algae
群體的藻

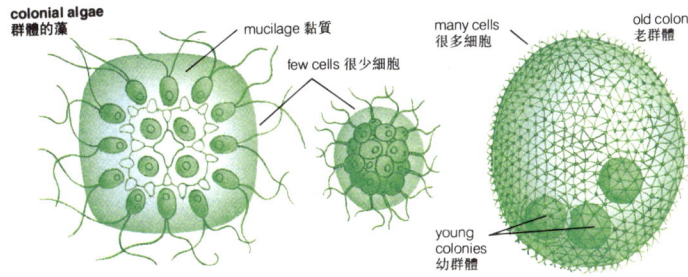

mucilage 黏質
few cells 很少細胞
many cells 很多細胞
old colony 老群體
young colonies 幼群體

chromatophore (*n*) the chloroplast (p. 32) of a green alga (↓), or, the pigment (p. 36)-containing body of a photosynthetic (p. 32) bacterium (p. 119).

paramylum (*n*) a polysaccharide (p. 30) made of glucose (p. 28) units. It is stored in grains in euglenoids (↓).

siphoneous (*adj*) of algae (p. 119) in which the plant body is not divided into cells, i.e. it is multinucleate (p. 163).

parenchymatous (*adj*) of multicellular (p. 119) algae (p. 119) whose cells divide in more than one direction.

filamentous (*adj*) of algae (p. 119) consisting of long threads of cells, e.g. *Spirogyra*.

coccoid (*adj*) of algae (p. 119) which are unicellular (p. 119) and non-motile (↓).

載色體(名) 綠藻類(↓)的葉綠體(第32頁)或光合(第32頁)細菌(第119頁)的含色素(第36頁)體。

裸藻澱粉(名) 由葡萄糖(第28頁)單元構成的多醣類(第30頁)。它在裸藻(↓)中呈粒狀貯存。

管狀(形) 指植物體不分裂為細胞的藻類(第119頁),即多核的(第163頁)。

薄壁組織的(形) 指多細胞(第119頁)藻類(第119頁)的細胞向多個方向分裂。

絲狀的(形) 指含有細胞長絲的藻類(第119頁)。如水綿。

球形的(形) 指單細胞(第119頁)及不能游動的(↓)藻類(第119頁)。

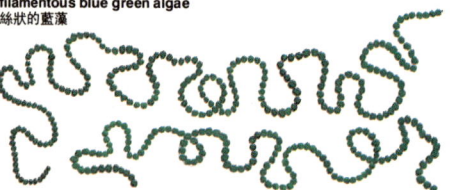

filamentous blue green algae
絲狀的藍藻

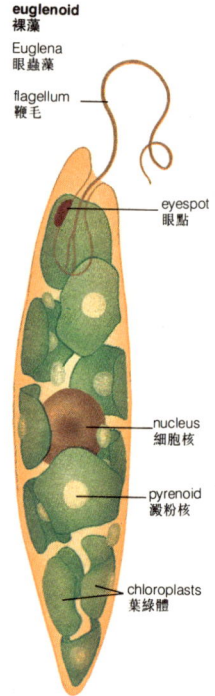

euglenoid
裸藻
Euglena 眼蟲藻
flagellum 鞭毛
eyespot 眼點
nucleus 細胞核
pyrenoid 澱粉核
chloroplasts 葉綠體

blue-green algae algae (p. 119) of the division (p. 134) Cyanophyta with prokaryotic (p. 16) cells. They are unicellular (p. 119) or multicellular (p. 119), lack flagella (↓) and have their own carotenoid (p. 37) photosynthetic (p. 32) pigments (p. 36).

red algae algae (p. 119) of the division (p. 134) Rhodophyta which have a red colour due to the presence of the pigments (p. 36) phycocyanin and phycoerythrin.

green algae algae (p. 119) of the division (p. 134) Chlorophyta. Green algae have chlorophyll *b* (p. 36), produce starch (p. 30), and have cellulose (p. 17) in their cell walls (p. 17).

euglenoid (*n*) an alga (p. 119) of the division (p. 134) Euglenophyta. Euglenoids are unicellular (p. 119), flagellate (↓) and have paramylum (↑) instead of starch (p. 30) as their main storage product.

藍藻 具原核(第16頁)細胞的藍藻植物門(第134頁)藻類(第119頁)。這些藻都是單細胞的(第119頁)或多細胞的(第119頁),沒有鞭毛(↓),有本身的類胡蘿蔔素(第37頁)和光合(第32頁)色素(第36頁)。

紅藻 屬紅藻植物門(第134頁)藻類(第119頁),因含藻藍蛋白及藻紅蛋白色素(第36頁)而呈紅色。

綠藻 是綠藻植物門(第134頁)的藻類(第119頁)。綠藻具葉綠素b(第36頁),能產生澱粉(第30頁),細胞壁(第17頁)含有纖維素(第17頁)。

裸藻;眼蟲藻(名) 屬裸藻植物門(第134頁)的藻類(第119頁)。裸藻是單細胞的(第119頁),有鞭毛(↓),並以裸藻澱粉(↑)而不是澱粉(第30頁)作為其主要貯存產物。

PLANT KINGDOM/ALGAE 植物界／藻類・121

flagellum (n) a long motile (↓) thread, consisting of a membrane (p. 18) enclosing a series of parallel (p. 110) microtubules (p. 21). Flagella (pl.) are found in the cells of unicellular (p. 119) motile algae (p. 119), e.g. *Euglena, Chlamydomonas*, and also in the male gametes (p. 61) of bryophytes (p. 122), pteridophytes (p. 126), and some gymnosperms (p. 128).
  flagellate (adj).
motile (adj) able to move, as in cells with flagella (↑).

鞭毛（名） 游動（↓）的長形絲狀體，由膜（第 18 頁）包着一系列平行的（第 110 頁）微管（第 21 頁）組成。單細胞（第 119 頁）能游動的藻類（第 119 頁），如裸藻、衣藻、苔蘚植物（第 122 頁）、蕨類植物（第 126 頁）及某些裸子植物（第 128 頁）的雄配子（第 61 頁）都有鞭毛。（形容詞為 flagellate）

能游動的（形） 能够運動，恰如有鞭毛（↑）的細胞。

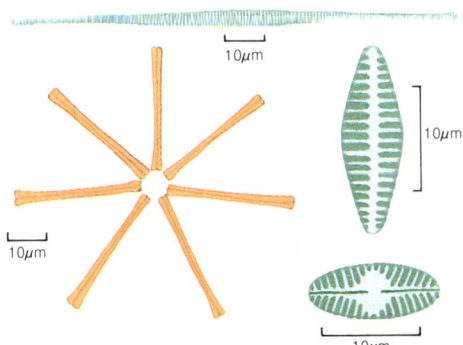

diatoms 矽藻

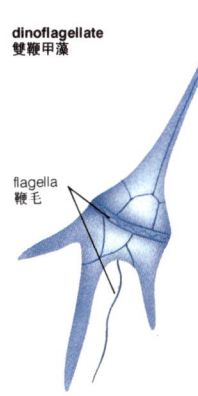

dinoflagellate 雙鞭甲藻
flagella 鞭毛

diatom (n) an alga (p. 119) of the division (p. 134) Bacillariophyta. Diatoms are mostly unicellular (p. 119), and their cell walls (p. 17) contain silicon. See also **siliceous skeleton**.
siliceous skeleton the silicon-containing cell wall (p. 17) of a diatom (↑).
dinoflagellate (n) a class (p. 134) of unicellular (p. 119) algae (p. 119), usually yellow in colour, which have two flagella (↑) and thick cell walls (p. 17) arranged in a characteristic pattern of plates. Dinoflagellates are a major component of marine phytoplankton (↓).
phytoplankton (n) the small plants, mostly diatoms (↑) and other unicellular (p. 119) algae (p. 119), which are found near the surface of oceans and lakes. Phytoplankton is one of the world's most important primary producers (p. 150).

矽藻；硅藻（名） 屬於矽藻植物門（第 134 頁）的藻類（第 119 頁）。矽藻多數是單細胞的（第 119 頁），其細胞壁（第 17 頁）含有矽質（硅質），參見**矽質骨骼**。

矽質骨骼 矽藻（↑）的含矽質的細胞壁（第 17 頁）。

雙鞭甲藻（名） 屬單細胞（第 119 頁）藻類（第 119 頁）的一綱（第 134 頁），常呈黃色，具有兩條鞭（↑）和厚的細胞壁（第 17 頁），排列成有特徵的板狀格局。雙鞭甲藻是海洋浮游植物（↓）的主要組成部分。

浮游植物（名） 微小的植物，多數是矽藻（↑）及其他單細胞的（第 119 頁）藻類（第 119 頁），海洋及湖泊的近水面處都可見到浮游植物。浮游植物是世界上最重要的初級生產者（第 150 頁）之一。

brown algae
(seaweeds)
褐藻（海藻）

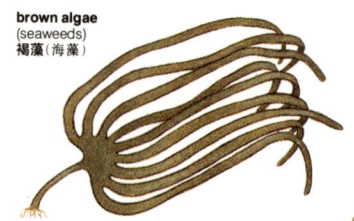

brown algae algae (p.119) of the division (p.134) Phaeophyceae, including many of the larger seaweeds (↓). Brown algae contain brown accessory pigments (p.36).

seaweed (n) the general name for any large, parenchymatous (p.120) alga (p.119) in the sea.

nonvascular (adj) of plants which do not have a vascular system (p.105). Nonvascular plants include most bryophytes (↓) and all algae (p.119).

vascular (adj) of plants with a vascular system (p.105). Vascular plants include all pteridophytes (p.126) and spermatophytes (p.128).

bryophyte (n) a plant of the division (p.134) Bryophyta, i.e. a moss (p.124) or a liverwort (↓). Bryophytes differ from other plants in having the gametophyte (p.65) as the main vegetative (p.60) stage. Most bryophytes have little or no vascular system (p.105), and live in wet, shady habitats (p.149).

thallus (n) a more or less undifferentiated (p.110) plant body, without distinct roots, stems and leaves, e.g. the gametophyte (p.65) of thalloid liverworts (↓), or the plant body of an alga (p.119). **thalloid** (adj), **thalli** (pl.).

prothallus (n) the gametophyte (p.65) of mosses (p.124), liverworts (↓) and pteridophytes (p.126). **prothalli** (pl.).

rhizoid (n) a thread-like cell which grows from the lower surface or base of a bryophyte (↑). Rhizoids have the function of roots.

sporogonium (n) the sporophyte (p.65) of a moss (p.124) or a liverwort (↓), consisting of a foot (↓), seta (↓) and capsule (↓).

foot (n) the base of the sporophyte (p.65) of a bryophyte (↑), which is the part that attaches it to the gametophyte (p.65).

褐藻　屬褐藻門（第134頁）的藻類（第119頁）包括許多較大型的海藻（↓）。褐藻含有褐色的輔助色素（第36頁）。

海藻（名）　為海上任何一種大型薄壁組織狀（第120頁）藻類（第119頁）的通稱。

無維管的（形）　不具維管系統（第105頁）的植物。無維管植物包括大多數苔蘚類植物（↓）及一切藻類植物（第119頁）。

維管的（形）　指具維管系統（第105頁）的植物。維管植物包括全部蕨類植物（第126頁）及種子植物（第128頁）。

苔蘚植物（名）　屬苔蘚植物門（第134頁）的植物，包括蘚類植物（第124頁）和苔類植物（↓）。它和其他植物不同之處在於有配子體（第65頁）作為主要營養（第60頁）階段。大多數苔蘚植物只具很少或不具維管系統（第105頁），生長在濕而陰的生境（第149頁）。

葉狀體（名）　一個近乎不分化的（第110頁）植物體沒有明顯的根、莖和葉。例如似葉狀體的苔類（↓）植物的配子體（第65頁）或藻類（第119頁）的植物體。（形容詞為thalloid，複數為thalli）

原葉體（名）　蘚類（第124頁）、苔類（↓）及蕨類（第126頁）的配子體（第65頁）。（複數為prothalli）

假根（名）　從苔蘚植物（↑）的下面或基部長出來的絲狀細胞。假根有根的功能。

孢子體（名）　蘚類（第124頁）或苔類（↓）的孢子體（第65頁），具有柄基（↓）、柄（↓）及蒴（↓）。

基部（名）　苔蘚植物（↑）孢子體（第65頁）的基部，使孢子體附在配子體（第65頁）的部分。

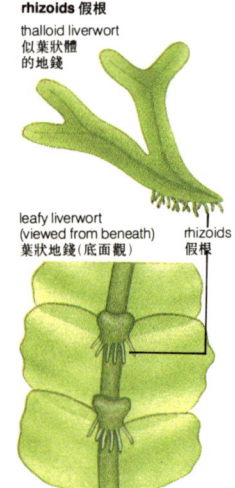

rhizoids 假根
thalloid liverwort
似葉狀體的地錢

leafy liverwort
(viewed from beneath)
葉狀地錢（底面觀）
rhizoids 假根

PLANT KINGDOM/BRYOPHYTES 植物界／苔蘚植物・123

seta (*n*) the stalk of the sporophyte (p. 65) of a bryophyte (↑).

capsule² (*n*) the spore (p. 66)-producing organ (p. 88) of the sporophyte (p. 65) of a bryophyte (↑), borne at the top of the seta (↑).

liverwort (*n*) one of the two groups of bryophytes (↑). Liverworts differ from mosses (p. 124) in having less differentiated (p. 110) cells in the gametophyte (p. 65), and in having elaters (p. 124) in the dehiscent (p. 84) capsule (↑). The gametophyte is either thalloid (↑) or leafy.

hepatic (*n*) = a liverwort (↑).

蒴柄（名） 苔蘚植物（↑）孢子體（第 65 頁）的柄。

蘚蒴（名） 苔蘚植物（↑）孢子體（第 65 頁）產生孢子（第 66 頁）的器官（第 88 頁），長在蒴柄（↑）頂端。

苔類植物（名） 苔蘚植物（↑）的兩大類之一。苔類與蘚類（第 124 頁）不同苔類配子體（第 65 頁）的細胞較少分化（第 110 頁），開裂的（第 84 頁）蒴（↑）有彈孢絲（第 124 頁），配子體為似葉狀體（↑）或葉狀。

地錢（名） 同苔類植物（↑）。

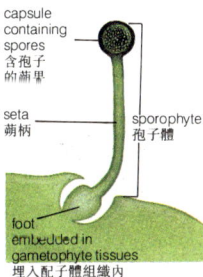

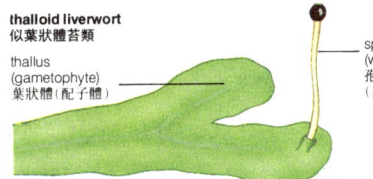

thalloid liverwort a liverwort (↑) in which the gametophyte (p. 65) is a flat, more or less undifferentiated (p. 110) thallus (↑). About 20% of liverwort species (p. 134) are thalloid.

似葉狀體苔類（名） 配子體（第 65 頁）扁平的苔類（↑），葉狀體（↑）或多或少未分化（第 110 頁），苔類的種（第 134 頁）約有 20% 是似葉狀體的。

leafy liverwort a liverwort (↑) in which the gametophyte (p. 65) has a simple stem, growing from the apex (p. 90) and bearing small leaves in rows along it. About 80% of liverwort species (p. 134) are leafy.

succubous (*adj*) of a kind of growth in leafy liverworts (↑), in which the front of each leaf lies underneath the leaf in front of it.

incubous (*adj*) of a kind of growth in leafy liverworts (↑), in which the front of each leaf lies on top of the leaf in front of it.

葉狀苔類 配子體（第 65 頁）有簡單的莖，從頂端（第 90 頁）長出，沿着莖有小葉排列成行生長的苔類（↑）。苔類的種（第 134 頁）約有 80% 是葉狀的。

蔽後式的（形） 指葉狀苔類（↑）生長的方式，每片葉的前緣位於其前面葉片的下方。

蔽前式的（形） 指葉狀苔類（↑）生長形式之一，每片葉的前緣位於其前面葉片的上方。

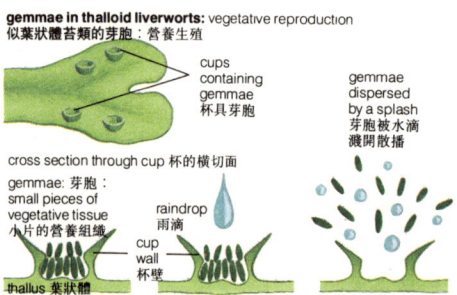

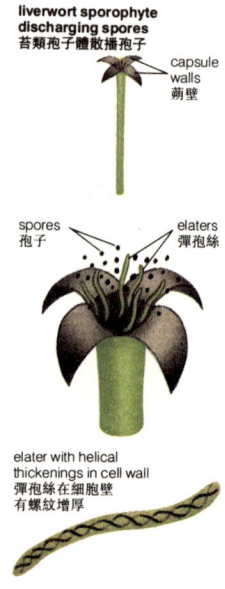

gemmae (*n.pl.*) small groups of green cells produced in cup-shaped structures on the surface of some thalloid liverworts (p. 123). Gemmae are dispersed (p. 84) by splashes of rain, and are a way of vegetative reproduction (p. 60). **gemma** (*sing.*).

elater (*n*) one of a bunch of long, thin cells in the capsule (p. 123) of the sporophyte (p. 65) of a liverwort (p. 123). Elaters have spiral (p. 98) thickening of the cell wall (p. 17). They alter their position with changes in humidity, and help with the dispersal (p. 84) of spores (p. 66) from the capsule.

moss (*n*) one of the two main groups of bryophytes (p. 122). Mosses differ from liverworts (p. 123) in having more differentiated (p. 110) cells in the gametophyte (p. 65). The gametophyte usually has a stem with leaves, and is often branched. The capsule (p. 123) of the sporophyte (p. 65) is also more differentiated in mosses, and the spores (p. 66) are released through a peristome (↓).

胞芽（名、複） 某些似葉狀體苔類（第 123 頁）表面杯狀構造內產生的小群綠色細胞。胞芽靠雨水濺開散播（第 84 頁），是營養體生殖（第 60 頁）的一種方式（單數為 gemma）

彈孢絲（名） 苔類植物（第 123 頁）孢子體（第 65 頁）的蘚蒴（第 123 頁）內的一束長形纖細胞。彈孢絲的細胞壁（第 17 頁）有螺旋（第 98 頁）增厚。彈孢絲隨濕度變化而改變位置，將孢子（第 66 頁）從蘚蒴內散播（第 84 頁）出去。

蘚類植物（名） 苔蘚植物（第 122 頁）的兩大類之一。蘚類與苔類（第 123 頁）不同，蘚類的配子體（第 65 頁）更加分化（第 110 頁）。配子體常有莖和葉，且常有分枝。孢子體（第 65 頁）的蘚蒴（第 123 頁）也更加分化。孢子（第 66 頁）通過蘚齒層（↓）放出。

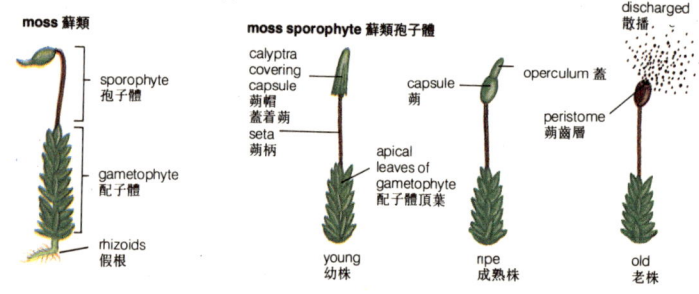

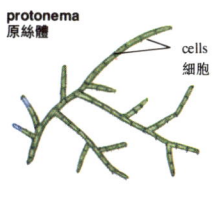

protonema 原絲體
cells 細胞

acrocarpous moss 頂生蘚類

pleurocarpous moss 側生蘚類

protonema (n) the young gametophyte (p. 65) of a moss (↑), in the early stages after the germination (p. 87) of the spore (p. 66).

pleurocarpous (adj) of mosses (↑) which have a stem with many branches, spreading across the ground. The reproductive (p. 59) organs (p. 88) are borne on short side-branches.

acrocarpous (adj) of mosses (↑) which have an upright stem, with the reproductive (p. 59) organs (p.88) at the apex (p.90).

paraphyses (n.pl.) small hairs one cell thick, often with a large round cell at the top, which grow among the antheridia (p. 65) in mosses (↑). They protect the antheridia and may provide them with photosynthetic (p. 32) products.

原絲體(名) 蘚類植物(↑)在孢子(第66頁)發芽(第87頁)後早期階段的幼配體(第65頁)。

側生蘚的(形) 指具有莖和許多分枝散佈在地面上，生殖(第59頁)器官(第88頁)長在短的側枝上的蘚類(↑)。

頂生蘚的(形) 指蘚類植物(↑)有一直立的莖，生殖(第59頁)器官(第88頁)生在莖頂(第90頁)上的蘚類(↑)。

側絲(名、複)細毛一個細胞厚，頂端常有一個大而圓的細胞，長在蘚類植物(↑)藏精器(第65頁)之間。側絲保護藏精器並可能向其提供光合作用(第32頁)的產物。

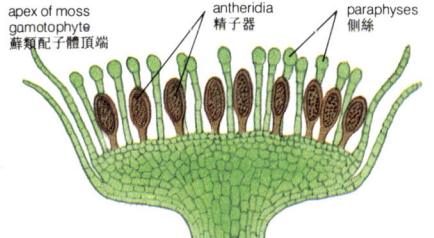

paraphyses 側絲
apex of moss gametophyte 蘚類配子體頂端
antheridia 精子器
paraphyses 側絲

calyptra (n) a hood of tissue (p. 88) produced from the wall of the archegonium (p. 65), especially in mosses (↑). The calyptra protects the young sporophyte (p. 65).

columella[1] (n) the tissue (p. 88) in the centre of the capsule (p. 123) of a moss (↑).

operculum (n) the lid which covers the pore (p. 19) at the apex (p. 90) of a moss (↑) capsule (p. 123). The operculum opens to allow spores (p. 66) to escape.

peristome (n) a set of tooth-like plates under the operculum (↑) of the capsule (p. 123) of a moss (↑). These plates are called peristome teeth. They control the release of spores (p. 66) into the air, by altering their position with changes in humidity.

蘚帽(名) 從藏卵器(第65頁)壁所產生組織(第88頁)的冠狀物，尤以蘚類植物(↑)為然。蘚帽保護幼嫩的孢子體(第65頁)。

蘚軸(名) 蘚類(↑)蘚(第123頁)中央的組織(第88頁)。

蘚蓋(名) 蘚類(↑)蘚(第123頁)頂端(第90頁)蓋着孔口(第19頁)的蓋。蘚蓋打開讓孢子(第66頁)逸出。

蘚齒層(名) 蘚類植物(↑)蘚(第123頁)蘚蓋(↑)的下一組齒狀板。這些板叫蘚齒，蘚齒隨濕度變化而改變其位置，控制孢子(第66頁)釋放入大氣。

126 · PLANT KINGDOM/PTERIDOPHYTES 植物界／蕨類植物

pteridophyte (*n*) a member of the division (p. 134) Pteridophyta, which includes the ferns (↓), clubmosses (↓) and horsetails (↓). In pteridophytes the sporophyte (p. 65) is the main vegetative (p. 60) stage of the life cycle (p. 64). The gametophyte (p. 65) is small, and is independent of the sporophyte, as in bryophytes (p. 122). Large tree-like forms of pteridophytes were common during the Carboniferous (p. 143) period, and were fossilized (p. 142) to form coal.

fern (*n*) a pteridophyte (↑) belonging to the order (p. 134) Filicales. Ferns have spirally (p. 98) arranged leaves, which are often pinnately (p. 98) compound (p. 98). Ferns are homosporous (p. 67), with the sporangia (p. 66) borne in sori (↓) on the abaxial (p. 97) surface of the leaf.

frond (*n*) the leaf of a fern (↑). Most ferns have pinnate (p. 98) or bipinnate (p. 98) fronds. The leaves of palms (p. 130) are also called fronds.

circinate (*adj*) coiled or rolled up, like the young frond (↑) of a fern (↑).

蕨類植物(名) 蕨類植物門(第134頁)的成員，包括真蕨(↓)、石松類(↓)及木賊類(↓)。在蕨類植物，孢子體(第65頁)是生活週期(第64頁)的主要營養(第60頁)階段。配子體(第65頁)細小，且獨立於孢子體，如同苔蘚植物(第122頁)。在石炭紀(第143頁)時期，巨樹狀蕨很普遍，現已化石化(第142頁)成煤。

真蕨(名) 屬真蕨目(第134頁)的蕨類植物(↑)。真蕨的螺旋狀(第98頁)排列的葉常為羽狀複葉(第98頁)。真蕨具同型孢子(第67頁)，孢子囊(第66頁)長在葉遠軸(第97頁)面的孢囊群內。

蕨葉；棕葉(名) 真蕨類(↑)的葉。大部分真蕨具有羽狀(第98頁)葉或二回羽狀(第98頁)葉。棕櫚(第130頁)的葉稱為棕葉。

拳捲的(形) 盤捲或包捲，如真蕨(↑)的幼蕨葉(↑)。

ferns 真蕨

frond 蕨葉

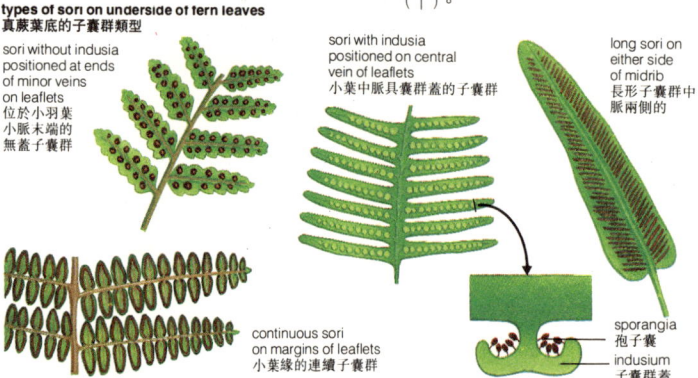

types of sori on underside of fern leaves
真蕨葉底的子囊群類型

sori without indusia positioned at ends of minor veins on leaflets
位於小羽葉小脈末端的無蓋子囊群

sori with indusia positioned on central vein of leaflets
小葉中脈具囊群蓋的子囊群

long sori on either side of midrib
長形子囊群中脈兩側的

continuous sori on margins of leaflets
小葉緣的連續子囊群

sporangia 孢子囊
indusium 子囊群蓋

sorus (*n*) an organ (p. 88) on the surface of the leaf of a fern (↑), in which sporangia (p. 66) are produced. Sori (*pl.*) have different shapes and are found on different parts of the leaf in different species (p. 134). Their function is to protect the sporangia.

子囊群(名) 真蕨(↑)葉面產生孢子囊(第66頁)的器官(第88頁)。子囊群有不同的形狀，不同種(第134頁)的葉的不同部分都有子囊群，其功能是保護孢子囊。

PLANT KINGDOM/PTERIDOPHYTES 植物界／蕨類植物・127

indusium (n) the flap of tissue (p. 88) in a sorus (↑), which protects the sporangia (p. 66).
indusia (pl.).
filmy fern a fern of the family (p. 134) Hymenophyllaceae. Filmy ferns have very delicate leaves, usually only one cell thick, and live in moist shady habitats (p. 149).
tree fern a fern (↑) of the family (p. 134) Cyatheaceae, in which the fronds (↑) grow from the top of a trunk (p. 92). The trunk in tree ferns consists partly of the woody bases of dead fronds. Most tree ferns are tropical (p. 162).

囊群蓋（名） 子囊群（↑）組織（第 88 頁）的蓋片，它保護著孢子囊（第 66 頁）。（複數為 indusia）。
膜蕨 屬膜蕨科（第 134 頁）的一種真蕨。膜蕨有極精緻的葉。通常僅一個細胞厚，長在陰濕的生境（第 149 頁）。
樹蕨 屬桫欏科（第 134 頁）的一種真蕨（↑）。蕨葉（↑）從樹幹（第 92 頁）頂端長出。樹蕨的幹莖有一部分是由死蕨葉木質基部組成。樹蕨多數是熱帶（第 162 頁）植物。

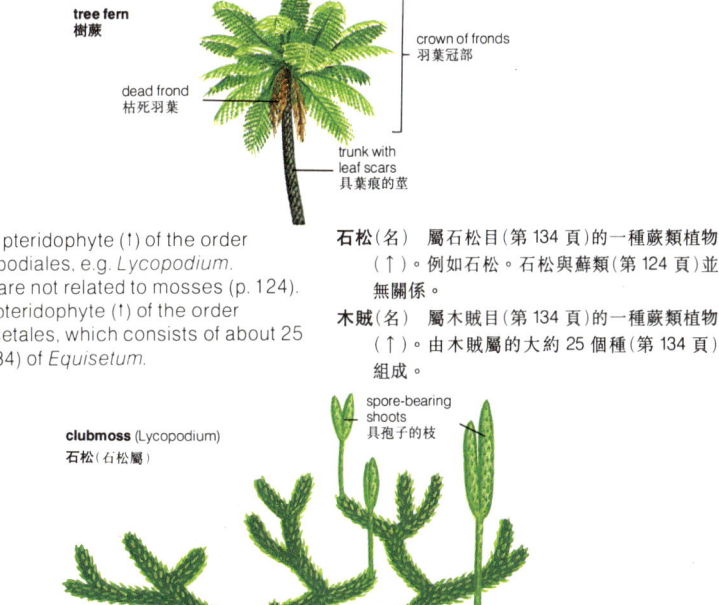

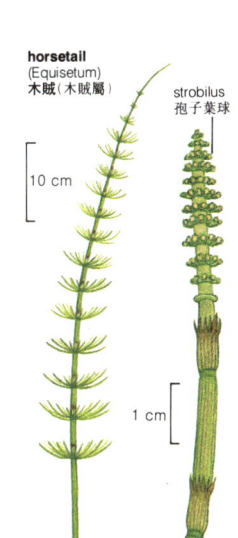

clubmoss (n) a pteridophyte (↑) of the order (p. 134) Lycopodiales, e.g. *Lycopodium*. Clubmosses are not related to mosses (p. 124).
horsetail (n) a pteridophyte (↑) of the order (p. 134) Equisetales, which consists of about 25 species (p. 134) of *Equisetum*.

石松（名） 屬石松目（第 134 頁）的一種蕨類植物（↑）。例如石松。石松與蘚類（第 124 頁）並無關係。
木賊（名） 屬木賊目（第 134 頁）的一種蕨類植物（↑）。由木賊屬的大約 25 個種（第 134 頁）組成。

cryptogam (*n*) a general name for all plants except gymnosperms (↓) and angiosperms (p. 130), in an old classification (p. 132) of the plant kingdom (p. 134). Cryptogams reproduce (p. 59) by spores (p. 66).

phanerogam (*n*) a general name for all seed plants (↓) in an old classification (p. 132) of the plant kingdom (p. 134), so called because the organs (p. 88) of reproduction (p. 59) are clearly visible. This name has been replaced by spermatophyte (↓).

spermatophyte (*n*) a member of the division (p. 134) Spermatophyta, or seed plants (↓). This division includes all angiosperms (p. 130) and gymnosperms (↓).

seed plant a plant which reproduces (p. 59) by seed, i.e. a spermatophyte (↑) or phanerogam (↑).

gymnosperm (*n*) a spermatophyte (↑) of the subdivision (p. 134) Gymnospermae. Gymnosperms differ from angiosperms (p. 130) in their unprotected ovules (p. 78), arrangement of reproductive (p. 59) organs (p. 88) in cones (p. 68), archegonia (p. 65) and wood without vessels (p. 107).

隱花植物(名) 植物界(第134頁)舊分類(第132頁)中除裸子植物(↓)及被子植物(第130頁)之外的一切植物的通稱。隱花植物以孢子(第66頁)繁殖(第59頁)。

顯花植物(名) 植物界(第134頁)舊分類(第132頁)中,一切種子植物(↓)的通稱,因其生殖(第59頁)器官(第88頁)明顯可見而得名。現已用種子植物(↓)代替此名稱。

種子植物(名) 是種子植物門(第134頁)的成員。英文亦稱 seed plant(↓)。此門植物包括全部被子植物(第130頁)和裸子植物(↓)。

種子植物 以種子繁殖(第59頁)的植物稱為種子植物(↑)或顯花植物(↑)。

裸子植物(名) 是種子植物(↑)亞門(第134頁)裸子植物。裸子植物與被子植物(第130頁)不同之處在於前者具有無保護的胚珠(第78頁),生殖(第59頁)器官(第88頁)安排在球果(第68頁),具有藏卵器(第65頁)以及木質部缺乏導管(第107頁)。

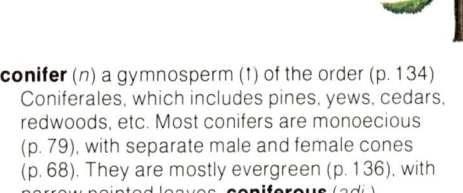

conifer 針葉樹

conifer (*n*) a gymnosperm (↑) of the order (p. 134) Coniferales, which includes pines, yews, cedars, redwoods, etc. Most conifers are monoecious (p. 79), with separate male and female cones (p. 68). They are mostly evergreen (p. 136), with narrow pointed leaves. **coniferous** (*adj.*).

針葉樹(名) 屬松柏目(第134頁)的一種裸子植物。包括松樹、紫杉、雪松、紅杉等。大多數針葉樹是雌雄同株(第79頁),雄球果與雌球果(第68頁)分開。大多數是常綠的(第136頁),並具有窄而尖的葉。(形容詞為 coniferous)

PLANT KINGDOM/GYMNOSPERMS 植物界／裸子植物・129

cycad 蘇鐵

cycad (n) a gymnosperm (↑) of the order (p. 134) Cycadales. Cycads are dioecious (p. 79), with motile (p. 121) male gametes (p. 61). They have palm- or fern-like leaves and are found mainly in tropical (p. 162) habitats (p. 149). The cycads are a primitive (p. 141) group, and there are many fossils (p. 142) dating from the Mesozoic (p. 143).

蘇鐵（名） 屬蘇鐵目（第 134 頁）的裸子植物（↑）。蘇鐵是雌雄異株（第 79 頁），具有游動的（第 121 頁）雄配子（第 61 頁），蘇鐵有棕櫚狀或真蕨狀的葉，主要長於熱帶（第 162 頁）生境（第 149 頁）。蘇鐵是一個原始的（第 141 頁）類群，現有許多是中生代（第 143 頁）的化石（第 142 頁）。

ginkgo 銀杏

Ginkgoales (n) an order (p. 134) of gymnosperms (↑) with only one living species, *Ginkgo biloba* (maidenhair tree), which is found in China. They have motile (p. 121) male gametes, like cycads (↑), but deciduous (p. 136) leaves.

Gnetales (n) a small order (p. 134) of gymnosperms (↑) made up of three genera (p. 134), *Ephedra*, *Gnetum*, and *Welwitschia*. Gnetales are similar to angiosperms (p. 130), because their wood contains vessels (p. 107) and their ovules (p. 78) do not have archegonia (p. 65).

銀杏目（名） 裸子植物（↑）一個目（第 134 頁），僅有一個生存的種。銀杏（銀杏樹）生長於中國。它們和蘇鐵（↑）一樣具有游動的（第 121 頁）雄配子，但會落葉（第 136 頁）。

買麻藤目（名） 屬裸子植物（↑）的一個小目（第 134 頁）。有三個屬（第 134 頁），即麻黃屬、買麻藤屬及千歲爹屬。買麻藤目近似被子植物（第 130 頁），因其木質部有導管（第 107 頁），胚珠（第 78 頁）不具藏卵器（第 65 頁）。

130 · PLANT KINGDOM/ANGIOSPERMS 植物界／被子植物

angiosperm (n) a spermatophyte (p. 128) of the subdivision (p. 134) Angiospermae. They differ from gymnosperms (p. 128) in that their ovules (p. 78) are protected in the ovary (p. 76) and in their possession of vessels (p. 107) in the xylem (p. 106). They also have double fertilization (p. 78) of the ovum (p. 61) and the endosperm (p. 86). There are more than 200 families (p. 134) and 250 000 species (p. 134) of angiosperms. They are divided into two classes (p. 134), the Monocotyledones (↓) and the larger Dicotyledones (↓).

flowering plant a plant which bears flowers. This term is usually used only for angiosperms (↑), but is sometimes used for some gymnosperms (p. 128) as well.

monocotyledon (n) an angiosperm (↑) of the class (p. 134) Monocotyledones. Their seeds have one cotyledon (p. 86). Monocotyledons do not have secondary thickening (p. 94) and most are small herbaceous (p. 136) plants with parallel (p. 110) venation (p. 97) and floral parts in whorls (p. 98) of three or multiples of three, e.g. grasses (↓), sedges (↓), orchids (↓), palms (↓). **monocot** (abbr.).

grass (n) a monocotyledon (↑) of the family (p. 134) Gramineae (sometimes called Poaceae). It is a very large family, and contains all the cultivated cereals (wheat, rice, etc.).

sedge (n) a monocotyledon (↑) of the genus (p. 134) Carex in the family (p. 134) Cyperaceae.

orchid (n) a monocotyledon (↑) of the family (p. 134) Orchidaceae. Most orchids are tropical (p. 162), and epiphytic (p. 137). The family is one of the largest in the plant kingdom (p. 134), with at least 17 000 species (p. 134).

palm (n) a monocotyledon (↑) of the family (p. 134) Palmae. Palms are the largest monocotyledons, and are found mostly in tropical (p. 162) forests (p. 158). They usually have a thick unbranched pachycaul (p. 94) trunk (p. 92), with rings on the surface. These rings mark the position of fallen leaves. Most palms have compound (p. 98) leaves, borne in a thick crown (p. 92) at the top of the trunk.

被子植物（名） 是種子植物（第128頁）中的一個亞門（第134頁），它和裸子植物（第128頁）的差別在於被子植物的胚珠（第78頁）包護在子房（第76頁）內，其木質部（第106頁）具有導管（第107頁）、卵（第61頁）及胚乳（第86頁）也具有受精作用（第78頁）。被子植物共有200多科（第134頁），250,000個種（第134頁），分為兩個綱（第134頁）：單子葉綱（↓）及較大的雙子葉綱（↓）。

有花植物 長有花朵的植物。此名通常只用於被子植物（↑），但有時亦用於某些裸子植物（第128頁）。

單子葉植物（名） 屬被子植物（↑）的單子葉植物綱（第134頁），其種子有一片子葉（第86頁）。單子葉植物沒有次生增厚（第94頁），多數是細小的草本（第136頁）植物，具平行（第110頁）脈序（第97頁）。花的各部以三或三的倍數成輪（第98頁）。例如禾草類（↓）、薹草類（↓）、蘭類（↓）和棕櫚類（↓）。單子葉植物英文縮寫為 monocot。

禾草（名） 屬禾本科（第134頁）的單子葉植物（↑），它是一個很大的科，包括全部栽培的禾穀類植物（小麥、水稻等）。

薹草（名） 屬於莎草科（第134頁）薹草屬（第134頁）的單子葉植物。

蘭（名） 屬蘭科（第134頁）的單子葉植物（↑）。蘭類大多是熱帶的（第162頁）附生（第137頁）植物。蘭是植物界（第134頁）最大的科之一，至少有17,000個種（第134頁）。

棕櫚（名） 屬棕櫚科（第134頁）的單子葉植物（↑），棕櫚是最大的單子葉植物，常見於熱帶（第162頁）森林（第158頁）。通常有一個粗而不分枝的粗短莖（第94頁）幹（第92頁），表面上有許多環圈，這些環圈標明已脫落的葉位置。棕櫚多數有複葉（第98頁），莖頂長著一個厚密的樹冠（第92頁）。

monocotyledons
單子葉植物

grass
禾本科

palm
棕櫚

orchid
蘭

dicotyledon (n) an angiosperm (↑) of the class (p. 134) Dicotyledones. The seeds of dicotyledons have two cotyledons (p. 86). Dicotyledons have secondary thickening (p. 94) in the stems. Most families (p. 134) and species (p. 134) of angiosperms are dicotyledons. **dicot** (abbr.).

雙子葉植物(名) 屬雙子葉植物綱(第134頁)的被子植物(↑)。雙子葉植物的種子有兩片子葉(第86頁), 莖有次生增厚(第94頁)。被子植物的大多數科(第134頁)及種(第134頁)都屬於雙子葉植物。其英文縮寫為 dicot。

dicotyledons 雙子葉植物
cactus 仙人掌
compocito 菊花
oak 橡樹
legume 豆科植物

cactus (n) a dicotyledon (↑) of the family (p. 134) Cactaceae, mainly found in hot dry areas in North and South America. Cacti (pl.) usually have thick succulent (p. 99) stems and groups of spines (p. 100) instead of leaves.

Leguminosae (n) a large family (p. 134) of dicotyledons (↑). The seeds of Leguminosae are borne in legumes (p. 84) or pods (p. 84). Many species (p. 134) of Leguminosae are important crop plants, e.g. beans, peas, clover, etc.

Compositae (n) a large family (p. 134) of dicotyledons (↑). Compositae have composite (p. 81) inflorescences (p. 80), as in daisies.

仙人掌(名) 仙人掌科(第134頁)的雙子葉植物(↑), 主要見於北美洲及南美洲的乾熱地區。仙人掌通常有厚的肉質(第99頁)莖及成群葉片變成的刺(第100頁)。

豆科(名) 雙子葉植物(↑)的一個大科(第134頁)。豆科的種子長在豆莢(第84頁)或莢果(第84頁)內;豆科的許多種(第134頁)是重要的農作物, 如菜豆、豌豆和三葉草等。

菊科(名) 雙子葉植物(↑)一個大科(第134頁)。菊科植物具有菊花(第81頁)花序(第80頁), 如雛菊。

CLASSIFICATION/GENERAL 分類／概述

classification (n) the naming of species (p.134) and their grouping into families (p.134), orders (p.134), divisions (p.134), etc. **classify** (v).

分類（名） 物種（第134頁）的命名及其分門別類為 科（第134頁）、目（第134頁）、門（第134頁）等。（動詞為 classify）

how a species is classified in the plant kingdom
植物界的一個物種怎樣分類

common oak 普通橡樹	*Quercus robur* 橡樹	species 種
oaks 橡樹	*Quercus* 櫟屬	genus 屬
beeches, chestnuts, oaks 山毛櫸、板栗、橡樹	Fagaceae 山毛櫸科	family 科
beeches, chestnuts, oaks, birches, alders, hazels, hornbeams 山毛櫸、板栗、橡樹、樺木、赤楊、榛、角木	Fagales 山毛櫸目	order 目
dicotyledons 雙子葉植物	Dicotyledones 雙子葉植物綱	class 綱
flowering plants 有花植物	Angiospermae 被子植物	subdivision 亞門
seed plants 種子植物	Spermatophyta 種子植物門	division 門
plants 植物	Plantae 植物界	kingdom 界
common name 普通名稱	Latin name 拉丁名稱	taxon 分類單元

nomenclature (n) the part of classification (↑) that involves the naming of species (p.134), families (p.134), orders (p.134), etc.

systematics (n) the part of classification (↑) that involves the arrangement of organisms (p.118) into related groups.

命名法（名） 為分類（↑）部分，包括對種（第134頁）、科（第134頁）、目（第134頁）等的命名。

系統；分類學（名） 為分類（↑）部分，包括將生物（第118頁）排列成有關的分類群。

CLASSIFICATION/TAXONOMY 分類／分類學

Linnaeus Carl Linnaeus, (1707-1778) was responsible for the modern system of naming plants and animals. This is the binomial (↓) system. His most famous work was *Species Plantarum*, published in 1753, in which he described all the plants then known to man.

binomial (*n*) the Latin name of a species (p. 134), consisting of two words. The first is the name of the genus (p. 134) to which the species belongs, and the second is the name which distinguishes the species from other species in the same genus. This system of naming species was introduced by Linnaeus (↑). **binomial** (*adj*).

authority (*n*) the name of the author who was the first to give a species (p. 134) or other taxon (↓) its name. In the case of species, the authority is given after the binomial (↑).

herbarium (*n*) a collection of dried pressed plants, used by taxonomists in the classification (↑) of plants.

type (*n*) the specimen of the individual (p. 135) plant from which a species (p. 134) was first described.

artificial key a way of identifying plants by steps. Each step involves a choice between at least two different characters, each of which leads to another choice of two characters, finally leading to the right identification.

taxonomy (*n*) the science of classification (↑) and the relationships of organisms. **taxonomic** (*adj*), **taxonomist** (*n*).

taxon (*n*) any taxonomic (↑) group, e.g. a species (p. 134), a family (p. 134). All the members of a taxon share similar characteristics, which are different from those of other groups. **taxa** (*pl.*).

characteristic (*adj*) of characters by which an organism or group of organisms can be recognized. For example, flowers are characteristic of angiosperms (p. 130), wood is characteristic of trees. **characteristic** (*n*).

character (*n*) any part or shape of an organism that makes it possible to classify (↑) the organism. Characters used in classification include the arrangement of the reproductive (p. 59) organs (p. 88), the shape of leaves, etc.

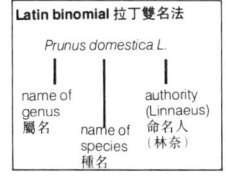

Latin binomial 拉丁雙名法
Prunus domestica L.
name of genus 屬名
name of species 種名
authority (Linnaeus) 命名人 (林奈)

林奈 (1707-1778) 他創建植物及動物的現代命名制，即雙名 (↓) 制。林奈最著名的著作是 1753 年出版的"植物種誌"，書中記述當時人們所知的全部植物。

雙名法；二名法（名）物種（第 134 頁）的拉丁名稱，由兩個詞組成，第一個詞為屬名（第 134 頁），是物種依歸的屬；第二個詞為種名，用以區別同一屬中的各個種。物種命名制是林奈（↑）創建的。（形容詞為 binomial）

命名人（名）作者之名，即第一個為一個種（第 134 頁）或其分類單元（↓）起名的人。命名人的名字放在物種雙名（↑）之後。

植物標本室（名）分類學家用於植物分類（↑）的壓乾植物標本收藏處。

模式標本（名）首次用於描述一個個體（第 135 頁）植物種（第 134 頁）的標本。

人為檢索表 按步驟鑑別植物的方式。每一步驟都包括至少在兩個不同的特徵之間作選擇，其中每一特徵都引向選擇兩個特徵中的另一特徵，最後引向準確的鑑別。

分類學（名）生物親緣關係及分類（↑）的學科。（形容詞為 taxonomic，名詞 taxonomist 意為分類學家）

分類單元（名）任一分類（↑）群。例如一個種（第 134 頁），一個科（第 134 頁）。同一個分類單元中的全部成員都有相似的特徵，使之和其他類群有所區別。（複數為 taxa）

特徵的（形）指一個生物體或一群生物具有可資識別的特徵。例如花是被子植物（第 130 頁）的特徵；木材是喬木的特徵。（名詞 characteristic 意為特徵、特性）

特徵（名）使生物體能被分類（↑）的任何部分或形狀。分類上所用的特徵包括生殖（第 59 頁）器官（第 88 頁）的排列、葉的形狀等等。

134 · CLASSIFICATION/TAXA 分類/分類單元

kingdom (*n*) the largest of all the taxa (p. 133). In older systems of classification (p. 132), there are only two kingdoms – plants and animals. In some newer systems, there are five – plants, animals, fungi (p. 163), bacteria (p. 119) and viruses (p. 118); before, the fungi, bacteria and viruses were included with the plants.

division (*n*) a major taxon (p. 133), e.g. bryophytes, which is made up of classes (↓). There are three divisions of land plants; these are bryophytes (p. 122), pteridophytes (p. 126) and spermatophytes (p. 128).

subdivision (*n*) a taxon (p. 133) within a division (↑).

class (*n*) a taxon (p. 133) consisting of orders (↓). Dicotyledons (p. 131) are a class.

order (*n*) a taxon (p. 133) consisting of families (↓). The Latin names of orders usually end with -ales.

family (*n*) a taxon (p. 133) consisting of related genera (↓). The Latin names of families usually end with -aceae.

tribe (*n*) a group of related genera (↓) within a family (↑).

genus (*n*) a group of related species (↓). The name of the genus is the first in a Latin binomial (p. 133). **genera** (*pl.*), **generic** (*adj*).

monotypic (*adj*) of a genus (↑) with only one species (↓), or a family (↑) with only one genus.

species (*n*) usually, the smallest unit of classification (p. 132). A species includes individuals (↓) which are alike and can breed (p. 59) with each other. Species are given Latin binomial (p. 133) names. They are sometimes divided into subspecies (↓) and varieties (↓) on the basis of small differences between populations (↓). **specific** (*adj*).

界(名) 全部分類單元(第133頁)中最大的分類階層。舊分類(第132頁)制只分植物和動物兩界。一些新的分類制則分植物、動物、真菌(第163頁)、細菌(第119頁)及病毒(第118頁)五界。過去是將真菌、細菌及病毒包括在植物界內。

門(名) 一個大分類單元(第133頁)，例如苔蘚植物門；門之下又分綱(↓)。陸生植物有三門，即苔蘚植物門(第122頁)、蕨類植物門(第126頁)及種子植物門(第128頁)。

亞門(名) 門(↑)之內的分類單元(第133頁)。

綱(名) 一個分類單元(第133頁)，下分許多目(↓)。雙子葉植物(第131頁)是一個綱。

目(名) 一個分類單元(第133頁)，下分許多科(↓)。目的拉丁名稱詞尾為 -ales。

科(名) 一個分類單元(第133頁)，下分許多有關屬(↓)。科的拉丁名稱詞尾為 -aceae。

族(名) 一科(↑)之內有關的屬(↓)的類群。

屬(名) 指一群有親緣關係的種(↓)，拉丁雙名法的(第133頁)第一個詞為屬名。(複數為 genera，形容詞為 generic)

單型的(形) 指一個屬(↑)只有一個種(↓)或一個科(↑)只有一個屬。

物種(名) 通常而言，種是最小的分類(第132頁)單元。一個種包括許多個體(↓)，這些個體很相似並且能彼此繁殖(第59頁)。種有拉丁雙名(第133頁)的名稱。有時還根據種群(↓)間的細小差異將種劃分為亞種(↓)及變種(↓)。(形容詞為 specific)

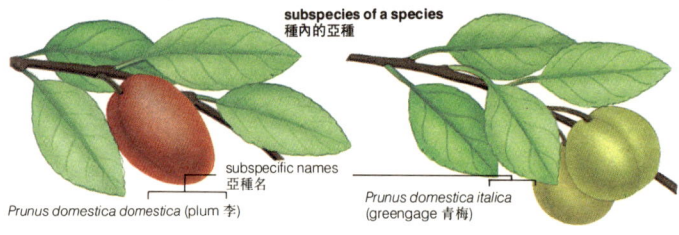

subspecies of a species 種內的亞種

subspecific names 亞種名

Prunus domestica domestica (plum 李)

Prunus domestica italica (greengage 青梅)

subspecies (*n*) a taxon (p.133) within a species (↑). Subspecies of a species differ in small ways. Although they can breed (p.59) with each other, they are usually found in different places, or different populations (↓). In the naming of subspecies, a third Latin name is put after the binomial (p.133). **subspecific** (*adj*)

variety (*n*) a taxonomic (p.133) group within a species (↑) or a subspecies (↑). The differences between varieties are small, and do not necessarily relate to differences in habitat (p.149) or place.

strain (*n*) a reproductively (p.59) isolated (p.142) population (↓) whose individuals (↓) have identical genotypes (p.41) over many generations (p.63) and show phenotypic (p.41) differences from other populations.

ecotype (*n*) a set of individuals (↓) or populations (↓) in a particular habitat (p.149) that differ phenotypically from members of the same species in other habitats.

cline (*n*) continuous or gradual variation (↓), within a population (↓), in some of the characters of a species (↑). This variation is related to gradual changes in ecological (p.149) conditions that occur, for instance, up the side of a mountain.

individual (*n*) a single organism.

variation (*n*) differences in the characteristics of individuals (↑) of a species (↑) or a population (↓).

infraspecific (*adj*) of variation (↑) between individuals (↑) of a species (↑).

polymorphism (*n*) the occurrence of two or more forms of a species (↑) in the same population (↓) or habitat (p.149). **polymorphic** (*adj*)

morph (*n*) one of the forms of a polymorphic (↑) species (↑).

population (*n*) a group of individuals (↑) of the same species (↑) living in the same place or area, close enough to breed (p.59) together.

endemic (*adj*) of taxa (p.133) that are found only in one particular place or area. **endemism** (*n*).

distribution (*n*) the whole geographical range in which a taxon (p.133) is found.

flora (*n*) the total of plant species (↑) in a particular region, country, continent, etc.

亞種（名） 一個種（↑）內的分類單元（第133頁），一個種的各個亞種之間稍微有些差異。雖然它們彼此能繁殖（第59頁），但常存在於不同的地方或不同的種群（↓）。亞種的命名是在雙名法（第133頁）後面加入第三個拉丁名。（形容詞為 subspecific）

變種（名） 種（↑）內或亞種（↑）內的一個分類（第133頁）群。變種之間的差異微小，不一定涉及生境（第149頁）或地點的差異。

品系（名） 生殖（第59頁）上隔離（第142頁）的種群（↓），它們的個體（↓）經歷許多世代（第63頁），已有相同的基因型（第41頁），並表現出和其他種群不同的表現型（第41頁）。

生態型（名） 在特定生境（第149頁）的一組個體（↓）或種群（↓），它們和生長在其他生境中的同種的成員在表型方面有差異。

生態群（名） 一個種群（↓）內，一個種（↑）的某些特徵方面出現連續的或逐漸的變異（↓）。這種變異和生態（第149頁）條件（例如在山的高坡上）出現的逐漸變化有關係。

個體（名） 單個的生物體。

變異（名） 一個種（↑）或一個種群（↓）的各個個體（↑）在特徵上的差異。

種以下的（形） 指一個種（↑）的個體（↑）間的變異（↑）。

多態性（名） 一個種（↑）在相同種群（↓）或生境（第149頁）出現兩個或多個形態。（形容詞為 polymorphic）

型（名） 多態的（↑）種（↑）的形態之一。

種群（名） 生長在相同地方或相同地區，足夠親近可以一起繁殖（第59頁）的一群同種（↑）個體（↑）。

特有的（形） 指僅見於一個特定地方或地區的分類單元（第133頁）。（名詞 endemism 意為特有分佈）

分佈（名） 一個分類單元（第133頁）所見的整個地理分佈範圍。

植物誌（名） 在一個特定地區、國家、大陸等的植物種（↑）的總和。

habit (*n*) the appearance of a plant, e.g. herb (↓), shrub (↓), tree (↓).
tree (*n*) a large, perennial (p. 117) plant with a woody trunk (p. 92), which usually bears branches.
sapling (*n*) a young tree (↑).
evergreen (*adj*) having green leaves throughout the year. **evergreen** (*n*).
deciduous (*adj*) (1) of plants which shed their leaves at least once a year and remain leafless for weeks or months; (2) of short-lived organs (p. 88) that are shed from a plant.
shrub (*n*) a small woody perennial (p. 117) plant, with branches from ground level upwards.

習性(名) 植物的外貌，如草本植物(↓)、灌木(↓)、喬木(↓)。
喬木(名) 高大的多年生(第117頁)植物，具有木質莖(第92頁)，常有分枝。
幼樹(名) 幼小的喬木(↑)。
常綠的(形) 指一整年都有綠葉的。(名詞為 evergreen)
落葉的；脫落的(形) (1)指植物每年至少落葉一次並保持數週或數月無葉；(2)指短命的器官(第88頁)從植物體上脫落。
灌木(名) 矮小的多年生(第117頁)木本植物，枝幹從地面處向上叢生。

two types of tree
喬木的兩種類型

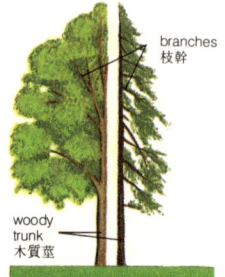

branches 枝幹
woody trunk 木質莖

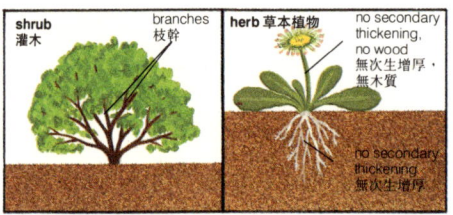

bush (*n*) a shrub (↑) with especially dense branching and foliage (p. 96).
herb (*n*) a small plant without wood in the stems or roots. **herbaceous** (*adj*).
climber (*n*) a plant with roots in the ground, which uses other plants to support itself. Climbers use tendrils (↓), adventitious roots (p. 89), or sucker-like discs to hold on to other plants, or sometimes twist around their stems.
tendril (*n*) a long, coiled, threadlike organ (p. 88), borne on the stems or leaves of many climbers (↑). Tendrils coil around the stems and branches of other plants, and help the climber to support itself. Tendrils are modified leaves, stems, leaflets (p. 98) or stipules (p. 99).
creeper (*n*) a plant which cannot support itself and spreads along the ground. Creepers usually have little or no secondary thickening (p. 94).
vine (*n*) (1) a climber (↑); (2) a member of the family (p. 134) Vitaceae.

叢枝灌木(名) 分枝及葉叢(第96頁)特別茂密的灌木(↑)。
草本植物(名) 莖及根都沒有木質的矮小植物。(形容詞為 herbaceous)
攀緣植物(名) 一種根長在地下，藉助其他植物支持本身的植物。攀緣植物用捲鬚(↓)、不定根(第89頁)或類吸盤固著在別的植物，或纏捲在該植物的莖上。
捲鬚(名) 一種細長捲曲的線狀器官(第88頁)，可長在許多種攀緣植物(↑)的莖或葉上。捲鬚纏捲在其他植物的莖及枝上，使攀緣植物得以自持。捲鬚為變態的葉、莖、小葉(第98頁)或托葉(第99頁)。
葡匐植物(名) 不能自持直立而沿地面散佈的一種植物。葡匐植物常有少量或者沒有次生增厚(第94頁)。
藤本植物(名) (1)攀緣植物(↑)；(2)屬葡萄科(第134頁)的植物。

climbers 攀緣植物
three examples of climbing plants
攀緣植物的三個例子

tendril 捲鬚
1 *Lathyrus*
羽扇豆
a plant that climbs by means of tendrils
利用捲鬚攀緣的植物

2 *Hedera*
常春藤
a plant that climbs by means of adventitious roots
利用不定根攀緣的植物
adventitious roots 不定根

3 *Convolvulus*
旋花
a plant that climbs by twisting growth
用纏繞生長而攀緣的植物

HABITS 習性 · 137

saprophyte
腐生植物
e.g. *Amanita muscari* (a fungus)
如蠅毒傘（一種真菌）

no photosynthetic tissue
無光合作用組織

dead organic matter 非生活有機物

saxicolous plant e.g. lichen 岩生植物如地衣

rock 岩石

xerophyte 旱生植物
e.g. cactus 如仙人掌

thick succulent stems with thick cuticle
具角質的厚肉質莖及厚角質葉

hot dry air
乾熱空氣

spines
針刺

dry desert sand
乾燥的沙漠沙土

liana (n) a woody perennial (p. 117) climber (↑). Lianas are very common in tropical (p. 162) rain forests (p. 158).

saprophyte (n) a plant which obtains all its nutrients (p. 111) from dead organic (p. 11) matter. Saprophytes are heterotrophic (p. 32). Many fungi (p. 163), bacteria (p. 119) and some vascular (p. 122) plants are saprophytes in the soil. They are important in the cycles of inorganic (p. 11) nutrients in ecosystems (p. 149). **saprophytic** (adj).

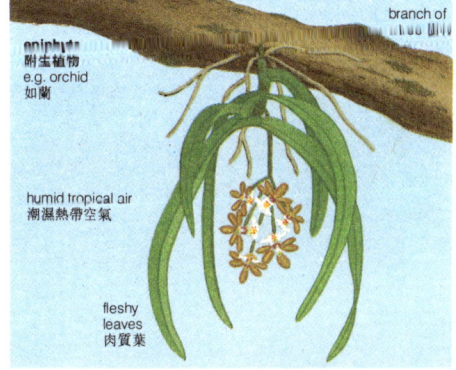

branch of
epiphyte
附生植物
e.g. orchid
如蘭

humid tropical air
潮濕熱帶空氣

fleshy leaves
肉質葉

epiphyte (n) a plant which grows on the stems and branches of other plants. Epiphytes have no roots in the ground. They are not parasitic (p. 144). **epiphytic** (adj).

epiphyll (n) a plant which grows on the leaves of other plants. Most epiphylls are tropical (p. 162) liverworts (p. 123) and lichens (p. 147).
epiphyllous (adj).

saxicolous (adj) of plants that live on rocks, e.g. lichens (p. 147).

xerophyte (n) a plant that lives in a desert or other dry habitat (p. 149). **xerophytic** (adj).

xeromorphic (adj) of plants with characteristics suited to very dry habitats (p. 149), such as deserts. Xeromorphic plants often have thick, succulent (p. 99) leaves with thick cuticles (p. 95) to prevent the loss of water.

籐本植物（名） 木質多年生（第117頁）攀緣植物（↑）。籐本植物在熱帶（第162頁）雨林（第158頁）非常普遍。

腐生植物（名） 從非生活有機（第11頁）取得其全部營養分（第111頁）的一種植物。腐生植物是異養（第32頁）植物。許多種真菌（第163頁）、細菌（第119頁）及某些維管（第122頁）植物都是長在土壤中的腐生植物。這些植物在生態系統（第149頁）的無機（第11頁）營養物循環上有重要意義。（形容詞為saprophytic）

附生植物（名） 長在另一種植物的莖及枝上的植物。附生植物的根不着地，它們不是寄生（第144頁）植物。（形容詞為epiphytic）

葉附生植物（名） 長在另一種植物葉面上的植物。大多數葉附生植物是熱帶的（第162頁）苔類植物（第123頁）及地衣（第147頁）。（形容詞為epiphyllous）

岩生的（形） 指長在岩石上的植物，如地衣（第147頁）。

旱生植物（名） 長在沙漠或其他乾燥生境（第149頁）的植物。（形容詞為xerophytic）

旱生結構的（形） 指具有適應極乾旱生境（第149頁）如沙漠的特徵的植物。旱生結構的植物常具有厚的肉質（第99頁）及厚角質（第95頁）的葉，以防止水分耗失。

halophyte (*n*) a plant which lives in salty habitats (p. 149). Halophytes are adapted (p. 141) for the uptake of water from concentrated solutions (p. 12). **halophytic** (*adj*).

cryophyte (*n*) a plant growing in very cold conditions, e.g. on ice or snow. Cryophytes are usually algae (p. 119) or bryophytes (p. 122).

cryptophyte (*n*) a plant with buds (p. 110) or shoot apices (p. 90) perennating (p. 117) underground or under water during unfavourable seasons, e.g. winter.

geophyte (*n*) a cryptophyte (↑) with buds (p. 110) or shoot apices (p. 90) perennating (p. 117) underground. Geophytes have rhizomes (p. 60), bulbs (p. 60) or corms (p. 60), etc. **geophytic** (*adj*).

鹽土植物（名） 適應在高鹽分土壤生境（第149頁）中生長的植物。鹽土植物能適應（第141頁）從高濃度溶液（第12頁）中吸收水分。（形容詞為 halophytic）

冰雪植物（名） 生在嚴寒條件下，例如在冰雪地上生長的植物。冰雪植物常為藻類（第119頁）或苔蘚植物（第122頁）。

隱芽植物（名） 在不良季節，如冬季能以芽（第110頁）或枝端（第90頁）多年生（第117頁）在地下或在水下的植物。

地下芽植物（名） 有芽（第110頁）或枝端（第90頁）多年生（第117頁）在地下的隱芽植物（↑）。地下芽植物有根狀莖（第60頁）、鱗莖（第60頁）、或球莖（第60頁）等。（形容詞為 geophytic）

geophyte
e.g. *Narcissus*
地下芽植物如水仙

bulb perennating underground during winter
球莖多年生在地下越冬

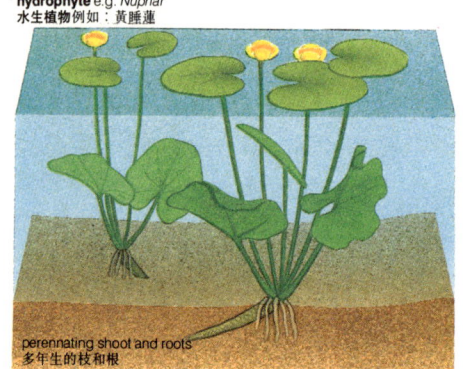

hydrophyte e.g. *Nuphar*
水生植物例如：黃睡蓮

perennating shoot and roots
多年生的枝和根

hydrophyte (*n*) a cryptophyte (↑) with buds (p. 110) or shoot apices (p. 90) perennating (p. 117) under water during unfavourable seasons.

hemicryptophyte (*n*) a plant with buds (p. 110) perennating (p. 117) at the surface of the soil.

chamaephyte (*n*) a woody or herbaceous (p. 136) plant less than 25 cm tall, with buds (p. 110) perennating (p. 117) above the ground.

phanerophyte (*n*) a woody plant with buds (p. 110) perennating (p. 117) more than 25 cm above the surface of the ground.

水生植物（名） 不良季節時能以芽（第110頁）及枝端（第90頁）多年生（第117頁）在水下的隱芽（↑）植物。

半隱芽植物（名） 有芽（第110頁）多年生（第117頁）在土壤表面的隱芽植物。

地上芽植物（名） 有芽（第110頁）多年生（第117頁）在離地面不到25厘米的木質或草本（第136頁）植物。

高位芽植物（名） 有芽（第110頁）多年生（第117頁）在離地面25厘米以上位置的木本植物。

evolution (*n*) the changes that occur in organisms over many generations (p. 63) and long periods of time. Evolution occurs by the natural selection (p. 140) of mutations (p. 54). **evolutionary** (*adj*), **evolve** (*v*).

進化（名） 生物體經過許多世代（第63頁）及漫長時間所發生的變化。進化是由於突變（第54頁）的自然選擇（第140頁）結果。（形容詞為 evolutionary，動詞為 evolve）

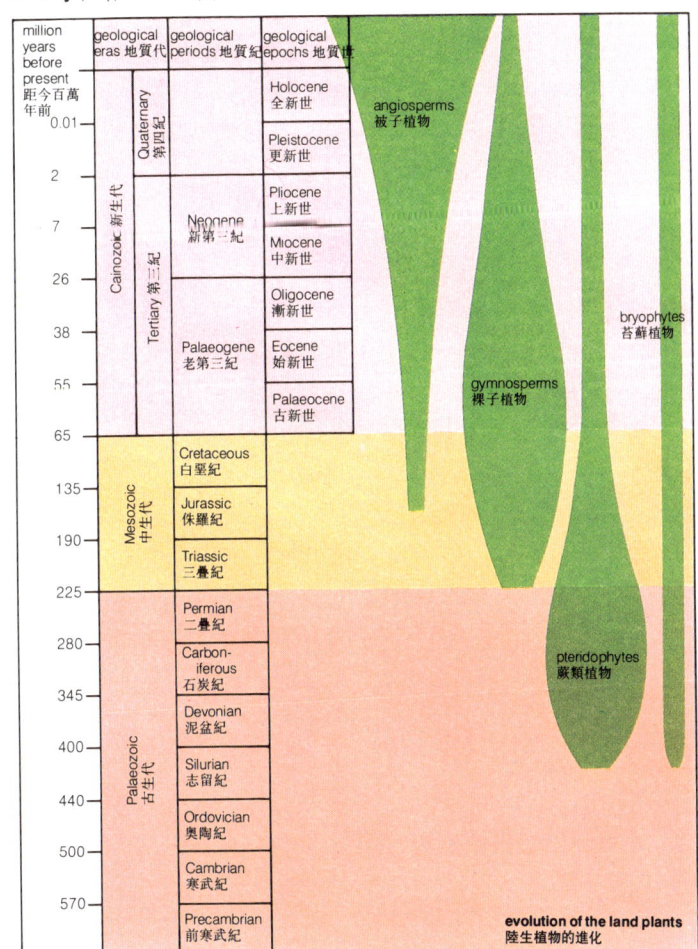

evolution of the land plants
陸生植物的進化

Lamarck Jean-Baptiste Lamarck (1744-1829) proposed that evolution (p.139) occurred as a result of the inheritance (p.41) of changes which happen to an organism during its lifetime. This is known as the Theory of Inheritance of Acquired Characters.

Darwin Charles Darwin (1809-1882) proposed that evolution (p.139) occurs by natural selection (↓) and the survival of the fittest (↓). His book on the subject, *On the Origin of Species*, was published in 1858. Darwin's theory of evolution is often known as Darwinism.

survival of the fittest the ability of some individuals (p.135) in a population (p.135) to live until after they have reproduced (p.59), because they have characters which enable them to survive for longer than those individuals which die before reproduction. This is natural selection (↓).

natural selection Darwin's (↑) theory of evolution (p.139), also called the survival of the fittest (↑). Natural selection is the selection by the environment (p.149) of the individuals (p.135) in a population (p.135) which are fittest. Only these individuals live until reproductive (p.59) age, and pass their genes (p.41) to future generations (p.63).

artificial selection the process by which man selects plant varieties (p.135) or individuals (p.135) with useful characters or traits (p.41), e.g. wheat plants with large seeds, in order to breed (p.59) them for his own purposes. Because of this, many cultivated plants appear to be very different from their wild ancestors.

neo-Darwinism (*n*) the theory of evolution (p.139) developed in the 20th century, after Darwin's (↑) death. It includes Darwin's theory of natural selection (↑) and the more recent knowledge of genetics (p.41) and inheritance (p.41) through chromosomes (p.46).

telome theory a theory of the evolution (p.139) of the organs (p.88) of plants, especially the branches and leaves, starting with very simple leafless primitive (↓) pteridophytes (p.126) with green stems.

拉馬克(1744-1829) 他提出，生物進化(第139頁)是由於生物體在其一生中對其所發生的種種變化獲得遺傳(第41頁)的結果。這是著名的"獲得性狀遺傳學說"。

達爾文(1809-1882) 他提出，生物進化(第139頁)是由於自然選擇(↓)和適者生存(↓)而發生的。他所著的《物種起源》於1858年出版。達爾文的進化論通常稱為達爾文主義。

適者生存 一個種群(第135頁)中的某些個體(第135頁)能夠生存直至完成生殖(第59頁)之後，是因為這些個體具有的性狀使他們能比那些在生殖之前死去的個體生存更長時間，這就是自然撰擇(↓)。

自然選擇 為達爾文(↑)的進化(第139頁)學說，亦稱適者生存(↑)。自然選擇是在一個種群(第135頁)中，那些最適應環境的個體(第135頁)對環境(第149頁)的選擇。只有這些個體能生存到生殖的年齡，並傳遞其基因(第41頁)給未來的後代(第63頁)。

人工選擇 人為選擇具有有用性狀或特質(第41頁)的植物變種(第135頁)或個體(第135頁)的過程。例如選擇大粒種的小麥，使人們能按自己的要求繁育(第59頁)這些小麥。因此許多栽培植物顯得和野生祖先十分不同。

新達爾文主義(名) 是達爾文(↑)逝世後，在二十世紀創立的進化(第139頁)學說。該學說包括達爾文的自然選擇(↑)學說、更近代的遺傳學(第41頁)知識以及通過染色體(第46頁)遺傳(第41頁)的知識。

頂枝學說 植物的各器官，(第88頁)尤其是枝及葉是從很簡單、無葉，具綠色莖的原始(↓)蕨類植物(第126頁)開始進化(第139頁)的學說。

neoteny (*n*) the condition in which some of the characters of an embryonic (p. 85) or young organism are found in the mature organism of reproductive (p. 59) age. Neoteny may have been important in the evolution (p. 139) of flowering plants (p. 130).

phylogeny (*n*) the evolutionary (p. 139) history of an organism or taxonomic (p. 133) group of organisms. **phylogenetic** (*adj*).

primitive (*adj*) of organisms which have the characteristics of early stages in evolutionary (p. 139) history. This can also apply to characters and organs (p. 88).

extinct (*adj*) of species (p. 134) which no longer exist.

extant (*adj*) of species (p. 134) which exist at present.

adaptation (*n*) a character or set of characters of an organism which help the organism to survive and reproduce (p. 59) in a particular habitat (p. 149). **adapt** (*v*), **adaptive** (*adj*).

adaptive radiation a process of evolution (p. 139) from one primitive (↑) species (p. 134) to more than one advanced species, each adapted (↑) to a particular niche (p. 153) or habitat (p. 149).

幼態成熟（名） 胚胎（第 85 頁）或幼嫩有機體的某些特徵，見於生殖（第 59 頁）年齡的成熟有機體的狀態。幼態成熟在有花植物（第 130 頁）的進化（第 139 頁）上也許具有重要意義。

系統發育（名） 一個生物或生物分類（第 133 頁）群的進化（第 139 頁）史。（形容詞為 phylogenetic）

原始的（形） 指在進化（第 139 頁）史中具有早期階段特徵的生物體。這個詞亦可用於描述各種特徵及器官（第 88 頁）。

滅絕的（形） 指已不生存的生物種（第 134 頁）。

現存的（形） 指現仍生存的生物種（第 134 頁）。

適應（名） 有助生物在特定生境（第 149 頁）中存活及生殖（第 59 頁）的生物體的一個性狀或一組性狀。（動詞為 adapt，形容詞為 adaptive）

適應輻射 從一個原始的（↑）的生物種（第 134 頁）到一個以上的進化物種的進化（第 139 頁）過程，其中的每個種都適應（↑）於一特定的生態區位（第 153 頁）或生境（第 149 頁）。

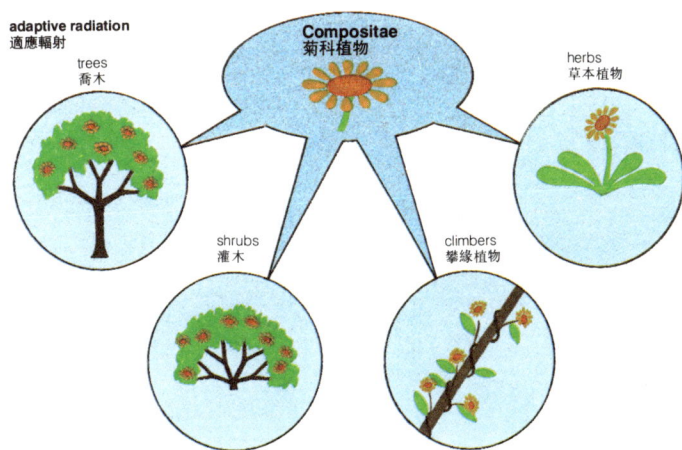

142 · EVOLUTION/SPECIATION, PALAEOBOTANY 進化／物種形成、古植物學

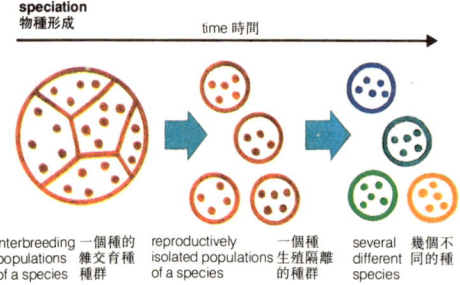

speciation 物種形成
time 時間
interbreeding populations of a species 一個種的雜交育種種群
reproductively isolated populations of a species 一個種生殖隔離的種群
several different species 幾個不同的種

sympatric 同域的
e.g. two species occuring in the same place
例如兩個種分佈在同一區域

allopatric 異區起源的
e.g. two species occuring in different places
例如兩個種分佈在不同區域

speciation (n) the evolutionary (p. 139) process in which new species (p. 134) are produced. **speciate** (v).

sympatric (adj) of two species (p. 134) which live in the same place, or of speciation (↑) in which two or more species evolve (p. 139) in the same place. **sympatry** (n).

allopatric (adj) of two or more species (p. 134) living in separate places, or of speciation (↑) in which different species evolve (p. 139) in different places. **allopatry** (n).

reproductive isolation a situation in which two individuals (p. 135) or populations (p. 135) cannot breed (p. 59) with each other.

fossil (n) the remains or marks left by dead organisms, converted to stone over geological time (↓). Fossils provide important clues to the history and evolution (p. 139) of living organisms. **fossilize** (v).

palaeobotany (n) the study of plants of the past, using fossils (↑).

palynology (n) the study of pollen (p. 74) remains and fossil (↑) pollen. This is an important way of studying the history of vegetation (p. 150). Pollen grains can last for thousands or millions of years, and it is possible to identify their genus (p. 134) or family (p. 134) by their shape and surface patterns.

pollen diagram a diagram showing the variation in the pollen (p. 74) content of sediments over long periods of time (often tens of thousands of years). Pollen diagrams are used to determine the history of vegetation (p. 150).

物種形成（名） 進化（第139頁）過程中產生新的物種（第134頁）。（動詞為 speciate）

同域的（形） 指兩個物種（第134頁）生活在同一區域，或指物種形成（↑）係由兩個或多個物種在同一區域演變（第139頁）而成。（名詞為 sympatry）

異區起源的（形） 指兩個或多個物種（第134頁）生活在隔離的區域；或指物種形成（↑）係由不同的種在不同區域演變（第139頁）而成。（名詞為 allopatry）

生殖隔離 兩個個體（第135頁）或種群（第135頁）彼此不能繁育（第59頁）的現象。

化石（名） 死亡生物體的遺骸或遺跡經過地質年代轉化為石（↓）。化石提供生存生物體的歷史及進化（第139頁）的重要線索。（動詞為 fossilize）

古植物學（名） 利用化石（↑）研究歷史上的植物的學科。

孢粉學（名） 研究花粉遺體或花粉（第74頁）化石（↑）的學科。這是研究植被（第150頁）歷史的重要途徑。花粉粒的壽命可長達數千或數百萬年，可通過花粉的形態及表面圖案鑑別出它的屬（第134頁）或科（第134頁）。

花粉式 顯示保存於沉積物中經過悠久的一段時間（往往幾萬年）的花粉（第74頁）變異的圖解。花粉式用來決定植被（第150頁）的歷史。

geological time the sequence of geological eras (↓) in the earth's history, measured in millions of years. Geological time began about 4½ thousand million years ago, when the earth was formed. *See also* page 139.

geological era a very long division of geological time (↑), lasting tens of millions of years, whose beginning and end are recognized by major changes in layers of rocks and fossils (↑) in the earth. Geological eras are divided into geological periods (↓) and epochs (↓).

geological period a major subdivision of a geological era (↑).

geological epoch a subdivision of a geological period (↑).

Palaeozoic (*n*) the geological era (↑) which ended about 225 million years ago, before the evolution (p. 139) of the angiosperms (p. 130). Pteridophytes (p. 126) were the dominant (p. 150) plants on land during this time.

Carboniferous (*n*) the geological period (↑) 345–280 million years ago, during the Palaeozoic (↑). The world's forests (p. 158) were dominated (p. 150) by tree-like pteridophytes (p. 126) which were fossilized (↑) to form coal.

Mesozoic (*n*) the geological era (↑) 225–65 million years ago, following the Palaeozoic (↑), with gymnosperms (p. 128) as the dominant (p. 150) plants on land. Angiosperms (p. 130) first appeared during the Mesozoic.

Cainozoic (*n*) the geological era (↑) which began 65 million years ago, after the Mesozoic (↑), and is still continuing. Angiosperms (p. 130) have been the dominant (p. 150) plants on land throughout this time. Also known as **Cenozoic**

Tertiary (*n*) one of two geological sub-eras (↑) into which the Cainozoic (↑) is divided. It began about 65 million years ago and ended about 2 million years ago.

Quaternary (*n*) the second geological sub-era (↑) of the Cainozoic (↑). It began with the Pleistocene (↓) and has lasted until the present.

Pleistocene (*n*) the geological epoch (↑) which began about 2 million years ago and ended 10 thousand years ago with the last Ice Age.

地質年代 地球歷史的地質代(↓)順序，以百萬年為計算單位。地質年代從距今約45億年前，地球形成時算起(參見第139頁)。

地質代 是地質年代(↑)中一段很漫長的分期，持續幾千萬年，它的開始和結束可由地球岩層及化石(↑)的重大變化來識別。地質代又劃分為地質紀(↓)及地質世(↓)。

地質紀 地質代(↑)的大次級劃分。

地質世 是地質紀(↑)的次級劃分。

古生代(名) 止於距今225百萬年之前，被子植物(第130頁)進化(第139頁)之前的地質代(↑)。在這時期是蕨類植物(第126頁)佔優勢(第150頁)。

石炭紀(名) 距今345–280百萬年前的古生代(↑)中的地質紀(↑)。當時地球上的森林(第158頁)是樹狀蕨類(第126頁)佔優勢(第150頁)。樹狀蕨類已礦化(↑)成煤了。

中生代(名) 繼古生代(↑)之後，距今225–65百萬年前的地質代(↑)。當時在陸上是裸子植物(第128頁)佔優勢(第150頁)。在中生代開始出現被子植物(第130頁)。

新生代(名) 在中生代(↑)之後從六千五百萬年前開始，現在仍延續著的地質代(↑)。這個時期被子植物(第130頁)已在陸上佔優勢(第150頁)。新生代亦稱 **Cenozoic**。

第三紀(名) 新生代(↑)劃分為兩個次級的地質紀(↑)之一。約始於六千五百萬年前，止於二百萬年前。

第四紀(名) 新生代(↑)的第二個地質次級紀(↑)，始於更新世(↓)延續至今。

更新世(名) 約始於二百萬年前，止於一萬年前最後一次冰期的地質世(↑)。

interaction (n) the process in which two or more organisms (p. 118), of the same or different species, act on each other, e.g. symbiosis (↓).
interact (v).

symbiosis (n) the state of two different organisms living closely together for much or all of their lives, e.g. the fungi (p. 163) and algae (p. 119) in lichens (p. 147). **symbiotic** (adj).

symbiont (n) an organism living in a symbiotic (↑) relationship with another organism.

mutualism (n) the kind of symbiosis (↑) in which each partner in the relationship gains something from the other.

commensalism (n) the kind of symbiosis (↑) in which the partners live in close association with each other and neither partner gains an obvious advantage.

pathogen (n) an organism (p. 118) which causes disease or illness in another organism. Many viruses (p. 118), fungi (p. 163) and bacteria (p. 119) are pathogens. **pathogenic** (adj).

infection (n) the entry or presence of a parasite (↓), pathogen (↑) or symbiont (↑) in the tissue (p. 88) of a host (↓). **infect** (v).

phytopathology (n) the study of the diseases of plants.

parasite (n) an organism which takes all its nutrients (p. 111) from the tissues (p. 88) of another organism, usually with harmful effects. Many fungi (p. 163) and bacteria (p. 119) are parasites, and so are a few flowering plants (p. 130), e.g. dodder, broomrape. **parasitic** (adj).

hemiparasite (n) a green plant whose roots grow into the tissues (p. 88) of another plant. Hemiparasites photosynthesize (p. 32), but take some of their nutrients (p. 111) and water from the other plant, e.g. mistletoe. Also known as semi-parasites.

vector² (n) an animal which carries parasites (↑) or pathogens (↑) from one organism to another.

host (n) the general name for an organism with a parasite (↑) in it, a plant with an epiphyte (p. 137) on it, or the larger partner in a symbiotic (↑) relationship.

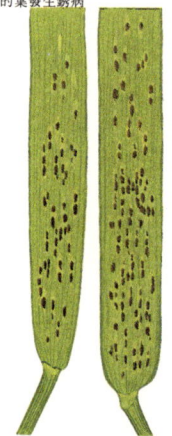
pathogens 病原體
e.g. basidiomycete fungus causing rust on wheat leaves
例如擔子菌使小麥的葉發生銹病

e.g. bacteria causing galls on apple tree stems
例如：細菌使蘋果樹莖長瘤

hemiparasite 半寄生植物
obtains nutrients and water from host
從寄主取得養分和水分

INTERACTIONS/MYCORRHIZAE 相互作用／菌根

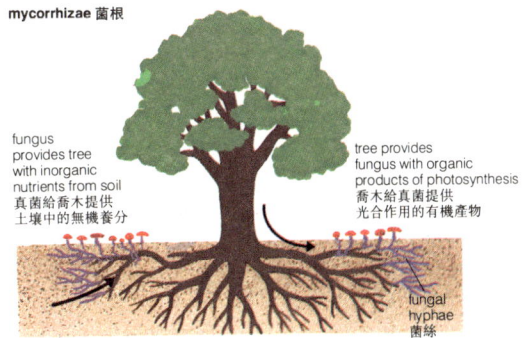

mycorrhizae 菌根

fungus provides tree with inorganic nutrients from soil
真菌給喬木提供土壤中的無機養分

tree provides fungus with organic products of photosynthesis
喬木給真菌提供光合作用的有機產物

fungal hyphae 菌絲

mycorrhiza (n) a symbiotic (↑) association between a fungus (p. 163) and the underground parts of a plant. Plants with mycorrhizae (pl.) often grow faster or larger than plants without, because the mycorrhizae provide extra nutrients (p. 111). In most cases, the plant provides the mycorrhiza with products of photosynthesis (p. 32). Mycorrhizae are common in most plant families (p. 134).

菌根（名） 一種真菌（第 163 頁）與植物的根或地下部分之間的共生（↑）體。有菌根的植物往往比沒有菌根的植物長得更快、更大，因為菌根能提供額外的營養素（第 111 頁）。在多數情況下，植物都會為菌根提供光合作用（第 32 頁）的產物。菌根常見於大部分植物科（第 134 頁）。

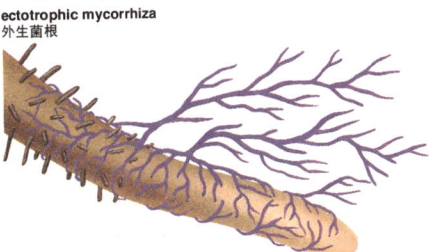

ectotrophic mycorrhiza 外生菌根

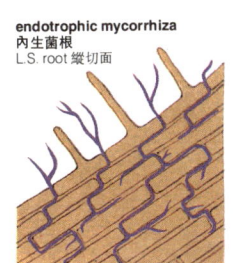

endotrophic mycorrhiza 內生菌根
L.S. root 縱切面

ectotrophic (adj) of mycorrhizae (↑) whose hyphae (p. 163) do not grow into the cells of the host (↑). Ectotrophic mycorrhizae form a sheath (p. 100) around the root of the host, and the mycelium (p. 163) also grows in the intercellular spaces (p. 95) of the root tissues (p. 88) See also **endotrophic** (↓).

endotrophic (adj) of mycorrhizae (↑) which do not form a sheath (p. 100) around the root of the host (↑). Endotrophic mycorrhizae usually grow into the cells of the host.

體外營養的（形） 指菌根（↑）的菌絲（第 163 頁）不長入寄主（↑）的細胞內。外生菌根形成鞘（第 100 頁）圍住寄主的根，而在根組織（第 88 頁）的細胞間隙（第 95 頁）中也長菌絲體（第 163 頁）。參見**內生的**（↓）。

內生的；體內營養的（形） 指不在寄主（↑）的根周圍形成鞘（第 100 頁）的菌根。內生菌根通常長入寄主（↑）的細胞內。

146 · INTERACTIONS/NITROGEN-FIXING BACTERIA 相互作用/固氮細菌

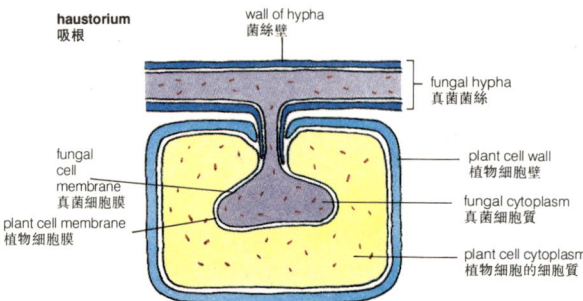

haustorium (*n*) the part of the hypha (p. 163) of a parasitic (p. 144) fungus (p. 163) which grows into the cell of the host (p. 144). **haustoria** (*pl.*).

nodule (*n*) a swelling on the root of a member of the family (p. 134) Leguminosae (p. 131), caused by symbiotic (p. 144) *Rhizobium* bacteria (p. 119). These bacteria are involved in nitrogen fixation (↓).

吸器 (名)　長入寄主 (第 144 頁) 細胞內的寄生 (第 144 頁) 真菌 (第 163 頁) 的菌絲 (第 163 頁) 部分。(複數為 haustoria)

根瘤 (名)　豆科植物 (第 131 頁) 科 (第 134 頁) 成員，根上由共生 (第 144 頁) 根瘤菌屬細菌 (第 119 頁) 引起的脹大部分。這些細菌參與固氮作用 (↓)。

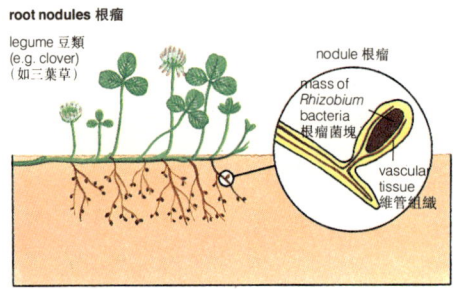

nitrogen fixation the process in which nitrogen in the air (N_2) is reduced (p. 11) by organisms to ammonia (p. 13). Only prokaryotic (p. 16) organisms such as blue-green algae (p. 120) and bacteria (p. 119) can carry this out. Some nitrogen-fixing organisms are found in symbiotic (p. 144) relationships, e.g. blue-green algae in lichens (↓) or *Rhizobium* bacteria in root nodules (↑).

gut flora the microorganisms (p. 118) found in the gut of animals. The gut flora assists the animal in breaking down its food.

固氮作用　空氣中的氮 (N_2) 被生物體還原 (第 11 頁) 成氨 (第 13 頁) 的過程。只有原核 (第 16 頁) 生物體，如藍藻 (第 120 頁) 及細菌 (第 119 頁) 能完成這種作用。共生 (第 144 頁) 關係中可見到某些固氮細菌，例如在地衣 (↓) 中共生的藍藻，或根瘤 (↑) 中的根瘤細菌。

腸道植物區系　動物腸道內的微生物 (第 118 頁)。腸道植物區系有助動物分解腸內的食物。

INTERACTIONS/LICHENS 相互作用／地衣 · 147

lichen (n) a symbiosis (p. 144) between a green or blue-green alga (p. 120) and a fungus (p. 163). Lichens are usually small plants, with a range of colour, which grow on rocks or as epiphytes (p. 137).

phycobiont (n) the algal (p. 119) symbiont (p. 144) in a lichen (↑).

mycobiont (n) the fungal (p. 163) partner in a lichen (↑).

fruticose (adj) of lichens (↑) whose habit (p.136) is shrub-like (p. 136).

foliose (adj) of lichens (↑) with a leafy thallus (p. 122). Foliose lichens have distinct upper and lower surfaces.

crustose (adj) of lichens (↑) whose thallus (p. 122) is closely pressed to the substrate (p. 154) or actually growing within it.

地衣（名）　綠藻或藍藻（第 120 頁）與真菌（第 163 頁）間的一種共生（第 144 頁）體，地衣通常是細小的植物，具有不同的顏色，長在岩石上，或成為附生物（第 137 頁）。

藻共生體（名）　地衣（↑）中的藻類（第 119 頁）共生體（第 144 頁）。

地衣共生菌（名）　與地衣（↑）結合的真菌（第 163 頁）。

灌木狀的（形）　指具有像灌木（第 136 頁）習性（第 136 頁）的地衣（↑）。

葉狀的（形）　指具葉片狀葉狀體（第 122 頁）的地衣（↑）。葉狀地衣分為明顯的上、下面。

殼狀的（形）　指葉狀體（第 122 頁）緊貼到基質上（第 154 頁）或實際上長在基質內的地衣（↑）。

lichens 地衣
crustose 殼狀的

foliose 葉狀的

fruticose 灌木狀的

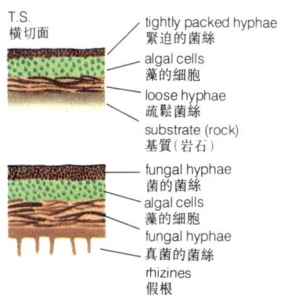
T.S. 橫切面
tightly packed hyphae 緊迫的菌絲
algal cells 藻的細胞
loose hyphae 疏鬆菌絲
substrate (rock) 基質（岩石）
fungal hyphae 菌的菌絲
algal cells 藻的細胞
fungal hyphae 真菌的菌絲
rhizines 假根

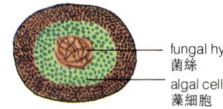
fungal hyphae 菌絲
algal cells 藻細胞

rhizine (n) a root-like bunch of hyphae (p. 163) growing from the bottom of the thallus (p. 122) of a lichen (↑).

apothecium (n) a cup-shaped structure which bears spores (p. 66), found in some lichens (↑).

perithecium (n) a flask-shaped hollow structure which bears spores (p. 66), in some lichens (↑). The perithecium opens through a pore (p. 19) in the surface of the thallus (p. 122).

假根（名）　地衣（↑）葉狀體（第 122 頁）的底部長出的根狀菌絲（第 163 頁）束。

子囊盤（名）　某些地衣（↑）中所長有孢子（第 66 頁）的杯狀構造。

子囊殼（名）　某些地衣（↑）長有孢子（第 66 頁）的瓶狀空心構造。子囊殼在葉狀體（第 122 頁）表面通過一個小孔（第 19 頁）開口。

obligate (adj) of organisms that can only live in one way. For instance, the fungi (p. 163) and algae (p. 119) in lichens (p. 147) are obligate symbionts (p. 144), being unable to live without each other, in most cases.

facultative (adj) of organisms which can live under several kinds of conditions. For instance, a facultative epiphyte (p. 137) is a plant which can grow either on the ground or on other plants.

toxin (n) a substance which is poisonous. Plants produce toxins, e.g. alkaloids (↓), to protect themselves from attack by herbivores (p. 153).
toxic (adj).

phytoalexin (n) a substance produced in some plants, which prevents attack by pathogenic (p. 144) or parasitic (p. 144) fungi (p. 163).

antibiotic (n) a substance that is harmful to bacteria (p. 119). Antibiotics are produced by many fungi (p. 163), e.g. penicillin is produced by several species of the fungus *Penicillium*.

tannins (n.pl.) a group of substances common in the outer tissues (p. 88) of many plants, which are bitter to the taste and are a defence against herbivores (p. 153). Tannins are used in the tanning of leather.

alkaloid (n) an organic (p. 11), nitrogen-containing compound, produced by many plants. Alkaloids are mostly toxic (↑) and often defend plants against attack by herbivores (p. 153).

allelopathy (n) discouragement by one plant of the growth of other plants around it, e.g. by toxins (↑) contained in the fallen leaves.
allelopathic (adj).

insectivorous (adj) of organisms which eat insects. Some plant species (p. 134) trap insects with sticky hairs (e.g. sundew), in bucket-shaped leaves (e.g. pitcher plants) or between hinged leaf-lobes (e.g. Venus fly-trap). Insectivorous plants obtain nutrients (p. 111) by secreting (p. 112) enzymes (p. 15) which break down the tissues (p. 88) and cells of the trapped insect. The insectivorous habit (p. 136) is an adaptation (p. 141) to habitats (↓) that are poor in nitrates (p. 13).

專性的（形） 指只能以一種方式生活的生物體。例如地衣（第 147 頁）中的真菌（第 163 頁）和藻類（第 119 頁）是專性的共生體（第 114 頁），在多數情況下要彼此依賴才能生活。

兼性的（形） 指能在幾種不同的條件下生活的生物體。例如兼性的附生植物（第 137 頁）能生長在地上或在其他植物體上。

毒素（名） 有毒的物質。植物能產生毒素，例如產生生物鹼（↓）以保護本身免受食草動物（第 153 頁）侵襲。（形容詞為 toxic）

植物抗毒素（名） 某些植物所產生的一種物質，能防止致病（第 144 頁）真菌或寄生（第 144 頁）真菌（第 163 頁）的侵襲。

抗生素（名） 對細菌（第 119 頁）有害的一種物質。抗生素由多種真菌（第 163 頁）生產的，如青黴素是由青黴菌屬的幾種真菌生產的。

丹寧（名、複） 常見於多種植物外層組織（第 88 頁）的一類物質，味苦，能防禦食草動物（第 153 頁）。

生物鹼（名） 許多植物所產生的含氮有機（第 11 頁）化合物。生物鹼多數有毒（↑），使植物能防禦食草動物（第 153 頁）的侵襲。

植物相剋作用（名） 指一株植物周圍的其他植物的生長受該株植物所阻礙，例如一株植物的落葉中含的毒素（↑）阻礙其周圍其他植物生長。（形容詞為 allelopathic）

食蟲的（形） 指食昆蟲的生物體。有些植物種（第 134 頁）會用黏毛（如茅膏菜）以桶狀葉（如瓶子草）或在鉸鏈狀葉裂之間（如維納斯捕蠅草）捕食昆蟲。食蟲植物能分泌（第 112 頁）酶素（第 15 頁）將所捕捉的昆蟲的組織（第 88 頁）和細胞分解，取得營養分（第 111 頁），這種食蟲習性（第 136 頁）是適應（第 141 頁）缺硝酸鹽（第 13 頁）生境（↓）的結果。

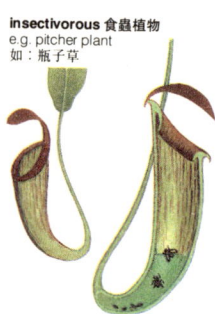

insectivorous 食蟲植物
e.g. pitcher plant
如：瓶子草

ecology (*n*) the study of organisms in relation to their environment (↓). **ecological** (*adj*), **ecologist** (*n*).

autecology (*n*) the ecology (↑) of a single species (p. 134) in a habitat (↓).

synecology (*n*) the ecology (↑) of all the organisms found in a habitat (↓) or an ecosystem (↓).

生態學(名) 研究生物與其環境(↓)之間相互關係的學科。(形容詞為 ecological，名詞 ecologist 意為生態學家)

個體生態學(名) 單獨一個種(第134頁)在生境(↓)中的生態學(↑)。

群體生態學(名) 有關一個生境(↓)或一個生態系(↓)全部生物的生態學(↑)。

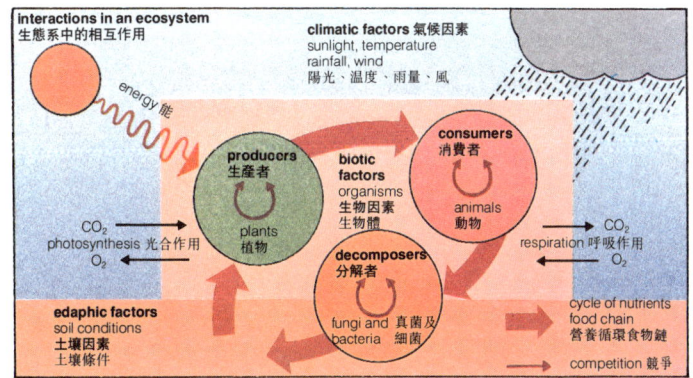

ecosystem (*n*) an ecological (↑) system in which organisms interact (p. 144) with each other and with their non-living environment (↓) and in which there is a more or less closed cycle of nutrients (p. 111).

biosphere (*n*) the parts of the earth in which organisms live, including the oceans, the land, the soil, and the atmosphere.

environment (*n*) the living and non-living surroundings of an organism, and the events which take place in those surroundings. **environmental** (*adj*).

habitat (*n*) the place or kind of place in which an organism, community (↓) or association (p. 150) is found, e.g. the habitat of an epiphyte (p. 137) is in the branches of trees, the habitat of algae (p. 119) is water.

community (*n*) the group of species (p. 134) of plants, animals, or both, living in the same habitat (↑) and interacting (p. 144) with each other.

生態系(名) 一個生態學(↑)的系統。在此系統中生物彼此間及與其非生活環境(↓)相互作用(第144頁)。在此生態系統內存在著或多或少封閉的營養素(第111頁)循環。

生物圈(名) 地球上有生物體生活的部分，包括海洋、陸地、土壤及大氣。

環境(名) 生物體周圍的生活部分和非生活部分以及在其中發生的事件。(形容詞為 environmental)

生境(名) 有生物個體、群落(↓)或群叢(第150頁)生活的地方或一類地方。例如附生植物(第137頁)的生境是喬木的枝；藻類植物(第119頁)的生境是水。

群落(名) 植物、動物種(第134頁)群或二者，這些種群生活在同一生境(↑)中，彼此相互作用(第144頁)。

association (n) a group of species (p. 134) that are usually found together, and require the same habitat (p. 149).

phytosociology (n) the study of the associations (↑) of plants.

dominant² (adj) of the most common or the largest species (p. 134) in a community (p. 149). **dominant** (n), **dominate** (v).

vegetation (n) the general term for all the plants in an ecosystem (p. 149).

primary vegetation vegetation (↑) which has not been disturbed or changed by man.

secondary vegetation vegetation (↑) growing in a place which has been disturbed by man, e.g. roadsides, old farmland, etc.

群叢(名) 通常在一起，並需要相同生境(第149頁)的一個種(第134頁)的群體。

植物社會學(名) 研究植物的群叢(↑)的一門學科。

優勢的(形) 指一個群落(第149頁)中最常見、最大的種(第134頁)。(名詞為 dominant，動詞為 dominate)

植被(名) 生態系(第149頁)中所生長的全部植物的總稱。

原生植被 沒有受到人類干擾或改變的植被(↑)。

次生植被 生長在一地已受人類干擾的植被(↑)，如路旁、老的農田等。

examples of regeneration 更新的實例

young grass leaves 幼嫩草葉 / eaten by herbivore 食草類啃過的 / new growth occurs and the grass regenerates 重新生長，草地更新

vegetation before burning 火燒前的植被 / vegetation burnt 植被被燒 / new plants grow and the vegetation regenerates 新植物生長，植被更新

regeneration² (n) the growth of new vegetation (↑) in a place where the old vegetation has been removed or damaged.

ecotone (n) the border between two habitats (p. 149) or types of vegetation (↑).

producer (n) an autotrophic (p. 32) organism in an ecosystem (p. 149), which produces organic (p. 11) matter using chemical energy or energy from light. Plants are the main producers in the biosphere (p. 149).

primary production the total amount of organic (p. 11) matter produced by the autotrophic (p. 32) organisms in an ecosystem (p. 149), using energy from sunlight.

primary productivity the amount of matter that can be synthesized (p. 13) by all the autotrophic (p. 32) organisms in a given area in a given time.

更新(名) 老的植被已廢掉或已破壞的地方生長新的植被。

群落交錯區(名) 兩個生境(第149頁)之間或兩個植被(↑)類型之間的邊界地帶。

生產者(名) 指生態系(第149頁)內的一種自養(第32頁)生物，此種生物利用化學能或陽光能量生產有機(第11頁)物質。植物是生物圈(第149頁)中的主要生產者。

初級生產量 自養(第32頁)生物體在生態系(第149頁)中利用陽光能量所生產有機(第11頁)物質的總量。

初級生產力 自養(第32頁)生物在一定面積和一定時間內所能合成(第13頁)的物質總量。

ecotone 群落交錯區

type of vegetation A / A型植被 ecotone 群落交錯區 type of vegetation B / B型植被

ECOLOGY/COLONIZATION 生態學／群落形成 · 151

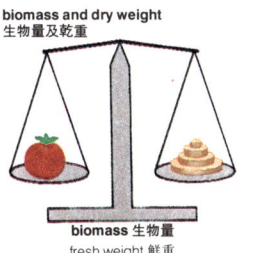
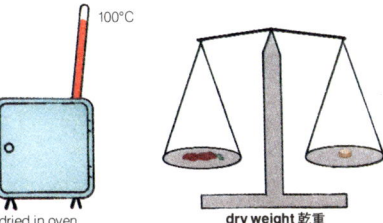

biomass and dry weight
生物量及乾重

biomass 生物量
fresh weight 鮮重
dried in oven
在烘箱內烘乾
dry weight 乾重
the weight after all water has evaporated
全部水分蒸發後的重量

biomass = weight of water content + dry weight
生物量 = 含水重 + 乾重

biomass (n) the weight of a living organism, or of all the organisms in an ecosystem (p. 149) or in a habitat (p. 149).

dry weight the weight of an organism, part of an organism, or the organisms in a habitat (p. 149) or in an ecosystem (p. 149), after drying. Because a large part of the biomass (↑) of most organisms is water, dry weight is usually small compared to biomass.

colonization (n) the arrival and germination (p. 87) of a seed on a substrate (p. 154), or the spread of plants to places where they have not grown before. Successful colonization depends on growth to reproductive (p. 59) age.

pioneer (n) a plant species (p. 134) that occurs in the early stages of succession (↓).

生物量(名)　在一個生態系(第 149 頁)或生境(第 149 頁)中單個個別生物體，或全部生物體的重量。

乾重　在一個生境(第 149 頁)或生態系(第 149 頁)中一個生物體、一個生物體的部分或許多生物體(第 149 頁)烘乾後的重量。因為大多數生物體的大部分生物量(↑)是水，因此乾重通常都比生物量小。

群落形成(名)　種子到達基質(第 154 頁)及萌發(第 87 頁)或者植物擴展到未長過這種植物的地方。群落形成的成功決定於生長到生殖(第 59 頁)年齡。

先鋒植物(名)　演替(↓)早期出現的植物種(第 134 頁)

succession 演替　　　time 時間

a pioneer species colonizes a habitat
先鋒種定居一個生境
pioneer plants grow and reproduce
先鋒植物生長和生殖
growth of plants alters edaphic and biotic factors and more species colonize
植物的生長改變了土壤及生物因素，使更多植物定居
climax community with many plant species. Conditions no longer suitable for pioneer species
有許多種的演替頂極群落，條件不再適合於先鋒種

succession (n) the process of development of vegetation (↑), involving changes of species (p. 134) and communities with time. Succession occurs because the growth of plants alters the biotic factors (p. 152) and edaphic factors (p. 154) of a habitat (p. 149), making possible the colonization (↑) of other species.

successional (adj).

演替(名)　植被(↑)發育的過程，涉及種(第 134 頁)及群落隨時間的變化。演替的發生是因為植物的生長改變了生境(第 149 頁)的生物因素(第 152 頁)及土壤因素(第 154 頁)，使其他種的群落形成(↑)成為可能。(形容詞為 successional)

152 · ECOLOGY/SUCCESSION 生態學/演替

open community a plant community (p. 149) in which the niches (↓) are unstable or 'empty', allowing entry into the community of new species (p. 134) from outside.

closed community a plant community (p. 149) in which the niches (↓) are stable and 'full', not allowing the entry of extra species (p. 134).

sere (n) a succession (p. 151) in a particular habitat (p. 149), e.g. a hydrosere is a succession in a shallow aquatic (p. 161) habitat, beginning with aquatic plants and ending with swamp forest (p. 158).

climax (n) the last stage in a succession (p. 151), after which there are no further great changes in the structure or the species (p. 134) in a habitat (p. 149).

biotic factors the effects of living organisms on an ecosystem (p. 149) and on each other, e.g. herbivores (↓) eating plants, or trees casting shade.

開闊群落　生態區位(↓)不穩定或未被佔據，容許外界的新種(第134頁)進入的植物群落(第149頁)。

密生群落　生態區位(↓)穩定且已被佔滿，不容許外種(第134頁)進入的植物群落(第149頁)。

演替系列(名)　在特定生境(第149頁)中的演替(第151頁)，例如水生演替系列是在淺水的(第161頁)生境的演替。以水生植物為始，沼澤林(第158頁)為終。

顛峰(名)　演替(第151頁)的最後階段，此後在一個生境(第149頁)中的結構或種(第134頁)不再有大的變化。

生物因素　活生物體對生態系(第149頁)的影響及彼此的影響，例如食草動物(↓)啃吃植物，或者喬木遮陰。

competition 競爭

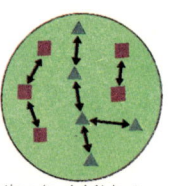

intraspecific between individuals of the same species in a habitat
種內競爭 生境中同種個體間的競爭

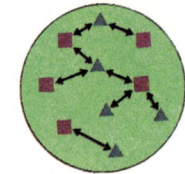

interspecific between individuals of different species in a habitat
種間競爭 生境中不同種的個體之間的競爭

competition (n) an interaction (p. 144) between two or more organisms in the same habitat (p. 149), when both, partly or wholly, share the same needs. Competition can be intraspecific (↓) or interspecific (↓). **compete** (v), **competitor** (n).

interspecific (adj) of competition (↑) between two species (p. 134).

intraspecific (adj) of competition (↑) between individuals (p. 135) of the same species (p. 134).

競爭(名)　在一個生境(第149頁)中兩個或多個生物之間在二者部分或全部地享有相同需求時的相互作用(第144頁)。競爭有種內的(↓)，或者種間的(↓)。(動詞為compete，名詞為 competitor)

種間的(形)　兩個種(第134頁)間的競爭(↑)。

種內的(形)　同種(第134頁)的個體(第135頁)之間的競爭(↑)。

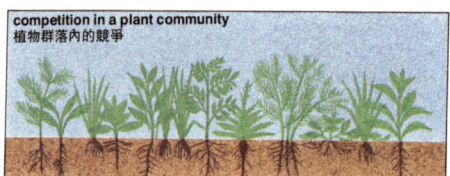

competition in a plant community
植物群落內的競爭

leaves compete for light
葉片競爭陽光

roots compete for nutrients
根競爭營養分

ECOLOGY/FOOD WEBS 生態學/食物網

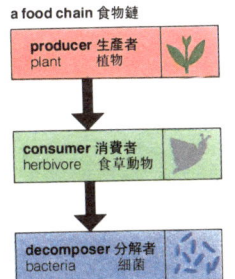

a food chain 食物鏈
producer 生產者 plant 植物
consumer 消費者 herbivore 食草動物
decomposer 分解者 bacteria 細菌

niche (n) the position and activities of an organism in its habitat (p.149). Each species (p.134) has its own niche, and competition (↑) occurs when the niches overlap.

herbivore (n) an animal that eats plants. **herbivory** (n), **herbivorous** (adj).

food chain the flow of energy and nutrients (p.111) from one group of organisms to another in an ecosystem (p.149), e.g. from producers (p.150) to consumers (p.154) to decomposers (p.157).

生態區位(名) 生物在其生境(第149頁)中的位置及活動。每個種(第134頁)都有本身的生態區位，生態區位重疊時即發生競爭(↑)。

食草動物(名) 以植物為食的動物。(名詞為 herbivory，形容詞為 herbivorous)

食物鏈 一個生態系(第149頁)中，從一群生物到另一群生物，例如從生產者(第150頁)到消費者(第154頁)到分解者(第157頁)的能量流及養分(第111頁)流。

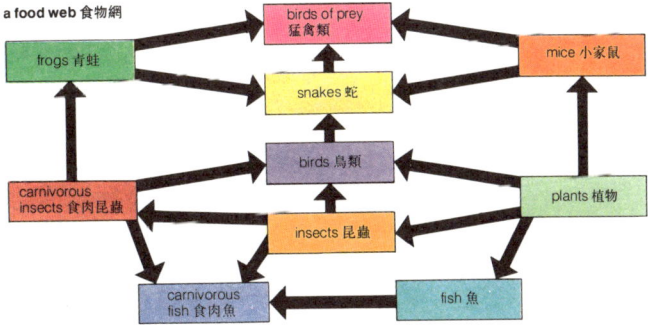

a food web 食物網

pyramid of available energy at the trophic levels of a food web
食物網食性層次的可用能量金字塔

trophic level 食性層次
higher order consumer (large carnivore) 高層消費者 (大的食肉類) — 4
secondary consumer (carnivore) 次級消費者 (食肉類) — 3
primary consumer (herbivore) 初級消費者 (食草類) — 2
producer 生產者 — 1

energy lost through respiration, heat radiation and other metabolic processes 能量通過呼吸、熱輻射及其他新陳代謝過程損失

energy available as food 可用作食物的能量

food web a set of interacting (p.144) food chains (↑), e.g. an animal may feed on several plant species (p.134), and the animal may be fed on by several animal species which in turn may be fed on by other animal species.

pyramid of numbers the number of organisms at each trophic level (↓) in a food web (↑) or food chain (↑). At each level, energy is lost through respiration (p.22) and other metabolic (p.14) processes, and there is less energy available for the next trophic level. For this reason, the number and biomass (p.151) of consumers (p.154) in an ecosystem (p.149) are less than the number and biomass of producers (p.150).

trophic level the position of an organism in a food chain (↑). The main trophic levels are those of producers (p.150), consumers (p.154) and decomposers (p.157).

食物網 一組相互作用(第144頁)的食物鏈(↑)。例如一種動物以幾個植物種(第134頁)為食，這種動物又成為其他幾種動物的食物，而這幾種動物又成為其他動物種的食物。

數量金字塔 在食物網(↑)或食物鏈(↑)中，每一個食性層次(↓)的生物數量。在每一個層次上都是通過呼吸作用(第22頁)及其他新陳代謝(第14頁)過程失去能量，下一級食性層次只有較少的能量可利用。因此，生態系(第149頁)中消費者(第154頁)的數量和生物量(第151頁)少於生產者(第150頁)的數量及生物量。

食性層次；營養級 一種生物在食物鏈(↑)中的位置。主要的食性層次是生產者(第150頁)、消費者(第154頁)及分解者(第157頁)。

consumer (*n*) a heterotrophic (p. 32) organism which eats other organisms, e.g. a herbivore (p. 153).

carbon cycle the pathway of the element (p. 8) carbon through ecosystems (p. 149). Carbon dioxide is fixed from the atmosphere by plants during photosynthesis (p. 32) and used for the synthesis of organic (p. 11) compounds. These are passed through the food web (p. 153) and metabolized (p. 14) by animals and decomposers (p. 157), and carbon dioxide is released back to the atmosphere by respiration (p. 22).

nitrogen cycle the cycle of the element (p. 8) nitrogen through ecosystems (p. 149). Organisms need nitrogen in order to synthesize (p. 13) amino acids (p. 56), proteins (p. 56) and other nitrogen-containing organic (p. 11) compounds. Nitrogen is taken up from the soil by plants in the form of nitrate (p. 13), converted to plant protein, and may then be passed to animals as protein. It is returned to the soil due to the death and decay (p. 157) of plants and animals, and by animal excretion (p. 112). Decomposers (p. 157) in the soil break down the organic nitrogen-containing compounds to inorganic (p. 11) compounds, e.g. nitrate and ammonia (p. 13), thus completing the cycle. Atmospheric nitrogen (N_2) enters the cycle as a result of nitrogen-fixing (p. 146) organisms and by oxidation (p. 11) in lightning in thunderstorms.

nitrifying bacteria bacteria (p. 119) in the soil which oxidize (p. 11) ammonia (NH_3) (p. 13) to nitrate (NO_3^-) (p. 13). This is one of the important stages in the nitrogen cycle (↑), which makes nitrate available for plants.

denitrifying bacteria bacteria (p. 119) in the soil which reduce (p. 11) nitrate (NO_3^-) (p. 13) to nitrite (NO_2^-) (p. 13) and molecular nitrogen (N_2).

edaphic factors the effects of the soil in an ecosystem (p. 149). Different soils have different structural and chemical characteristics, and different plant species (p. 134) are adapted (p. 141) for growth on particular types of soil.

substrate[2] (*n*) general term for the soil or the surface on which an organism is growing.

消費者（名）　指吃其他生物的異養生物（第32頁），例如食草動物（第153頁）。

碳循環　碳元素（第8頁）經過生態系（第149頁）的途徑。植物在行光合作用（第32頁）時將大氣中的二氧化碳固定，用於合成有機（第11頁）化合物。這些有機化合物通過食物網（第153頁），為動物及分解者（第157頁）新陳代謝（第14頁），由呼吸作用（第22頁）將二氧化碳釋放回大氣。

氮循環　氮元素（第8頁）通過生態系（第149頁）的循環。生物需要氮來合成（第13頁）氨基酸（第56頁）、蛋白質（第56頁）及其他含氮有機（第11頁）化合物。植物從土壤中吸收硝酸鹽（第13頁）形式的氮，轉化為植物蛋白質，然後進入動物體成為動物蛋白質。最後，由於植物及動物的死亡和腐化（第157頁）及動物的排泄（第112頁）而回歸土壤中。土壤中的分解者（第157頁）再將含氮有機化合物分解成無機（第11頁）化合物，如硝酸鹽及氨（第13頁），至此完成循環。大氣中的氮（N_2）是通過固氮（第146頁）生物體及雷暴時閃電的氧化作用（第11頁）進入循環。

硝化細菌　土壤中能將氨（NH_3）（第13頁）氧化（第11頁）成硝酸鹽（NO_3^-）（第13頁）的細菌（第119頁）。這是氮循環（↑）的重要階段之一，此階段使硝酸鹽能被植物利用。

反硝化細菌　土壤中能將硝酸鹽（NO_3^-）（第13頁）還原（第11頁）成亞硝酸鹽（NO_2^-）（第13頁）及分子氮（N_2）的細菌（第119頁）。

土壤因素　生態系（第149頁）中土壤的種種影響。不同土壤有不同的結構和化學特性，不同的植物種（第134頁）適應（第141頁）在特定類型的土壤生長。

基質（名）　供生物生長的土壤或表層的通稱。

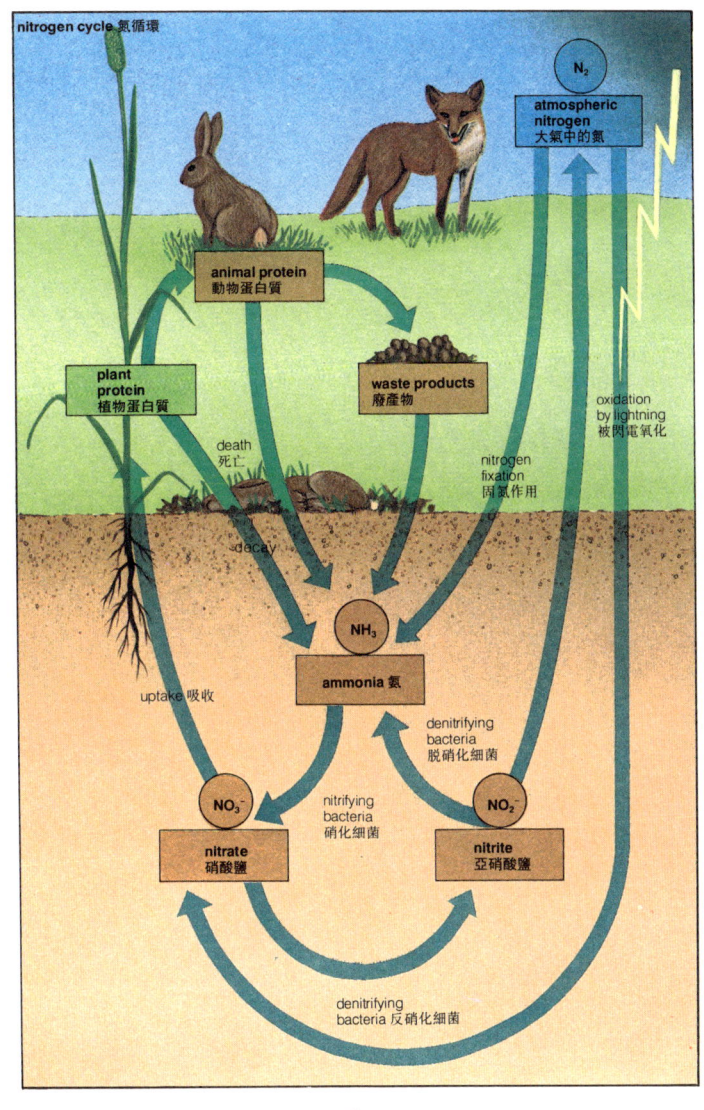

soil profile the sequence of different layers of material in the soil. The layers, or horizons (↓), differ in chemical composition and thickness. The top layers are usually organic (p. 11), derived from the litter (↓), and the layers underneath are inorganic (p. 11), derived from the rock beneath them. Soils in different places have their own characteristic soil profiles, depending on the climate and the type of rock on which they occur.

土壤剖面 土壤中不同物質層的順序,土壤的每一層或各層層位(↓)的化學組成及厚度都不同。上面的幾層通常為枯枝落葉層(↓)產生的有機(第11頁)質土,下面幾層則為其下方岩石產生的無機(第11頁)質層。不同地區的土壤各有其特徵性的土壤剖面,這有賴於氣候及其來源的岩石類型。

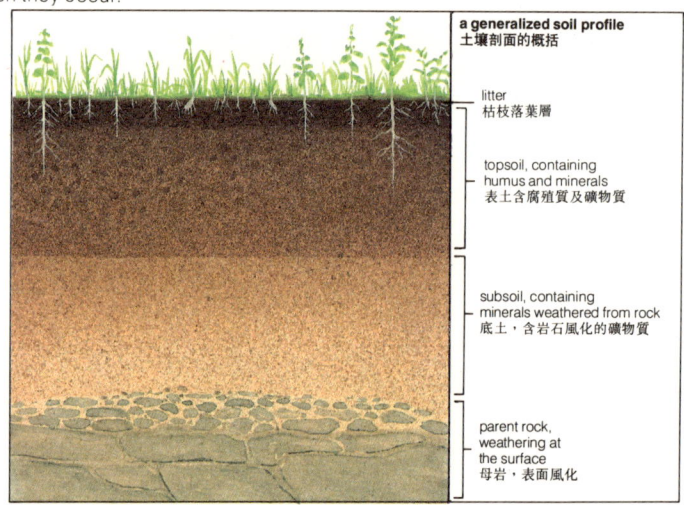

a generalized soil profile
土壤剖面的概括

- litter 枯枝落葉層
- topsoil, containing humus and minerals 表土含腐殖質及礦物質
- subsoil, containing minerals weathered from rock 底土,含岩石風化的礦物質
- parent rock, weathering at the surface 母岩,表面風化

topsoil (*n*) the general term for the upper, organic (p. 11) horizons (↓) in a soil profile (↑).

subsoil (*n*) the general term for the lower, inorganic (p. 11) horizons (↓) in a soil profile (↑).

horizon (*n*) a layer in a soil profile (↑). Different soil profiles can be compared by examining the structure, thickness and chemical composition of their horizons.

humus (*n*) the layer of organic (p. 11) matter at the top of a soil profile (↑). Humus is the habitat (p. 149) of most decomposers (↓).

litter (*n*) dead plant and animal material on the surface of the ground, above the humus (↑) layer.

表土(名) 土壤剖面(↑)上層,即有機(第11頁)層位(↓)的通稱。

底土(名) 土壤剖面(↑)下層,即無機(第11頁)層位(↓)的通稱。

層位(名) 土壤剖面(↑)的一層。不同的土壤剖面可以通過檢查其各個層位的結構、厚度及化學組成作比較。

腐殖質(名) 土壤剖面(↑)表層的有機(第11頁)物層。腐殖質是大多數分解者(↓)的生境(第149頁)。

枯枝落葉層(名) 腐殖(↑)層上面的枯死植物及動物死屍層。

ECOLOGY/SOILS 生態學／土壤 · 157

peat (*n*) a kind of litter (↑) layer found in certain very wet or waterlogged habitats (p. 149), such as bogs, which decomposes (↓) very slowly, often under very acid conditions. Layers of peat may be several metres thick.

mor (*n*) a very acid humus (↑) which hardly mixes with the inorganic (p. 11) soil underneath it.

mull (*n*) humus (↑) which is well-mixed with the inorganic (p. 11) soil.

calcareous (*adj*) of substrates (p. 154) which contain calcium carbonate, $CaCO_3$, e.g. soils on limestone or chalk.

calcicole (*n*) a plant which grows only or mainly on calcareous (↑) soil. **calcicolous** (*adj*).

calcifuge (*n*) a plant which grows only or mainly on non-calcareous (↑) soil. **calcifugous** (*adj*).

decomposer (*n*) an organism which breaks down organic (p. 11) matter, releasing carbon dioxide and inorganic (p. 11) compounds, e.g. nitrates (p. 13), phosphates (p. 13), ammonia (p. 13). The main decomposers are bacteria (p. 119) and fungi (p. 163). **decomposition** (*n*), **decompose** (*v*).

decay (*n*) the process of rotting and decomposition (↑) which takes place after the death of an organism. Decay involves the breakdown of organic (p. 11) compounds in the organism, by saprophytic (p. 137) bacteria (p. 119) and fungi (p. 163). It is an important part of the cycle of nutrients (p. 111) and energy in ecosystems (p. 149). **decay** (*v*).

rhizosphere (*n*) the general name given by some ecologists to the parts of the biosphere (p. 149) in which roots grow.

泥炭土（名） 在極濕或積水的生境（第149頁）中，例如在沼澤地所見的一種枯枝落葉層（↑），其分解（↓）非常緩慢，常常是在酸性很大的條件下。泥炭土的各層可厚達數米。

酸性有機質（名） 酸性很大的腐殖質（↑），很難和其下面的無機（第11頁）質土壤混合。

腐熟腐殖質（名） 能與無機（第11頁）質土壤很好混和的腐殖質（↑）。

鈣質的（形） 指含碳酸鈣（$CaCO_3$）的基質（第154頁）。例如石灰岩或白堊質的土壤。

喜鈣植物（名） 只在或主要在鈣質（↑）土壤生長的植物。（形容詞為 calcicolous）

避鈣植物（名） 只在或主要在非鈣質（↑）土壤生長的植物。（形容詞為 calcifugous）

分解者（名） 能分解有機（第11頁）物，釋放二氧化碳及無機（第11頁）化合物，例如硝酸鹽（第13頁）、磷酸鹽（第13頁）、氨（第13頁）的生物體。主要分解者是細菌（第119頁）及真菌（第163頁）。（名詞為 decomposition，動詞為 decompose）

腐化（名） 生物死後發生的腐爛及分解過程（↑）。腐化和生物體內的有機（第11頁）化合物被腐生（第137頁）細菌（第119頁）及真菌（第163頁）分解有關。這是生態系（第149頁）中養分（第111頁）及能量循環的一個重要組成部分。（形容詞為 decay）

根圍（名） 生態學家對生物圈（第149頁）中植物根生長的周圍土壤部分的通稱。

158 · ECOLOGY/FORESTS 生態學／森林

forest (*n*) a habitat (p. 149) or type of vegetation (p. 150) in which large trees are dominant (p. 150), forming a dense canopy (↓).
canopy (*n*) the top layer of a forest (↑), consisting of the crowns (p. 92) of trees.
understorey (*n*) the part of a forest (↑) or woodland (↓) underneath the canopy (↑), consisting of shrubs (p. 136), saplings (p. 136) and herbs (p. 136).

森林（名） 高大樹木佔優勢（第150頁）形成濃密樹冠層（↓）的植被（第150頁）生境（第149頁）或類型。
樹冠層（名） 森林（↑）的頂層，由喬木的樹冠（第92頁）組成。
下層林木（名） 森林（↑）或林地（↓）樹冠層（↑）的下面部分，由灌木（第136頁）、幼樹（第136頁）及草本植物（第136頁）組成。

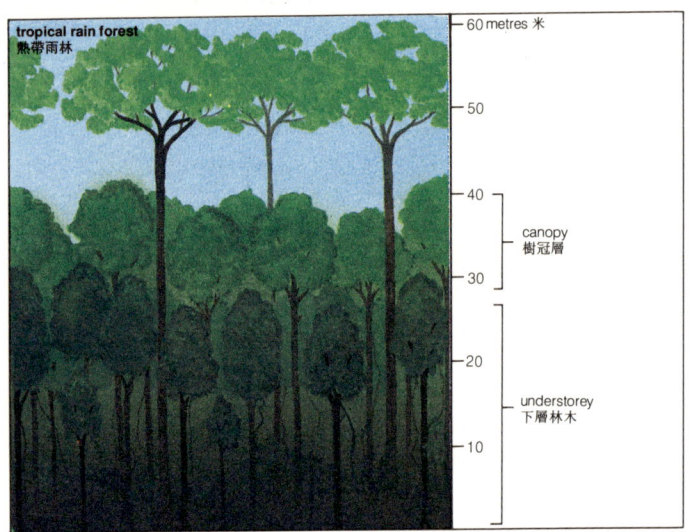

rain forest a wet forest (↑), where there is heavy rain during most months of the year. Most rain forests are tropical (p. 162), although some are found in mild rainy temperate (p. 162) regions. Tropical rain forests usually have very tall trees and a very large number of plant species (p. 134).
jungle (*n*) thick, secondary vegetation (p. 150) in wet tropical (p. 162) regions.
montane forest a forest (↑) on a mountain. Montane forests have smaller trees than lowland forests, and the trees get smaller towards the tree line (↓).

雨林 潮濕的森林（↑），一年中大多數月份雨量充沛。雨林大多數是熱帶（第162頁）森林，但有些是溫和多雨的溫帶（第162頁）雨林。熱帶雨林常有極高大的樹木和大量的植物種（第134頁）。
熱帶植叢（名） 潮濕熱帶（第162頁）地區的稠密次生植被（第150頁）。
山地森林 山地上的森林（↑）。山地森林的樹木比低地森林矮小，靠近森林木線（↓）的樹木更矮小。

ECOLOGY/FORESTS 生態學／森林・**159**

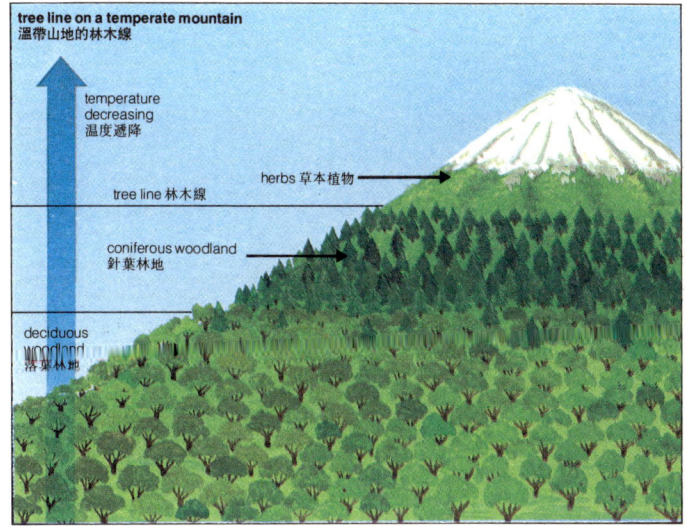

tree line the level on the side of a mountain above which trees do not grow. Vegetation (p. 150) at higher levels is herbaceous (p. 136) or absent altogether.

woodland (n) vegetation (p. 150) in which trees are dominant (p. 150). The trees in woodland are smaller and more spaced apart than those in forest (↑).

林木線　在山坡上的層面，在此層面高度以上不長樹木。在更高高度的植被(第150頁)是草本植物(第136頁)，或幾乎沒有植被。

林地(名)　以喬木佔優勢(第150頁)的植被(第150頁)。林地的喬木比森林(↑)的矮小，而且比較稀落。

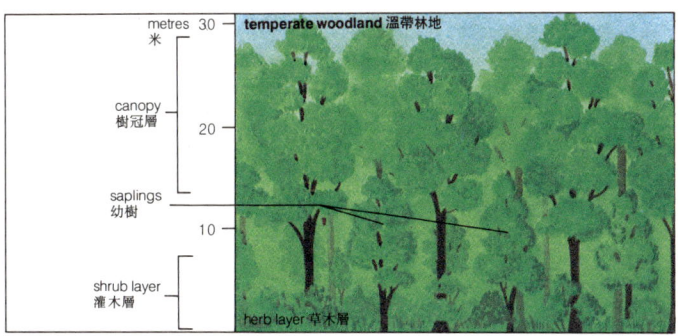

160 · ECOLOGY/SCRUB, GRASSLANDS 生態學／密灌叢、草地

scrub landscape dominated by shrubs
灌叢灌木佔優勢的景觀

scrub (*n*) vegetation (p. 150) in which shrubs (p. 136) and small trees are dominant (p. 150).
grassland (*n*) vegetation (p. 150) in which grasses (p. 130) are dominant (p. 150), e.g. prairies (↓), savanna (↓).

密灌叢(名) 以灌木(第 136 頁)及小樹佔優勢(第 150 頁)的植被(第 150 頁)。
草本植被區(名) 以禾草類(第 130 頁)佔優勢(第 150 頁)的植被(第 150 頁)。例如高草原(↓)、稀樹乾草原(↓)。

grassland landscape dominated by grasses
草本植被區以禾草類佔優勢的景觀

prairie (*n*) a North American grassland (↑).
savanna (*n*) a tropical (p. 162) grassland (↑).
steppes (*n.pl.*) the grasslands (↑) of temperate (p. 162) Asia.
sward (*n*) an area of vegetation (p. 150) consisting mainly of grasses (p. 130).

高草原(名) 北美洲的草本植被區(↑)。
稀樹乾草原(名) 熱帶的(第 162 頁)草本植被區(↑)。
乾草原(名、複) 亞洲溫帶(第 162 頁)區的草本植被區(↑)。
草地(名) 主要由禾草類(第 130 頁)組成的植被(第 150 頁)區。

ECOLOGY/AQUATIC HABITATS 生態學／水生生境 · 161

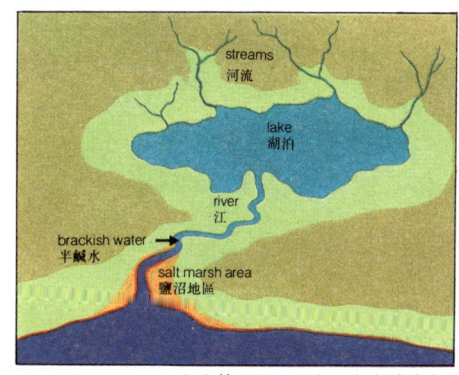

aquatic (*adj*) of organisms that live in water, or of underwater habitats (p. 149).
freshwater (*adj*) of aquatic (↑) habitats (p. 149) in which the concentration (p. 12) of dissolved (p. 12) ions (p. 9) is very low, e.g. rivers, streams, lakes.
eutrophic (*adj*) of habitats (p. 149) rich in nutrients (p. 111).
oligotrophic (*adj*) of habitats (p. 149) poor in nutrients (p. 111).
eutrophication (*n*) a process which can occur in rivers and shallow lakes when the addition of extra nutrients (p. 111), e.g. from fertilizers, causes heavy growth of algae (p. 119). When the algae die, their decay (p. 157) by bacteria (p. 119) reduces the concentration (p. 12) of oxygen in the water, so that aerobic (p. 22) organisms may not survive.
brackish water water in which the concentration (p. 12) of dissolved (p. 12) ions (p. 9) is more than in freshwater (↑) habitats (p. 149) but less than in seawater.
salt marsh a coastal habitat (p. 149) with a wet substrate (p. 154) containing a high concentration (p. 12) of dissolved (p. 12) salts, due to flooding by very high tides. The vegetation (p. 150) of salt marshes is mostly herbaceous (p. 136).
littoral (*adj*) of habitats (p. 149) between the high and low tide marks on the sea shore.

水生的（形）　指生活在水域或水下生境（第149頁）的生物。
淡水的（形）　指溶解的（第12頁）離子（第9頁）濃度（第12頁）非常低的水生（↑）生境（第149頁）。例如江河及湖泊。
富營養的（形）　指富含營養素（第111頁）的生境（第149頁）。
寡營養的（形）　指營養（第111頁）貧乏的生境（第149頁）。
富營養化（名）　江河、淺湖中加入過多營養素（第111頁）時發生的過程。例如施肥引致藻類（第119頁）大量生長，當藻類死亡，被細菌（第119頁）腐化（第157頁），造成水體的氧濃度（第12頁）降低，使需氧（第22頁）生物無法生存。
半鹹水　溶解（第12頁）的離子（第9頁）濃度（第12頁）比淡水（↑）生境（第149頁）高，但比海水低的水體。
鹽沼　帶有濕的基質（第154頁），溶解（第12頁）鹽類濃度（第12頁）高的海岸生境（第149頁）。這是很高潮汐泛濫引起的。鹽沼的植被（第150頁）大多數是草本植物（第136頁）。
潮汐區帶（形）　指位於海岸高潮痕及低潮痕之間的生境（第149頁）。

162 · ECOLOGY/CLIMATES 生態學／氣候

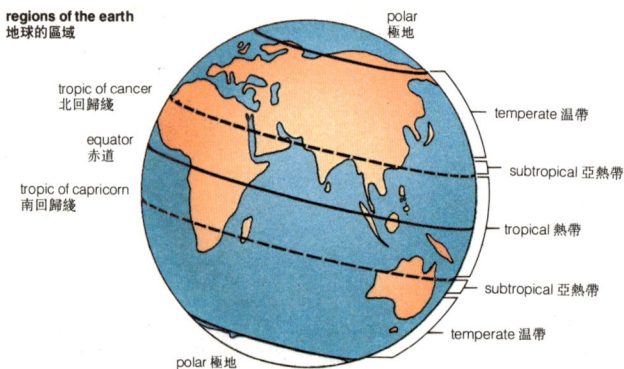

regions of the earth 地球的區域
tropic of cancer 北回歸線
equator 赤道
tropic of capricorn 南回歸線
polar 極地
temperate 溫帶
subtropical 亞熱帶
tropical 熱帶
subtropical 亞熱帶
temperate 溫帶
polar 極地

climatic factors the effects of temperature, sunlight, rainfall, wind, etc. on ecosystems (p. 149).

tropical (adj) of the regions of the world where the average monthly temperature changes little during the year, and the length of the day changes only slightly at different times of year.

subtropical (adj) of the regions of the world between the tropical (↑) and temperate (↓) regions.

temperate (adj) of the regions of the world which have warm summers with long days and cool winters with long nights.

polar (adj) of the very cold regions of the world close to the north and south poles, where the sun does not rise in midwinter and does not set in midsummer. Hardly any plants survive in polar regions.

phenology (n) the study of organisms and their activities in relation to the seasons of the year.

microclimate (n) the climate of a small or limited space, e.g. the surface of the soil, or under the canopy (p. 158) of a forest (p. 158).

quadrat (n) a square area marked out by an ecologist (p. 149) for counting and sampling organisms in a habitat (p. 149).

transect (n) a long rectangular area or a set of quadrats in a line marked out by an ecologist (p. 149) for counting and sampling organisms in one or several habitats (p. 149).

氣候因素 指溫度、陽光、雨量、風等對生態系（第149頁）的影響。

熱帶的(形) 指一年之中月平均氣溫變化小，在一年中的不同時期日長變化輕微的世界區域。

亞熱帶的(形) 指介於熱帶(↑)與溫帶(↓)之間的世界區域。

溫帶的(形) 指夏季暖和，日長而冬季涼爽，夜長的世界區域。

極地的(形) 指地球上接近北極和南極的嚴寒區域，在極地區域仲冬時節太陽不升起，仲夏時節太陽不落，難有任何植物生長。

物候學(名) 研究生物及其與年中各個季節有關活動的學科。

小氣候(名) 小空間或有限空間的氣候。例如土壤表面，或者森林(第158頁)的樹冠層(第158頁)下面的氣候。

樣方(名) 生態學家(第149頁)為計數一個生境(第149頁)中的生物和取樣而劃分出的一個正方面積。

狹樣區(名) 生態學家(第149頁)為計數一個或幾個生境(第149頁)中的生物及取樣而劃分出的一個狹長的矩形面積或一組樣方。

fungi (*n.pl.*) the large group of organisms, sometimes regarded as a separate kingdom (p.134), which are set apart from other plants because they are heterotrophic (p.32), lack chlorophyll (p.36), and often have chitin (↓) in their cell walls (p.17). Most fungi consist of thread-like hyphae (↓), which together form the mycelium (↓), although some, like yeast (p.164), are unicellular (p.119). Fungi reproduce (p.59) by spores (p.66). They are important as decomposers (p.157) in ecosystems (p.149), and many are parasitic (p.144). **fungus** (*sing.*), **fungal** (*adj*).

mycology (*n*) the study of fungi (↑). **mycologist** (*n*).

真菌(名、複) 生物中的大類群,有時被作為一個獨立的生物界(第134頁)和其他植物分開,因為真菌是異養(第32頁)的,缺乏葉綠素(第36頁),而其細胞壁(第17頁)中常含有幾丁質(↓)。大多數真菌是由絲狀菌絲(↓)組成並聯成菌絲體(↓),只有一些真菌如酵母(第164頁)是單細胞的(第119頁)。真菌是以孢子(第66頁)生殖(第59頁)。真菌是生態系(第149頁)中重要的分解者(第157頁),而有許多是寄生的(第144頁)真菌。(單數為fungus,形容詞為fungal)

真菌學(名) 研究真菌(↑)的學科。(名詞mycologist意為真菌學家)

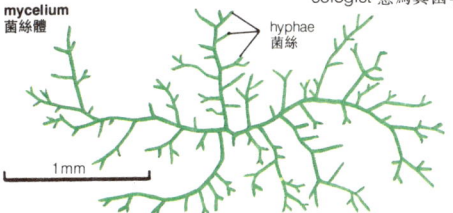

mycelium 菌絲體
hyphae 菌絲
1mm

mycelium (*n*) the vegetative (p.60) part of a fungus (↑), consisting of many hyphae (↓). **mycelia** (*pl.*).

hypha (*n*) a thread-like multinucleate (↓) tube with a cell wall (p.17), the organ (p.88) of vegetative (p.60) growth (p.109) in most fungi (↑). Hyphae increase in length by growth at their tips and give rise to new hyphae by side-branching. **hyphae** (*pl.*).

chitin (*n*) a polysaccharide (p.30) containing nitrogen. This is the main substance in most fungal (↑) cell walls (p.17). Chitin also occurs in insects.

multinucleate (*adj*) of cells with many nuclei (p.19), as in the hyphae (↑) of fungi (↑).

septum (*n*) a wall across a hypha (↑). The number of nuclei (p.19) between each septum varies from one or two, as in Basidiomycetes (p.165). to many, as in other groups. **septa** (*pl.*).

aseptate (*adj*) of hyphae (↑) without septa (↑). e.g. in the Phycomycetes (p.164).

菌絲體(名) 許多菌絲(↓)組成真菌(↑)的營養(第60頁)體。(複數為mycelia)

菌絲(名) 具細胞壁(第17頁)的絲狀多核(↓)管,為大部分真菌(↑)的營養(第60頁)生長(第109頁)器官(第88頁),菌絲以其尖端生長增加其長度,並以側枝長成新菌絲。(複數為hyphae)

幾丁質(名) 含氮的多醣類(第30頁),是大多數真菌(↑)細胞壁(第17頁)的主要物質。昆蟲體內也有幾丁質。

多核的(形) 指含有許多個核(第19頁)的細胞,如同真菌(↑)的菌絲(↑)。

隔膜(名) 橫貫菌絲(↑)的壁。各個隔膜之間的核(第19頁)數相差一或二個,如在擔子菌綱(第165頁)所見,以至相差多個如其他類中所見。(複數為septa)

無隔膜的(形) 指沒有隔膜(↑)的菌絲(↑),如在藻菌綱(第164頁)所見。

FUNGI/PHYCOMYCETES, ASCOMYCETES 真菌／藻菌綱、子囊菌綱

Phycomycetes (n) a group of simple, aseptate (p. 163) fungi (p. 163), living mainly in damp conditions; the hyphae (p. 163) of Phycomycetes are not usually organized into mycelia (p. 163).

mildew (n) a disease of plants caused by a fungus (p. 163) growing on the surface. There are two common types of mildew, downy and powdery, which are produced by different kinds of fungi.

mould (n) the general name for a fungal (p. 163) growth on a surface.

藻菌綱(名) 一類簡單的無隔膜(第 163 頁)真菌(第 163 頁)，主要生長在潮濕的條件下，藻菌綱的菌絲(第 163 頁)通常不組成菌絲體(第 163 頁)。

黴(名) 真菌(第 163 頁)生長在表面引致的植物病，有兩種常見的黴，即霜黴病及白粉病，這是由不同類真菌引起的病害。

黴菌(名) 生長在表面上的真菌(第 163 頁)的通稱。

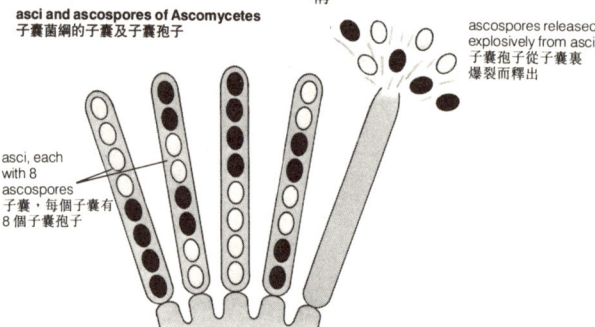

asci and ascospores of Ascomycetes
子囊菌綱的子囊及子囊孢子

ascospores released explosively from asci
子囊孢子從子囊裏爆裂而釋出

asci, each with 8 ascospores
子囊，每個子囊有 8 個子囊孢子

Ascomycetes (n) a large group of fungi (p. 163), recognized by their production of asci (↓) and ascospores (↓), e.g. yeast (↓).

ascospore (n) the haploid (p. 50) spore (p. 66) of an ascomycete (↑), formed by meiosis (p. 49) immediately after nuclear fusion (p. 61). Ascospores are contained within asci (↓), from which they are violently ejected when ripe (p. 83).

ascus (n) the reproductive (p. 59) organ (p. 88) of an ascomycete (↑), usually containing 8 ascospores (↑). Asci (pl.) are usually long and thin, with the ascospores arranged in a row.

yeast (n) kind of ascomycete (↑) fungus (p. 163). Yeasts, e.g. *Saccharomyces*, are unicellular (p. 119) and do not produce hyphae (p. 163) or mycelia (p. 163). Yeast cells can reproduce by budding. Yeasts are used by man for baking and brewing.

子囊菌綱(名) 一大類真菌(第 163 頁)，可通過其產生子囊(↓)及子囊孢子(↓)識別出，例如酵母(↓)。

子囊孢子(名) 子囊菌(↑)的單倍體(第 50 頁)孢子(第 66 頁)核融合(第 61 頁)後立即行減數分裂(第 49 頁)形成。子囊孢子藏在子囊(↓)內，當它成熟(第 83 頁)時猛烈地彈射出子囊孢子。

子囊(名) 子囊菌(↑)的生殖(第 59 頁)器官(第 88 頁)，通常含有 8 個子囊孢子(↑)。子囊為細長形有排成一列的子囊孢子。

酵母(名) 子囊菌綱(↑)的真菌(第 163 頁)類。酵母，例如酵母屬為單細胞的(第 119 頁)真菌，酵母不產生菌絲(第 163 頁)或菌絲體(第 163 頁)。酵母細胞能靠出芽生殖。人們使用酵母來烤製麵包及釀酒。

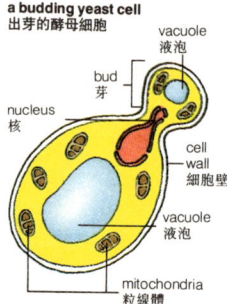

a budding yeast cell
出芽的酵母細胞

vacuole 液泡
bud 芽
nucleus 核
cell wall 細胞壁
vacuole 液泡
mitochondria 粒線體

FUNGI/BASIDIOMYCETES 真菌／擔子菌綱・165

Deuteromycetes (*n*) a group of fungi (p. 163) that only reproduce (p. 59) asexually (p. 59). They are common and some, e.g. *Penicillium*, are useful to man. Also known as Fungi Imperfecti (↓).
Fungi Imperfecti = Deuteromycetes (↑).

半知菌綱（名） 僅行無性（第59頁）生殖（第59頁）的一類真菌（第163頁）。這類真菌很常見，其中有些真菌如青黴屬菌對人類很有用，半知菌亦稱不完全菌類（↓）。

不完全菌類 同半知菌綱（↑）。

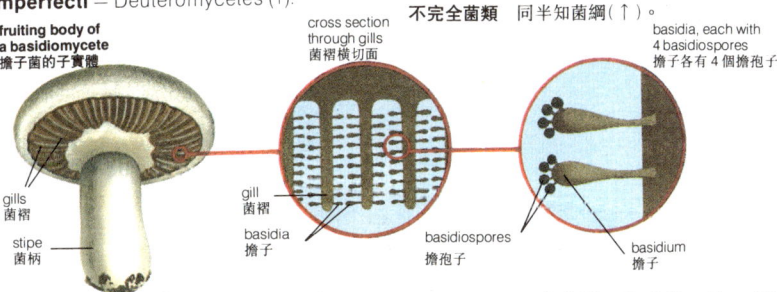

fruiting body of a basidiomycete 擔子菌的子實體
gills 菌褶
stipe 菌柄
cross section through gills 菌褶橫切面
gill 菌褶
basidia 擔子
basidiospores 擔孢子
basidia, each with 4 basidiospores 擔子各有4個擔孢子
basidium 擔子

Basidiomycetes (*n*) a group of fungi (p. 163), recognized by their production of spores (p. 66) externally on a basidium (↓). Mushrooms (↓) and toadstools (↓) are the fruiting bodies (↓) of basidiomycetes.

basidium (*n*) the reproductive (p. 59) organ (p. 88) of a basidiomycete (↑). A basidium consists of one or four cells, bearing four basidiospores (↓) on short stalks. **basidia** (*pl*.).

basidiospore (*n*) the haploid (p. 50) spore (p. 66) of a basidiomycete (↑), produced on a basidium (↑).

fruiting body = fruit.

mushroom (*n*) the name for the reproductive (p. 59) structure in basidiomycete (↑) fungi (p. 163) of the family (p. 134) Agaricaceae.

toadstool (*n*) the fruiting body (↑) of a basidiomycete (↑) fungus (p. 163), consisting of a stalk and a cap. The cap has gills (p. 166) on the underside, on which the spores (p. 66) are produced. Toadstools are often poisonous.

擔子菌綱（名） 真菌（第163頁）的一類，可從其孢子（第66頁）生在擔子（↓）外面識別出。菇類（↓）及毒蕈（↓）都是擔子菌綱的子實體（↓）。

擔子（名） 擔子菌（↑）的生殖（第59頁）器官（第88頁）。一個擔子由一或四個細胞組成，短柄上長著四個擔孢子（↓）。（複數為 basidia）

擔孢子（名） 擔子菌（↑）的長在擔子（↑）上單倍體（第50頁）孢子（第66頁）。

子實體 同果實。

菇類（名） 擔子菌（↑）真菌（第163頁）的一個科（第134頁），傘菌科的生殖（第59頁）結構的名稱。

毒蕈（名） 擔子菌（↑）真菌（第163頁）的子實體（↑），由菌柄和菌帽組成。菌帽下面有菌褶（第166頁），孢子（第66頁）在其中產生。毒蕈往往有毒。

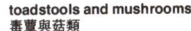
toadstools and mushrooms 毒蕈與菇類

166 · FUNGI/BASIDIOMYCETES 真菌／擔子菌綱

bracket fungus
簇狀菌

bracket fungus a basidiomycete (p. 165) fungus growing in the wood of living or dead trees, producing large flat-topped bracket-like fruiting bodies (p. 165) on the side of the host (p. 144).

stipe (n) a stalk, especially of a mushroom (p. 165) or toadstool (p. 165), or of a large seaweed (p. 122).

gill (n) a flat, vertically-positioned structure on the underside of the cap of a mushroom (p. 165) or toadstool (p. 165). The cap has many gills, radiating from the centre. The gills bear basidia (p. 165) on their surfaces.

dikaryon (n) a stage in the life cycle (p. 64) of many basidiomycetes (p. 165) when all the cells contain two haploid (p. 50) nuclei (p. 19). Each nucleus is derived from a different parent. **dikaryotic** (adj).

plasmogamy (n) fusion (p. 61) of the cytoplasm (p. 18) of two cells from different parents. This is the beginning of sexual (p. 59) reproduction (p. 59) in fungi (p. 163).

karyogamy (n) the fusion (p. 61) of two nuclei (p. 19) after plasmogamy (↑). In some fungi (p. 163), e.g. Basidiomycetes (p. 165), karyogamy takes place many cell divisions (p. 45) after plasmogamy. Between plasmogamy and karyogamy, each cell is a dikaryon (↑).

簇狀菌　一種生長在活的或枯死樹木上的擔子菌（第 165 頁）真菌，在寄主的（第 144 頁）側面部位生長大的平頂簇狀子實體（第 165 頁）。

菌柄（名）　菌類的柄，特別是菇類（第 165 頁）或毒蕈（第 165 頁）的柄，或者大海藻（第 122 頁）的柄。

菌褶（名）　在菇類（第 165 頁）或毒蕈（第 165 頁）菌帽下面的扁平豎直位置構造。菌帽有許多從中央放射出的菌褶，菌褶的表面長著擔子（第 165 頁）。

雙核體（名）　許多擔子菌綱（第 165 頁）生活史（第 64 頁）的一個階段，這時全部細胞都有兩個單倍體（第 50 頁）的核（第 19 頁）。各個核是由不同的親代所產生。（形容詞為 dikaryotic）

胞質配合（名）　不同親代所產生的兩個細胞的細胞質（第 18 頁）融合（第 61 頁）。這是真菌（第 163 頁）有性（第 59 頁）生殖（第 59 頁）的開始。

核配合（名）　胞質融合（↑）後兩個細胞核（第 19 頁）的融合（第 61 頁），在某些真菌（第 163 頁）例如擔子菌綱（第 165 頁）在胞質配合之後核配合，進行大量的細胞分裂（第 45 頁）。在胞質配合和核配合之間，每個細胞都是一個雙核體（↑）。

FUNGI/BASIDIOMYCETES 真菌／擔子菌綱・167

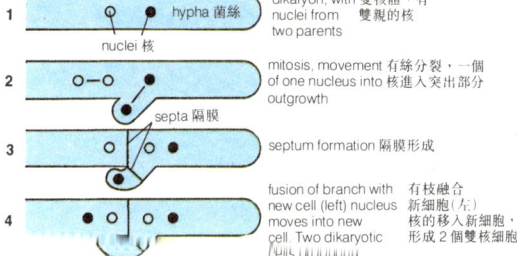

clamp connection a small looped branch of a hypha (p. 163) which grows at the time of cell division (p. 45) and septum (p. 163) formation in the dikaryon (↑) of a basidiomycete (p. 165).

dolipore septum a complex pore (p. 19) in the septum (p. 163) of a basidiomycete (p. 165) hypha (p. 163).

鎖狀連合　細胞分裂(第45頁)時長出的細小菌絲(第163頁)環圈狀分枝，而隔膜(第163頁)是在擔子菌(第165頁)雙核體(↑)中形成。

陷孔隔　擔子菌(第165頁)菌絲(第163頁)的隔膜(第163頁)的複合孔(第19頁)。

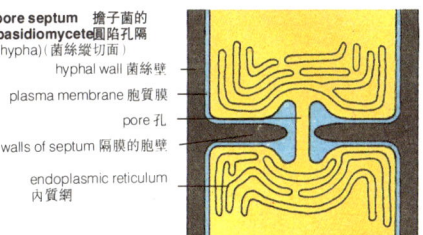

rust (n) a parasitic (p. 144) basidiomycete (p. 165) fungus (p. 163) of the order (p. 134) Uredinales. Rusts cause dark spots on the stems and leaves of plants. Some rusts, e.g. *Puccinia graminis* (cereal rust) are economically serious parasites.

uredospore (n) a type of vegetative (p. 60) spore (p. 66) produced by rust (↑) fungi (p. 163). Uredospores are dikaryotic (↑).

teleutospore (n) a type of thick-walled resting spore (p. 66) produced by rust (↑) fungi (p. 163). The teleutospore is the basidium (p. 165), eventually producing basidiospores (p. 165).

銹病菌(名)　真菌一個目(第134頁)銹菌目中的寄生(第144頁)擔子菌(第165頁)真菌(第163頁)。銹病菌引起植物的莖及葉生黑斑。某些銹病菌，例如禾穀柄銹屬(禾穀類銹菌)是造成嚴重經濟損失的寄生菌。

夏孢子(名)　銹病菌(↑)真菌(第163頁)產生的一種營養(第60頁)孢子(第66頁)，夏孢子是雙核的。(↑)。

冬孢子(名)　銹病菌(↑)真菌(第163頁)產生的一種厚壁休眠孢子(第66頁)。冬孢子是擔子(第165頁)，最後產生擔孢子(第165頁)。

Zygomycetes (n) a group of fungi (p. 163) which produce non-motile (p. 121) spores (p. 66) in sporangia (p. 66).

homothallic (adj) of zygomycete (↑) species (p. 134) which exist in a single physiological (p. 111) form. Zygospores (↓) can be produced as a result of sexual (p. 59) fusion (p. 61) between identical mycelia (p. 163) growing together.

heterothallic (adj) of zygomycete (↑) species (p. 134) which exist in two different forms that appear identical but have different physiologies (p. 111). Zygospores (↓) are only produced if the two forms are growing together.

接合菌綱(名) 在孢子囊(第66頁)內產生不能游動(第121頁)孢子(第66頁)的一類真菌(第163頁)。

同宗配合的(形) 指存在於單個的生理學(第111頁)形態接合菌(↑)的種(第134頁)。接合孢子(↓)可在完全相同的菌絲體(第163頁)之間生長在一起,作為有性(第59頁)融合(第61頁)的結果而形成。

異宗配合的(形) 指接合菌(↑)的種(第134頁)以兩種不同形態存在,這兩種形態看似相同,但生理學(第111頁)上都是不同的。接合孢子(↓)只能在兩種形態一起生長時才產生。

zygospore formation
接合孢子形成

2 zygomycete mycelia 2個接合菌的菌絲

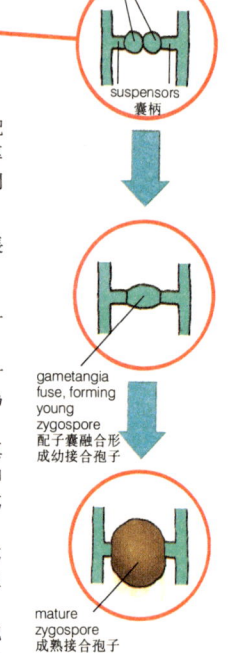

gametangia 配子囊

suspensors 囊柄

gametangia fuse, forming young zygospore 配子囊融合形成幼接合孢子

mature zygospore 成熟接合孢子

zygospore (n) thick-walled spore (p. 66) produced by zygomycete (↑) fungi (p. 163) after the fusion (p. 61) of gametangia (p. 65). Zygospores can rest for long periods of time before starting to grow.

suspensor[2] (n) the short branch of a hypha (p. 163) bearing a gametangium (p. 65) in a zygomycete (↑).

sporangiospore (n) an asexual (p. 59) spore (p. 66) produced in a sporangium (p. 66).

conidium (n) an asexual (p. 59) spore (p. 66) produced at the tip or the side of a hypha (p. 163). **conidia** (pl.).

Oomycetes (n) a group of fungi (p. 163) recognized by their zoospores (↓), which have two flagella (p. 121). Oomycetes are often aquatic (p. 161) or parasitic (p. 144).

zoospore (n) a motile (p. 121) flagellate (p. 121) spore (p. 66), found in many aquatic (p. 161) fungi (p. 163), e.g. Oomycetes (↑).

chlamydospore (n) asexual (p. 59), thick-walled structure containing food reserves, which can survive periods when hyphae (p. 163) cannot grow.

接合孢子(名) 接合菌(↑)真菌(第163頁)在配子囊(第65頁)融合(第61頁)之後產生的厚壁孢子(第66頁)。接合孢子可以經長時間休眠周期之後才開始生長。

囊柄(名) 接合菌(↑)配子囊(第65頁)中生長的菌絲(第163頁)的短分枝。

孢囊孢子(名) 在孢子囊(第66頁)內產生的一種無性(第59頁)孢子(第66頁)。

分生孢子 菌絲(第163頁)頂端或側面產生的一種無性(第59頁)孢子(第66頁)。(複數為conidia)

卵菌綱(名) 可從其游動孢子(↓)識別的一類真菌(第163頁),具有雙鞭毛(第121頁)。卵菌通常是水生的(第161頁)或寄生的(第144頁)。

游動孢子(名) 能游動(第121頁)、有鞭毛(第121頁)的孢子,見於許多水生(第161頁)真菌類(第163頁),如卵菌綱(↑)。

厚垣孢子(名) 有厚壁結構的無性(第59頁)孢子,內貯有食物,當菌絲(第163頁)不能生長時,可生存一定時間。

FUNGI/CHYTRIDIOMYCETES, MYXOMYCETES 真菌／壺菌綱、黏菌綱・169

Chytridiomycetes (*n*) a group of aquatic (p. 161) and soil-dwelling fungi (p. 163), commonly unicellular (p. 119), which produce zoospores (↑).

oogonium (*n*) reproductive (p. 59) organ (p. 88) in some fungi (p. 163) and algae (p. 119), which produces female gametes (p. 61), or oospheres (↓). Oogonia (*pl.*) are multinucleate (p. 163).

oosphere (*n*) the female gamete (p. 61) produced in an oogonium (↑).

oospore (*n*) a dormant (p. 117), thick-walled zygote (p. 61), formed by the fertilization (p. 62) of an oosphere (↑).

columella (*n*) the central part of a sporangium (p. 66) in some fungi (p. 163), e.g. the order (p. 134) Mucorales.

Myxomycetes (*n*) the taxonomic (p. 133) group containing the true slime moulds (↓), also known as the acellular (↓) slime moulds.

壺菌綱（名） 一類水生（第161頁）真菌及土居真菌（第163頁），常為單細胞（第119頁），能產生游動孢子（↑）。

卵囊（名） 某些真菌（第163頁）及藻類（第119頁）產生雌配子（第61頁）或卵球（↓）的生殖（第59頁）器官（第88頁），卵囊是多細胞的（第163頁）。

卵球（名） 卵囊（↑）產生的雌配子（第61頁）。

卵孢子（名） 由卵球（↑）受精（第62頁）形成的休眠（第117頁）厚壁合子（第61頁）。

蒂軸（名） 某些真菌（第163頁）如毛黴目（第134頁）中孢子囊（第66頁）的中心部分。

黏菌綱（名） 含有真正黏菌（↓）的分類（第133頁）群，亦稱非細胞（↓）黏菌。

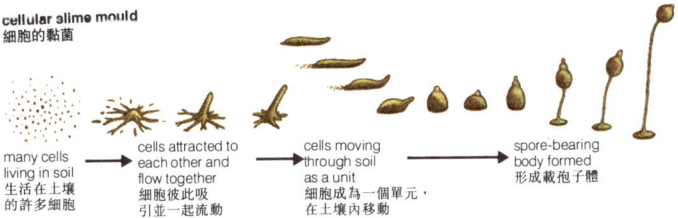

cellular slime mould
細胞的黏菌

many cells living in soil
生活在土壤的許多細胞

cells attracted to each other and flow together
細胞彼此吸引並一起流動

cells moving through soil as a unit
細胞成為一個單元，在土壤內移動

spore-bearing body formed
形成載孢子體

slime moulds a group of heterotrophic (p. 32) soil-dwelling organisms. They can be either acellular (↓) or cellular (↓). In acellular slime moulds, or Myxomycetes (↑), the organism is a plasmodium (↓). In cellular slime moulds, the cells are amoeba-like and single when feeding; when starved, they flow together to form a single sporulating (p. 66) structure.

plasmodium (*n*) a jelly-like multinucleate (p. 163) mass of protoplasm (p. 16) surrounded by a membrane (p. 18). This is the vegetative (p. 60) stage of an acellular (↓) slime mould (↑). A plasmodium can move through the soil.

acellular (*adj*) not made of cells, like the multinucleate (p. 163) plasmodium (↑) of a myxomycete (↑) slime mould (↑).

cellular (*adj*) made of cells.

黏菌 一群異養（第32頁）的土居有機種，可以是非細胞狀的（↓），或細胞狀的（↓）。在非細胞型的黏菌或黏菌綱（↑），此類有機體是原質團（↓）。在細胞狀黏菌，其細胞為阿米巴狀，當飼飽時為單體，當飢餓時它們流在一起，形成單一的孢子形成（第66頁）結構。

原質團（名） 為一層膜（第18頁）包著的凍膠狀多核（第163頁）原生質（第16頁）團，這是非細胞狀（↓）黏菌（↑）的營養（第60頁）階段。一個原質團能在土壤內移動。

非細胞的（形） 指不是細胞形成的，像黏菌綱（↑）黏菌（↑）的多核（第163頁）原質團（↑）。

細胞狀的（形） 形容由細胞形成的。

structure (*n*) (1) the three-dimensional arrangement of the parts of a substance or organism, e.g. the structure of a molecule (p. 9) is the arrangement of its atoms (p. 8), the structure of a plant is the arrangement of its tissues (p. 88) and organs (p. 88); (2) any object with a definite shape or arrangement of parts, e.g. a molecule, a cell, a trunk (p. 92) of a tree.

function (*n*) the part played by a structure or a system, e.g. the function of chloroplasts (p.32) is photosynthesis (p. 32), the function of photosynthesis (p. 32) is to produce carbohydrates (p. 28). **function** (*v*). **functional** (*adj*).

unit (*n*) (1) a single structure or object, which repeated many times forms a whole functioning object, e.g. the units of a nucleic acid (p. 51) are nucleotides (p. 52), the units of a population (p. 135) are individuals (p. 135); (2) a standard measurement, e.g. a metre, a kilogram.

sequence (*n*) (1) the order in which units are arranged one after another in a line, e.g. nucleotides (p. 52) in a nucleic acid (p. 51), amino acids (p. 56) in a protein (p. 56); (2) the order in which a set of chemical reactions (p. 11) in a metabolic pathway (p. 14) take place.

specialized (*adj*) of organisms and structures which are adapted for living in a particular habitat (p. 149) or shaped for a particular function, e.g. epiphytes (p. 137) are specialized for living on the branches of trees, leaves are specialized for photosynthesis (p. 32). **specialize** (*v*). **specialization** (*n*).

stimulus (*n*) an effect of the environment (p. 149) that causes, activates or quickens a process in an organism. A stimulus can be continuous, e.g. gravity, which causes the downward growth of roots, or periodic, e.g. light, which activates photosynthesis (p. 32), or sudden and occasional, e.g. damage, which activates the growth of callus (p. 108) tissue (p. 88) in plants.

modification (*n*) a minor change in structure or function, e.g. a bulb (p. 60) is an evolutionary (p. 139) modification of stem and leaves. **modify** (*v*).

結構（名）　(1)一個物質或有機體的各部分的三維空間排列。例如分子（第 9 頁）結構是組成分子的原子（第 8 頁）的排列，植物的結構是植物組織（第 88 頁）和器官（第 88 頁）的排列；(2) 具有確定形狀的任何物體或部件如分子、細胞、樹幹（第 92 頁）的排列。

功能（名）　一個結構或系統所起的作用。例如葉綠體（第 32 頁）的功能是光合作用（第 32 頁），光合作用（第 32 頁）的功能是製造碳水化合物（第 28 頁）。（動詞為 function，形容詞為 functional）

單元；單位（名）　(1)一個單元的結構或物體，它重複多次形成一個功能完整的物體。例如核酸（第 51 頁）的單元是核苷酸（第 52 頁），種群（第 135 頁）的單元是個體（第 135 頁）；(2) 標準度量單位例如米、公斤。

順序；序列（名）　(1)許多單元一個接一個排成一列的次序，例如核酸（第 51 頁）分子中的許多核苷酸（第 52 頁），蛋白質（第 56 頁）分子中的許多氨基酸（第 56 頁）都有排列順序；(2) 在一個代謝途徑（第 14 頁）中發生一組化學反應（第 11 頁）的次序。

特化的（形）　指各種有機體及結構適應於生活一種特定生境（第 149 頁），或適應於特定功能而成某種形狀。例如附生植物（第 137 頁）特化以適應於生長在樹枝上，葉片特化以適應光合作用（第 32 頁）。（動詞為 specialize，名詞為 specialization）

刺激（名）　環境（第 149 頁）的影響引起、激活或加速有機體內的過程。刺激有連續性的，例如地心吸力使根向下生長；呈週期性的，例如光照激活光合作用（第 32 頁）；或者是突然和偶然的，例如損傷激活植物的癒傷（第 108 頁）組織（第 88 頁）的生長。

變態；誘發變異（名）　結構或功能的細小變化，例如鱗莖（第 60 頁）是莖和葉的進化（第 139 頁）變態。（動詞為 modify）

GENERAL AND TECHNICAL WORDS IN BOTANY 植物學一般詞彙及技術詞彙 · 171

mechanism (n) the way in which a process takes place, e.g. the mechanism of a chemical reaction (p. 11), or the way in which a functional unit works, e.g. the mechanism of an enzyme (p. 15).

medium (n) a solid or liquid substrate (p. 15), containing all the materials necessary for growth, used by biologists for the cultivation of organisms such as bacteria (p. 119), fungi (p. 163) and algae (p. 119), and also for the growth of tissue cultures (p. 69).

light microscope an instrument that uses light rays passing through a system of lenses to magnify small objects. The light microscope can be used for observing the arrangement of cells and tissues (p. 88) and the larger structures inside cells. It is not powerful enough for the observation of small details of cell structure. **microscopy** (n).

electron microscope a powerful instrument that uses electrons (p. 8) instead of light rays to magnify very small objects. The electron microscope can magnify more than 100 000 times, and can be used for observing very small details of cell structure.

stain (n) any one of many dye substances used in microscopy (↑) to show up particular parts of cells or tissues (p. 88).

機制；機理(名) 一個過程發生的方式，例如化學反應(第11頁)的機制，或者一個功能單元的工作方式，例如酶(第15頁)的機制。

培養基(名) 含有生長所需一切物質的固體或液體受質(第15頁)。生物學家用之於培養生物體如細菌(第119頁)、真菌(第163頁)及藻類(第119頁)及組織培養(第69頁)。

光學顯微鏡 讓光線透過一個透鏡組將細小物體放大的儀器。光學顯微鏡可用於觀察細胞及組織(第88頁)的排列，以及細胞內較大的結構，但用於觀察細胞結構的微小細部，其放大倍數不足夠。(名詞為 microscopy)

電子顯微鏡 用電子(第8頁)代替光線以放大極細微物體的強力儀器，電子顯微鏡可放大十萬倍以上，可用於觀察細胞內極微小的細部。

染色劑(名) 顯微術(↑)中任一種用於顯示細胞或組織(第88頁)特定部分的多種染料物質之一。

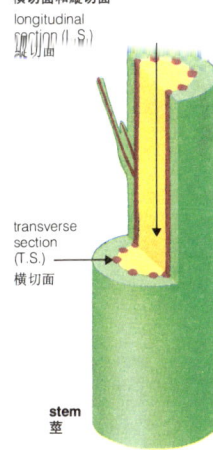

transverse sections and longitudinal sections
橫切面和縱切面

longitudinal section (L.S.)
縱切面

transverse section (T.S.)
橫切面

stem
莖

longitudinal section (L.S.)
縱切面

transverse section (T.S.)
橫切面

root tip
根尖

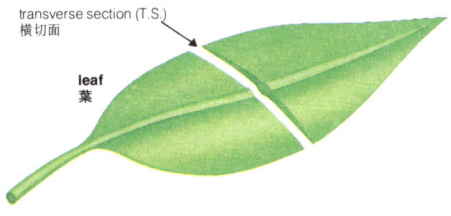

transverse section (T.S.)
橫切面

leaf
葉

transverse section a cut made across an organ (p. 88) or tissue (p. 88), at right angles to the main direction of growth. T.S. (abbr.).

longitudinal section a cut made through or along an organ (p. 88) or tissue (p. 88), in the same direction as the main direction of growth. L.S. (abbr.).

橫切面 和生長主方向成直角橫切一個器官(第88頁)或組織(第88頁)的截面。(英文縮寫為 T.S.)

縱切面 與生長主方向同向切過一個器官(第88頁)或組織(第88頁)的截面。(英文縮寫為 L.S.)

mode (*n*) the most frequent (↓) value or class in a set of measurements or sample (↓).

mean (*n*) the arithmetic average of a set of measurements, defined by the equation

$$\bar{x} = \frac{x_1 + x_2 + x_3 + \ldots + x_n}{n},$$

where $\bar{x}$ is the mean, $x_1, x_2, x_3, \ldots, x_n$ are the individual measurements, and n is the number of measurements.

眾數(名) 一組測量或樣本(↓)中最頻(↓)出現的數據或分組。

平均值(名) 一組測量數據的算術平均值，由下式確定：

$$\bar{X} = \frac{X_1 + X_2 + X_3 + \cdots + Xn}{n}$$

其中 $\bar{X}$ 為平均值，$X_1、X_2、X_3 \cdots Xn$ 為個別的測量數據，n 為測量數據的個數。

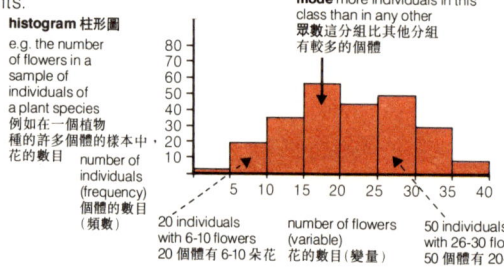

histogram (*n*) a way of showing the frequency (↓) with which different values of a variable (↓) occur in a sample (↓). The variable is divided into classes, and the frequency of each class is represented by the height of the bars on the chart.

normal distribution a symmetrical (p. 71) curve, showing the frequency (↓) of different values of a variable (↓) in a whole population (p. 135), i.e. the largest possible sample (↓). The mean (↑) and the mode (↑) are equal to each other in a normal distribution. Many biological variables are normally distributed.

柱形圖(名) 一個變量(↓)之不同值在一個樣本(↓)中出現的頻數(↓)的方式。將變量劃分為若干分組，用圖上柱的高度代表每一組的頻數。

正態分佈 表示在一個完整種群(第135頁)中一個變量(↓)的不同數據出現的頻數，即最大的可能樣本(↓)的一條對稱(第71頁)曲線，在一個正態分佈中平均值(↑)與眾數(↑)彼此相等。許多生物學的變量都是正態分佈的。

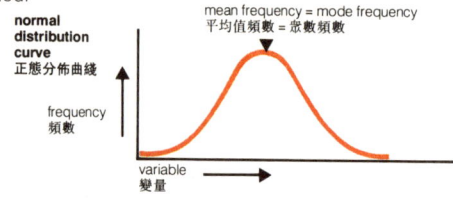

frequency (*n*) (1) a measure of how often an event takes place; (2) the number of times a particular class or value of a variable (↓) is recorded or observed in a sample (↓).

頻數(名) (1)事件發生頻繁程度的量度；(2)在一個樣本(↓)中一個特定分組或變量(↓)值被記錄或觀察到的次數。

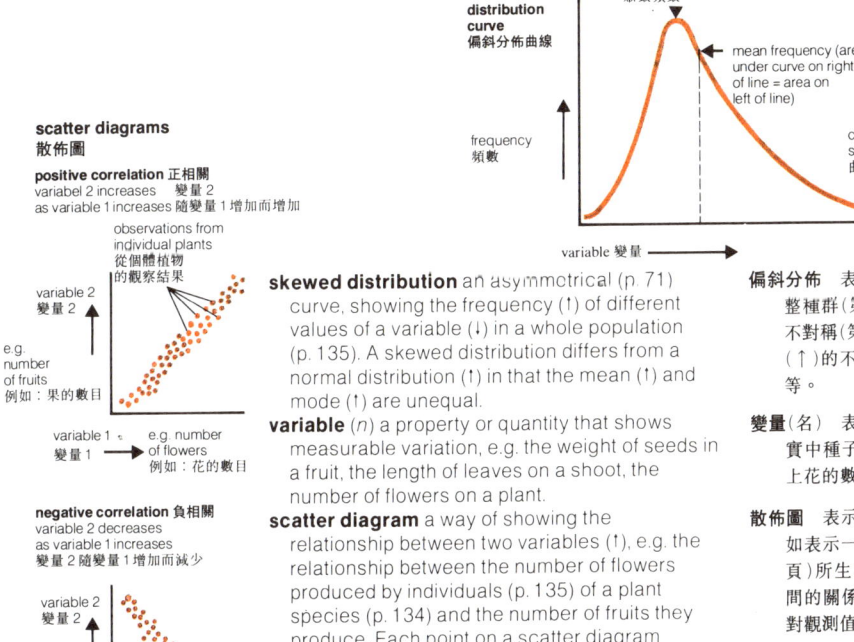

skewed distribution an asymmetrical (p. 71) curve, showing the frequency (↑) of different values of a variable (↓) in a whole population (p. 135). A skewed distribution differs from a normal distribution (↑) in that the mean (↑) and mode (↑) are unequal.

variable (n) a property or quantity that shows measurable variation, e.g. the weight of seeds in a fruit, the length of leaves on a shoot, the number of flowers on a plant.

scatter diagram a way of showing the relationship between two variables (↑), e.g. the relationship between the number of flowers produced by individuals (p. 135) of a plant species (p. 134) and the number of fruits they produce. Each point on a scatter diagram represents a pair of observations, one of each variable.

correlation (n) the process of determining whether the variation (p. 135) of one variable (↑) is related to the variation of another. If one variable increases at the same time as the other, they are said to be positively correlated. If one decreases as the other increases they are said to be negatively correlated. The correlation between two variables can be shown on a scatter diagram (↑).

sample (n) a small piece or part of a larger structure, area, or group, used by a scientist to measure or describe the properties of the larger object, e.g. quadrats (p. 162) are samples of vegetation (p. 150); dried plants in a herbarium (p. 133) are samples of species (p. 134).

偏斜分佈　表示一個變量(↓)的不同值在一個完整種群(第135頁)中出現的頻數(↑)的一條不對稱(第71頁)曲線。偏斜分佈和正態分佈(↑)的不同在於平均值(↑)與眾數(↑)不相等。

變量(名)　表示可測量變量的性質或量，例如果實中種子的重量、枝上葉的長度、一株植物上花的數目。

散佈圖　表示兩個變量(↑)間的關係的方式。例如表示一個植物種(第134頁)各個體(第135頁)所生花的數目與它們所形成果實數目之間的關係的方式。散佈圖上的每個點代表一對觀測值每個變量之一。

相關(名)　決定一個變量的變差是否和另一變量(↑)的變差(第135頁)有關係的方法。如果一個變量隨另一個變量的增加而增加，則為正相關，如果隨另一變量的增加而減少則為負相關。兩個變量之間的相關性可在散佈圖(↑)上顯示出來。

樣本(名)　科學家用於計量或描述較大物體性質的一個較大的結構、面積或組別的一小塊或小部分，例如樣方(第162頁)是植被(第150頁)的樣本；植物標本室(第133頁)的已乾植物是植物種(第134頁)的樣本。

Understanding botanical words
理解意義的植物學用詞彙

Many botanical and other scientific words or parts of words are derived from the Latin and Greek languages. The following pages contain some of the commoner word parts in the form of **prefixes**, that is, word parts that are added to the front of another word or word part to alter or specify its meaning. Many of the word parts in the list are not only used as prefixes, but also in the middle or at the end of words (sometimes in a slightly altered form); usually they cannot be used on their own. For instance, the prefix 'phyto-' (meaning: concerning plants) also appears as the word ending '-phyte' (meaning: plant), but not as a separate word.

Prefixes describing numbers or quantities are taken from Greek or Latin words. The following table shows the common prefixes from these two languages.

植物學及其他科學用的詞彙或構詞成分有許多是源於拉丁語及希臘語。以下各頁收列一些較常用的**詞頭**形式構詞成分，即加在另一個詞或構詞成分前面以改變或規定其詞義的構詞成分。表中所列的構詞成分有許多不僅用作詞頭，而且也用作詞的中間或詞尾（有時稍改變詞形），但通常不能單獨使用。例如詞頭 phyto-（義為有關植物的）亦出現於詞尾" -phyte "（表植物之義），但不能用作一個獨立的詞。

表示數字或數量的詞頭都源自希臘詞或拉丁詞。下表為源自這兩種語言的常用詞頭。

	GREEK PREFIX 希臘詞頭	LATIN PREFIX 拉丁詞頭	PREFIX 詞頭	MEANING 意義	詞源
1 一	mono-	uni-	hemi-	half 半	Gr 希臘
2 二	di-	bi-	semi-	half 半	L 拉丁
3 三	tri-	ter-	poly-	many 多；聚	Gr 希臘
4 四	tetra-	quad-	multi-	many 多	L 拉丁
5 五	penta-	quinq-	omni-	all 全；總	L 拉丁
6 六	hexa-	sex-	dupli-	twice 兩倍；兩次	L 拉丁
7 七	hepta-	sept-	tripli-	three times 三倍；三次	L 希臘
8 八	octo-	oct-	hypo-	less, under 次；較少；在下	Gr 希臘
9 九	nona-	novem-	hyper-	more, over 超；過；高；在上	Gr 希臘
10 十；分	deca-	deci-	sub-	under 亞；次；在下、較低	L 拉丁
100 百；厘	hecta-	cente-	super-	over 過、超、高於	L 拉丁
1000 千；毫	kilo-	milli-	iso-	same, equal, identical 相同、相等、同一	Gr 希臘

APPENDIX ONE/UNDERSTANDING BOTANICAL WORDS 附錄一／理解意義的植物學用詞彙・175

PREFIXES

a-	without, not, lacking, e.g. *a*sexual, not sexual; *a*symmetrical, without symmetry.
ab-	from, away, e.g. *ab*axial, the side of a leaf facing away from the stem.
ad-	to, towards, e.g. *ad*axial, the side of a leaf facing towards the stem.
allo-	different, differing, other, e.g. *allo*polyploid, a polyploid resulting from the fusion of two different nuclei; *allo*patric, of species occurring in different regions.
an-	same prefix as **a-**, used before words beginning with a vowel or the letter h, e.g. *an*aerobic, not aerobic.
andro-	male, e.g. *andro*ecium, the male parts of a flower.
anti-	against, opposite, e.g. *anti*biotic, a substance that acts against living organisms (especially bacteria); *anti*podal, the cells at the opposite end of the embryo sac from the micropyle.
apo-	from, away from, without, e.g. *apo*gamy, reproduction without sexual fusion; *apo*carpous, of flowers with carpels separate from one another.
auto-	caused by or originating in itself, e.g. *auto*polyploid, a polyploid resulting from an increase in the number of sets of chromosomes within a nucleus; *auto*troph, an organism producing its own food.
bi-	two, twice, double, e.g. *bi*nomial, the Latin name of a species, consisting of two words; *bi*ennial, a plant with a two-year life cycle.
bio-	life, living, e.g. *bio*logy, the study of living things.
caul(i)-	relating to stems, e.g. *cauli*florous, having flowers growing directly from the stem.
chromo-	colour, coloured, e.g. *chromo*plast, a plastid containing pigments; *chromo*somes, so called because they become deeply coloured when stained for microscopy.
cleisto-	closed, without an opening, e.g. *cleisto*gamy, self-pollination before the flower opens.
co-	together, with, associated, e.g. *co*enzyme, a substance (not a substrate) that is necessary for the functioning of an enzyme.
crypto-	hidden, e.g. *crypto*phyte, a plant whose perennating organs are underground; *crypto*gam, a plant whose reproductive organs are very small or hidden.
cyto-	relating to cells, e.g. *cyto*logy, the study of cells; *cyto*plasm, the parts of the cell outside the nucleus.

詞頭

	無；不；缺。例如 *a*sexual 無性的；*a*symmetrical 不對稱的。
	遠；離。例如 *ab*axial 遠軸的，指葉的一面朝向遠離莖的。
	近向；走向。例如 *ad*axial 近軸的，指葉的一面朝向接近莖的。
	異；別；他。例如 *allo*polyploid 異源多倍體，由兩個不同細胞核融合產生的多倍體；*allo*patric 異區起源的，物種出現在不同區域。
	同詞頭 **a-** 加在以母音或 h 字母為首的詞之前。例如 *an*aerobic 厭氧的，即不需氧的。
	雄性。例如 *andro*ecium 雄蕊群，花的雄性部。
	抗；反。例如 *anti*biotic 抗生素，一種對抗活有機體作用(尤其是細菌)的物質；*anti*podal 反足的，指細胞位於胚囊珠孔相對的一端。
	自；離；無。例如 *apo*gamy 無配子生殖，指無性融合的生殖方式；*apo*carpous 離心皮的，指與心皮彼此分離。
	自發或同源。例如 *auto*polyploid 同源多倍體，指多倍體產生於細胞核內染色體組數目的增加；*auto*troph 自養生物，指本身能製造食物的有機體。
	二；兩次；兩倍。例如 *bi*nomial 二名的，指物種的拉丁名由兩個詞組成；*bi*ennial 二年生的，指植物的生活週期為兩年。
	生命；活的。例如 *bio*logy 生物學，生物學科。
	與莖相關的。例如 *cauli*florous 莖花的，指花直接生於莖上的。
	色；有色。例如 *chromo*plast 有色體，含色素的質體；*chromo*somes 染色體，因在染色作顯微檢查時顏色變深而得名。
	閉合；無開口。例如 *cleisto*gamy 閉花受精，即在開花之前自花受精。
	共；與；關聯。例如 *co*enzyme 輔酶，為酶起作用所必須有的物質(但不是受質)。
	隱藏。例如 *crypto*phyte 隱芽植物，多年生器官長在地下的植物；*crypto*gam 隱花植物，生殖器官很小或隱藏起來的一種植物。
	細胞的；與細胞有關的。例如 *cyto*logy 細胞學，細胞學科；*cyto*plasm 細胞質，細胞核外的細胞部分。

di-	two, twice, double, e.g. *di*saccharide, a carbohydrate consisting of two sugar molecules (monosaccharides); *di*cotyledon, a plant with two cotyledons in the seed.	雙；二次；兩倍。例如 *di*saccharide 雙醣，由兩個單醣分子組成的碳水化合物；*di*cotyledon 雙子葉植物，種子有兩片子葉的植物。	
ecto-	outside, e.g. *ecto*trophic mycorrhizae grow outside the cells of the host root.	外；在外面。例如 *ecto*trophic mycorrhizae 外生菌根生長在寄主根部細胞之外。	
endo-	inside, inner, e.g. *endo*carp, the inner layer of the fruit wall; *endo*trophic mycorrhiza, with hyphae growing into the cells of the host root.	內；在內面；內部。例如 *endo*carp 內果皮，果壁的內層；*endo*trophic mycorrhiza 內生菌根的菌絲生在寄主根部細胞內。	
epi-	on, upon, above, outer, e.g. *epi*carp, the outer layer of the fruit wall; *epi*phyte, a plant growing on another plant; *epi*geal germination, when the cotyledons emerge above the ground.	上；在上面；在外。例如 *epi*carp 外果皮，果殼外層；*epi*phyte 附生植物，生在另一種植物上的植物；*epi*geal germination 出土萌發，指子葉長出土表之上時。	
eu-	good, normal, e.g. *eu*trophic, a habitat well-supplied or rich in nutrients.	良好；正常。例如 *eu*trophic 富營養的，指營養素供應良好或豐富的生境。	
ex-	without, e.g. *ex*albuminous, without endosperm; *ex*stipulate, without stipules.	無。例如 *ex*albuminous 無胚乳的，即沒有胚乳的；*ex*stipulate 無托葉的，即不具有托葉。	
extra-	outside, beyond, separate from, e.g. *extra*floral, away from the flower.	外面；在外；分離。例如 *extra*floral 花外的，指不在花的。	
flavo-	yellow, e.g. *flavo*protein, one of a group of yellow-coloured proteins.	黃色。例如 *flavo*protein 黃素蛋白，黃色蛋白類之一。	
gam(o)-	joining together, fusion, e.g. *gamo*petalous, having fused petals.	聯合；融合。例如 *gamo*petalous 合瓣的，花瓣部合生的。	
gymno-	naked, exposed, e.g. *gymno*sperm, a plant in which the seed is not enclosed in an ovary.	裸的；露的。例如 *gymno*sperm 裸子植物，種子不包裹在子房的植物。	
gyno-	female, e.g. *gyno*ecium, the female parts of a flower.	雌性的。例如 *gyno*ecium 雌蕊群，花的雌性部。	
halo-	salt, salty, e.g. *halo*phyte, a plant that grows in salty habitats.	鹽；含鹽的。例如 *halo*phyte 鹽生植物，生於含鹽生境的植物。	
hemi-	half, partly, e.g. *hemi*parasite, a parasite that produces some of its own food.	半；部分。例如 *hemi*parasite 半寄生植物，能生產本身所需營養的植物。	
hetero-	different, other, e.g. *hetero*zygous, with different alleles at the same locus on homologous chromosomes; *hetero*troph, an organism that obtains food other than from itself.	異；其他。例如 *hetero*zygous 異型結合的，指同源染色體同位點具不同等位基因；*hetero*troph 異養生物，非靠本身獲得食物的生物體。	
homo-	same, similar, e.g. *homo*logous chromosomes have the same sequence of loci; *homo*sporous plants produce spores all of the same size.	相同；相似。例如 *homo*logous chromosomes 同源染色體具相同的位點序；*homo*sporous 同型孢子植物所生孢子大小相等。	
hydro-	relating to water, e.g. *hydro*phyte, a plant with perennating organs under water; *hydro*lysis, a chemical reaction involving the addition of water molecules and the breakdown of organic molecules.	水的。例如 *hydro*phyte 水生植物，具水下多年生器官的植物；*hydro*lysis 水解作用，和加入水分子及有機分子的分解相關的化學反應。	
hyper-	more, above, very, e.g. *hyper*tonic, a more concentrated solution.	過；超；極。例如 *hyper*tonic 高滲的，一種更濃的溶液。	
hypo-	less, below, under, e.g. *hypo*tonic, a less concentrated solution; *hypo*gynous, a flower in which the corolla, calyx and anthers arise below the gynoecium.	少；在下；低。例如 *hypo*tonic 低滲的，濃度更低的溶液；*hypo*gynous 下位的，指花的花冠、花萼及花藥在雌蕊下長出。	

infra-	below, under, e.g. *infra*specific, variation below the level of species.	在下；在…以下。例如 *infra*specific 種以下的，指種分類級以下的變異。	
inter-	between, e.g. *inter*specific, competition between species.	之間。例如 *inter*specific 種間的，指種間的競爭。	
intra-	within, e.g. *intra*specific, competition within a species.	內部的。例如 *intra*specific 種內的，指種內的競爭。	
iso-	identical, e.g. *iso*gamy, the fusion of morphologically identical gametes.	等同。例如 *iso*gamy 同配生殖，指形態相同的配子的融合。	
lepto-	thin, slender, e.g. *lepto*tene, the stage in the first meiotic prophase when the chromosomes appear as thin threads.	瘦小；細。例如 *lepto*tene 細絲期，指第一次減數分裂的前期階段，此時染色體呈細絲狀。	
macro-	large, great, long, e.g. *macro*molecule, a large molecule composed of many smaller molecular units.	大；巨大；長。例如 *macro*molecule 大分子，係由許多較小的分子單元組成一個大的分子。	
mega-	(1) great, large, e.g. *mega*spore, the larger of the two kinds of spore produced by heterosporous plants; (2) one million times.	(1)大；巨。例如 *mega*spore 大孢子，異型孢子植物所生兩孢子之較大者；(2)百萬次(倍)。	
meso-	middle, between, e.g. *meso*phyll, the layer of tissue between the palisade and lower epidermis in a leaf; *meso*carp, the middle layer of the pericarp of a fruit.	中間；居間。例如 *meso*phyll 葉肉，為葉內位於柵欄組織與下表皮之間的組織層；*meso*carp 中果皮，指果實的果皮的中間一層。	
micro-	small, very small, e.g. *micro*scope, an instrument used for observing very small objects; *micro*spore, the smaller of the two kinds of spore produced by heterosporous plants.	小；微。例如 *micro*scope 顯微鏡，用於觀察極微小物體的儀器；*micro*spore 小孢子，指異型孢子植物所生兩種孢子之較小者。	
mono-	one, once, single, e.g. *mono*cotyledon, a plant with one cotyledon in the seed; *mono*carpic, a plant that produces fruit once in its lifetime.	一；一次；單一。例如 *mono*cotyledon 單子葉植物，種子只具一片子葉的植物；*mono*carpic 結果一次的，一生只結一次果的植物。	
morph(o)-	shape, relating to shape, e.g. *morph*ology, the study of shape.	形狀；形態。例如 *morph*ology 形態學，植物形態的學科。	
multi-	many, e.g. *multi*nucleate, of cells with many nuclei.	多。例如 *multi*nucleate 多核的，細胞具有多個核。	
myco-	relating to fungi, e.g. *myco*logy, the study of fungi.	真菌的。例如 *myco*logy 真菌學，真菌的學科。	
neo-	new, e.g. *neo*Darwinism, the science of evolution developed after Darwin, including the more recently-discovered principles of genetics.	新。例如 *neo*Darwinism 新達爾文主義，達爾文之後提出的進化科學，其中包括許多現代發現的遺傳學原理。	
oligo-	few, e.g. *oligo*trophic, a habitat with few nutrients or low fertility; *oligo*saccharide, a carbohydrate consisting of a few monosaccharide units.	寡；少。例如 *oligo*trophic 寡養的，指生境的營養或肥力低；*oligo*saccharide 寡糖，由幾個單醣單元組成的碳水化合物。	
ortho-	upright, correct, e.g. *ortho*tropic, an upright axis.	直；正確。例如 *ortho*tropic 直生的，直立的軸。	
pachy-	thick, fat, e.g. *pachy*tene, the stage in the first meiotic prophase when the chromosomes become short and thick.	厚；肥。例如 *pachy*tene 粗絲期，第一次減數分裂前期，此時染色體變粗短。	
palaeo-	old, ancient, e.g. *palaeo*botany, the study of fossil plants.	古；古代。例如 *palaeo*botany 古植物學，化石植物的學科。	
pent(a)-	five, e.g. *pent*ose, a monosaccharide with five carbon atoms.	五。例如 *pent*ose 戊醣，具五個碳原子的單醣。	
peri-	around, on the surface, e.g. *peri*anth, the parts of the flower around the reproductive parts; *peri*carp, the wall of the fruit.	周圍；在表面。例如 *peri*anth 花被，生殖部周圍的花部；*peri*carp 果皮，果實的外壁。	
photo-	relating to light, e.g. *photo*synthesis, the production of carbohydrates using energy from light; *photo*tropism, curved growth towards light.	光的。例如 *photo*synthesis 光合作用，指利用陽光能生產碳水化合物的作用；*photo*tropism 向光性，向著陽光彎曲生長。	

phyco-	concerning algae, e.g. *phyco*biont, the algal partner in a lichen symbiosis.	有關藻類的。例如 *phyco*biont 藻共生體，指參加地衣共生的藻類。
phyll(o)-	relating to leaves, e.g. *phyllo*taxy, the way in which leaves are arranged.	與葉相關的。例如 *phyllo*taxy 葉序，指葉排列的方式。
phyto-	concerning plants, e.g. *phyto*chemistry, the chemistry of plants.	有關植物的。例如 *phyto*chemistry 植物化學。
poly-	many, e.g. *poly*peptide, a molecule with many peptide bonds.	多。例如 *poly*peptide 多肽，指具多肽鍵的分子。
rhiz(o)-	relating to roots, rootlike organs or other underground plant parts, e.g. *rhiz*oid, the 'roots' of bryophytes; *rhiz*ome, an underground stem.	和根、根狀器官及其他地下部分有關的。例如，*rhiz*oid 假根，為苔蘚植物的根；*rhiz*ome 根狀莖，是一種長在地下的莖。
sapro-	concerning decay, e.g. *sapro*phyte, a plant that lives on decaying organic matter.	和腐爛有關的。例如 *sapro*phyte 腐生植物，生於腐爛有機物質上的植物。
schiz(o)-	splitting, dividing, e.g. *schizo*carp, a fruit that splits into the separate carpels when ripe.	裂；分開。例如 *schizo*carp 分果，成熟時裂成分開心皮的果實。
schler(o)-	hard, rigid, e.g. *scler*enchyma, a hard, supporting plant tissue.	硬；堅。例如 *scler*enchyma 厚壁組織，為支持植物的堅硬組織。
semi-	half, partly, e.g. *semi*permeable, of membranes that allow the passage of some molecules but not others.	半；部分。例如 *semi*permeable 半透性的，指一種膜只能讓某些分子通過。
sub-	under, below, somewhat, e.g. *sub*species, a taxon below the level of species; *sub*acute, a leaf apex that is somewhat acute.	下；在下；稍。例如 *sub*species 亞種，為種級以下的分類單元。*sub*acute 稍尖，指葉的頂端略尖。
sym-	together, united, e.g. *sym*biosis, two different organisms living with and depending on each other.	共；合；連。例如 *sym*biosis 共生現象，指兩種不同的生物生活在一起並互相依賴。
syn-	together, united, e.g. *syn*carpous, of ovaries in which the carpels are united.	合；同；連。例如 *syn*carpous 合生心皮的，指子房的心皮連在一起。
tetra-	four, e.g. *tetra*ploid, having four sets of homologous chromosomes.	四。例如 *tetra*ploid 四倍體，有四組同源染色體。
tri-	three, e.g. *tri*ose, a monosaccharide with three carbon atoms; *tri*ploid, having three sets of homologous chromosomes.	三。例如 *tri*ose 丙醣，為含三個碳原子的單醣；*tri*ploid 三倍體；具有三組同源染色體。
uni-	one, once, single, e.g. *uni*cellular, an organism consisting of one cell.	一；一次；單。例如 *uni*cellular 單細胞的，指由一個細胞組成的有機體。
xero-	dry, e.g. *xero*phyte, a plant that occurs in dry habitats.	乾。例如 *xero*phyte 旱生植物，指生長於乾旱生境的植物。

International System of Units (SI)
國際單位制

PREFIXS 詞頭

PREFIX 詞頭	FACTOR 因數	SIGN 記號
milli- 毫	$\times 10^{-3}$	m
micro- 微	$\times 10^{-6}$	μ
nano- 納(毫微)	$\times 10^{-9}$	n
pico- 皮(微微)	$\times 10^{-12}$	p

PREFIX 詞頭	FACTOR 因數	SIGN 記號
kilo- 千	$\times 10^{3}$	k
mega- 兆	$\times 10^{6}$	M
giga- 吉	$\times 10^{9}$	G
tera- 太	$\times 10^{12}$	T

BASIC UNITS 基本單位

UNIT	單位	SYMBOL 符號	MEASUREMENT	量
metre	米	m	length	長度
kilogram	千克(公斤)	kg	mass	質量
second	秒	s	time	時間
ampere	安培	A	electric current	電流
kelvin	開爾文	K	temeprature	溫度
mole	摩爾	mol	amount of substance	物質的量

DERIVED UNITS 導出單位

UNIT	單位	SYMBOL 符號	MEASUREMENT	量
newton	牛頓	N	force	力
joule	焦耳	J	energy, work	能量、功
hertz	赫茲	Hz	frequency	頻率
pascal	帕斯卡	Pa	pressure	壓強、壓力
coulomb	庫倫	C	quantity of electric charge	電荷量
volt	伏特	V	electrical potential	電位
ohm	歐姆	Ω	electrical resistance	電阻

SOME MULTIPLES OF SI UNITS HAVING SPECIAL NAMES　某些有專門名稱的國際制單位倍數量

UNIT	單位	SYMBOL 符號	DEFINITION 定義	MEASUREMENT	量
angstrom	埃	Å	10^{-10} m = 10^{-1} nm	length	長度
micron	微米	μm	10^{-6} m	length	長度
litre	升	l	10^{-3} m^3 = dm^3	volume	容積
tonne	噸	t	10^3 kg	mass	質量
dyne	達因	dyn	10^{-5} N	force	力
bar	巴	bar	10^5 Pa	pressure	壓力

SOME NON-SI UNITS　某些國際制單位

UNIT	單位	SYMBOL 符號	DEFINITION 定義	MEASUREMENT	量
atm	大氣壓	atm	101325 Pa, 帕斯卡 1.01325 ba 巴	pressure	壓力
degree Celsius	攝氏度	°C	K($t_C = t_K - 273$)	temperature	溫度
million years	百萬年	Ma, m.y.	10^6 years 年	time	時間
billion (US) years	十億年(美國)	Ga	10^9 years 年	time	時間

INDEX 索引

abaxial /æb'æksɪəl/ 遠軸的 97
aberration /ˌæbə'reʃən/ 畸變 41
abscisic acid /æb'sɪsɪk 'æsɪd/ 脫落酸 113
abscission /æb'sɪʒən/ 脫落 114
absorption spectrum /əb'sɔrpʃən 'spɛktrəm/ 吸收光譜 38
accessory pigment /æk'sɛsərɪ 'pɪgmənt/ 輔助色素 36
acellular /e'sɛljələ/ 非細胞的 169
achene /e'kin/ 瘦果 84
acid /'æsɪd/ 酸 11
acrocarpous /ˌækrə'kɑrpəs/ 頂生蒴的 125
actinomorphic /ˌæktɪə'mɔrfɪk/ 輻射對稱的 71
action spectrum /'ækʃən 'spɛktrəm/ 作用光譜 38
active site /'æktɪv saɪt/ 活性部分 15
active transport /'æktɪv 'trænsport/ 主動運輸 101
adaptation /ˌædəp'teʃən/ 適應 141
adaptive radiation /ə'dæptɪv ˌredɪ'eʃən/ 適應輻射 141
adaxial /æd'æksɪəl/ 近軸的 97
adenine /'ædənɪn/ 腺嘌呤 52
ADP /e di pi/ 二磷酸腺苷 26
adventitious root /ˌædvɛn'tɪʃəs rut/ 不定根 89
aerenchyma /eərɛnkɪmə/ 通氣組織 92
aerial root /e'ɪrɪəl rut/ 氣生根 89
aerobic /ˌeə'robɪk/ 需氧的；好氧的 22
aestivation /'ɛstəˌveʃən/ 花被捲疊式 72
agamospermy /ˌægəməspɚm/ 無融合結籽 59
aggregation /ˌægrɪ'geʃən/ 族聚 119
air layering /ɛr 'leɚ.ɪŋ/ 空中壓條 69
albumen /æl'bjumən/ 胚乳 86
alcohol /'ælkəˌhɔl/ 醇 24
aldose /'ældos/ 醛醣；醛糖 29
aleurone layer /ə'ljuron 'leɚ/ 糊粉層 86
algae /'ældʒi/ 藻類 119
alkaloid /'ælkəˌlɔɪd/ 生物鹼 148
alleles /ə'lilz/ 對偶基因；等位基因 43
allelopathy /ə'lilə'pæθɪ/ 植物相剋作用 148
allogamy /ə'lagəmɪ/ 異體受精 62
allopatric /ˌælə'pætrɪk/ 異區起源的 142
allopolyploid /ˌælə'pɑlɪlɔɪd/ 異源多倍體 50
alternate /'æltɚnɪt/ 互生的 98
alternation of generations /ˌɔltɚ'neʃən ɑv ˌdʒɛnə'reʃənz/ 世代交替 64

amino acid /ə'mino 'æsɪd/ 氨基酸 56
ammonia /ə'monjə/ 氨 13
amylase /'æmɪˌles/ 澱粉酶 30
amylopectin /ˌæmɪlə'pɛktɪn/ 支鏈澱粉 30
amyloplast /ə'mɪləplæst/ 造粉體；澱粉體 19
amylose /'æmɪˌlos/ 直鏈澱粉 30
anaerobic /ˌænˌeə'robɪk/ 厭氧的；缺氧的 22
anaphase /'ænəˌfez/ 後期 45
anatomy /ə'nætəmɪ/ 解剖學 88
anatropous /ə'nætrəpəs/ 倒生的 78
androecium /æn'drɪʃɪəm/ 雄蕊群 73
andromonoecious /ˌændrəmə'niʃəs/ 雄花兩性花同株的 79
anemophily /ˌænə'mɑfəlɪ/ 風媒 75
angiosperm /'ændʒɪoˌspɚm/ 被子植物 130
anisogamous /ˌænaɪ'sɑgəməs/ 異配生殖的 61
annual /'ænjʊəl/ 一年生的 117
anther /'ænθɚ/ 花藥 73
antheridium /ˌænθə'rɪdɪəm/ 藏精器 65
antherozoid /ˌænθərə'zoɪd/ 游動精子 65
anthesis /æn'θɪsɪs/ 開花期 73
antibiotic /ˌæntɪbaɪ'ɑtɪk/ 抗生素 148
anticlinal /ˌæntɪ'klaɪnl̩/ 垂周的 110
anticodon /ˌæntɪˌkodɑn/ 反密碼子 53
antipodal cells /æn'tɪpədl̩ sɛlz/ 反足細胞 78
apetalous /e'pɛtləs/ 無瓣的 71
apex /'epɛks/ 尖端；頂端 90
apical dominance /'æpɪkl̩ 'dɑmənəns/ 頂端優勢 114
apocarpous /ˌæpə'kɑrpəs/ 離心皮的 76
apogamy /ə'pɑgəmɪ/ 無配子生殖 59
apomixis /ˌæpə'mɪksɪs/ 無融合生殖 59
apoplast /'æpə'plæst/ 非原質體 102
apospory /'æpəˌsporɪ/ 無孢子生殖 59
apothecium /ˌæpə'θɪʃɪəm/ 子囊盤 147
aquatic /ə'kwætɪk/ 水生的 161
aqueous /'ekwɪəs/ 水的 12
archegonium /ˌɑrkɪ'gonɪəm/ 藏卵器 65
architecture /'ɑrkəˌtɛktʃɚ/ 結構 92
aril /'ærɪl/ 假種皮 86
armed /ɑrmd/ 具刺的 100
aromatic /ˌærə'mætɪk/ 芳香的 31

artificial key /ˌɑrtəˈfɪʃəl ki/ 人為檢索表　133
artificial selection /ˌɑrtəˈfɪʃəl səˈlɛkʃən/ 人工選擇　140
Ascomycetes /ˌæskəmaɪˈsits/ 子囊菌綱　164
ascospore /ˈæskəˌspor/ 子囊孢子　164
ascus /ˈæskəs/ 子囊　164
aseptate /əˈsɛptɪk/ 無隔膜的　163
asexual /eˈsɛkʃʊəl/ 無性的　59
association /əˈsosɪˈeʃən/ 群叢　150
asymmetrical /ˌesɪˈmɛtrɪkl/ 不對稱的　71
atom /ˈætəm/ 原子　8
ATP /e ti pi/ 三磷酸腺苷　26
auricle /ˈɔrɪkl/ 葉耳　100
autecology /ˌɔtəˈkɑlədʒɪ/ 個體生態學　149
authority /əˈθɔrətɪ/ 命名人　133
autogamy /ɔˈtɑgəmɪ/ 自花受粉　62
autopolyploid /ˈɔtoplɔɪd/ 同源多倍體　50
autotrophic /ˌɔtəˈtrɑfɪk/ 自養的；自營的　32
auxin /ˈɔksɪn/ 植物生長素　112
awn /ɔn/ 芒　82
axil /ˈæksɪl/ 葉腋　96
axile /ˈæksɪl/ 中軸的　77
axis /ˈæksɪs/ 軸　92

bacillus /bəˈsɪləs/ 桿菌　119
bacteria /bækˈtɪrɪə/ 細菌　119
bacteriophage /bækˈtɪrɪəˌfedʒ/ 噬菌體　118
bark /bɑrk/ 樹皮　94
base¹ /bes/ 鹼　11
base² /bes/ 鹼基　52
Basidiomycetes /bəˌsɪdɪəmaɪˈsits/ 擔子菌綱　165
basidiospore /bəˈsɪdɪəˌspor/ 擔孢子　165
basidium /bəˈsɪdɪəm/ 擔子　165
basifixed /ˈbesəˌfɪkst/ 基生的　73
berry /ˈbɛrɪ/ 漿果　83
biennial /baɪˈɛnɪəl/ 二年生的　117
binary fission /ˈbaɪnərɪ ˈfɪʃən/ 二分裂　45
binomial /baɪˈnomɪəl/ 雙名法；二名法　133
biomass /ˈbaɪoˌmæs/ 生物量　151
biosphere /ˈbaɪəˌsfɪr/ 生物圈　149
biotic factors /baɪˈɑtɪk ˈfæktɚz/ 生物因素　152
bipinnate /baɪˈpɪnet/ 二回羽狀的　98

bisexual /baɪˈsɛkʃʊəl/ 兩性的　62
bivalent /baɪˈvelənt/ 二價染色體　47
blade /bled/ 葉片　96
blue-green algae /ˈblu ˈgrin ˈældʒi/ 藍藻　120
bole /bol/ 樹幹　92
bracket fungus /ˈbrækɪt ˈfʌŋgəs/ 簷狀菌　166
brackish water /ˈbrækɪʃ ˈwɑtɚ/ 半鹹水　161
bract /brækt/ 苞葉　99
bracteole /ˈbræktɪol/ 小苞片　99
branch /bræntʃ/ 枝條　92
breed /brid/ 繁育　59
brown algae /braun ˈældʒi/ 褐藻　122
bryophyte /ˈbraɪəˌfaɪt/ 苔蘚植物　122
bud /bʌd/ 芽　110
bulb /bʌlb/ 鱗莖　60
bulbil /ˈbʌlbəl/ 珠芽　60
bundle sheath /ˈbʌndl ˈʃiθ/ 維管束鞘　106
bush /bʊʃ/ 叢枝灌木　136
buttress /ˈbʌtrɪs/ 板狀根　92

C₃ pathway /siθri ˈpæθˌwe/ 三碳途徑　35
C₄ pathway /si for ˈpæθˌwe/ 四碳途徑　35
Cactus /ˈkæktəs/ 仙人掌　131
Cainozoic /ˌkaɪnəˈzoɪk/ 新生代　143
calcareous /kælˈkɛrɪəs/ 鈣質的　157
calcicole /ˈkælsəˌkol/ 喜鈣植物　157
calcifuge /ˈkælsəˌfjudʒ/ 避鈣的　157
callose /ˈkælos/ 胼胝質　108
callus¹ /ˈkæləs/ 愈傷組織　69
callus² /ˈkæləs/ 胼胝體　108
Calvin cycle /ˈkælvɪn ˈsaɪkl/ 卡爾文循環　34
calyptra /kəˈlɪptrə/ 蒴帽　125
calyx /ˈkelɪks/ 花萼　70
cambium /ˈkæmbɪəm/ 形成層　108
campylotropous /ˌkæmpɪlətrəpəs/ 彎生的　78
canopy /ˈkænəpɪ/ 樹冠層　158
capitate /ˈkæpəˌtet/ 頭狀的　81
capitulum /kəˈpɪtʃʊləm/ 頭狀花序　81
capsule¹ /ˈkæpsl/ 蒴果　84
capsule² /ˈkæpsl/ 蘚蒴　123
carbohydrate /ˈkɑrboˈhaɪdret/ 碳水化合物；醣　28

carbon cycle /ˈkɑrbən ˈsaɪkl/ 碳循環　154
Carboniferous /ˌkɑrbəˈnɪfərəs/ 石炭紀　143
carotene /ˈkærəˌtin/ 胡蘿蔔素　37
carotenoids /kəˈrɑtɪˌnɔɪdz/ 類胡蘿蔔素　37
carpel /ˈkɑrpl/ 心皮　75
casparian strip /ˈkæspəriən strɪp/ 卡氏帶　89
catalysis /kəˈtæləsɪs/ 催化作用　15
catalyst /ˈkætl̩ɪst/ 催化劑　15
catkin /ˈkætkɪn/ 柔荑花序　80
cauliflorous /ˌkɔləˈflɔrəs/ 莖花的　82
cell /sɛl/ 細胞　16
cell division /sɛl dəˈvɪʒən/ 細胞分裂　45
cell membrane /sɛl ˈmɛmbren/ 細胞膜　18
cellular /ˈsɛljələ/ 細胞狀的　169
cellulose /ˈsɛljəˌlos/ 纖維素　17
cell wall /sɛl wɔl/ 細胞壁　17
Cenozoic, /ˌsinəˈzoɪk/ see Cainozoic 新生代　143
centriole /ˈsɛntrɪˌol/ 中心粒　46
centromere /ˈsɛntrəˌmɪr/ 著絲點　46
centrosome /ˈsɛntrəˌsom/ 中心體　46
chalaza /kəˈlezə/ 合點　78
chamaephyte /kəˈmifaɪt/ 地上芽植物　138
character /ˈkærɪktə/ 特徵　133
characteristic /ˌkærɪktəˈrɪstɪk/ 特徵的　133
chartaceous /kɑrˈteʃəs/ 堅紙質的　99
chemiosmosis /ˌkɛmɑzˈmosɪs/ 化學滲透　24
chemotropism /kɪˈmɑtrəpɪzm/ 向化性　115
chiasmata /kaɪˈæzmətə/ 交叉　47
chimaera /kaɪˈmɪrə/ 嵌合體　44
chitin /ˈkaɪtɪn/ 幾丁質　163
chlamydospore /kləˈmɪdəˌspor/ 厚垣孢子　168
chlorophylls /ˈklɔrəˌfɪlz/ 葉綠素　36
chloroplast /ˈklɔrəˌplæst/ 葉綠體　32
chloroplast envelope /ˈklɔrəˌplæst ˈɛnvəˌlop/ 葉綠體被膜　32
chromatid /ˈkromətɪd/ 染色分體　46
chromatophore /ˈkromətəˌfor/ 載色體　120
chromoplast /ˈkroməˌplæst/ 有色體　19
chromosome /ˈkroməˌsom/ 染色體　46
Chytridiomycetes /kaɪˌtrɪdɪəmaɪˈsitəs/ 壺菌綱　169
circinate /ˈsɜsn̩ˌet/ 拳捲的；渦狀捲的　126

citric acid cycle /ˈsɪtrɪk ˈæsɪd ˈsaɪkl/ 檸檬酸循環　24
clamp connection /klæmp kəˈnɛkʃən/ 鎖狀連合　167
class /klæs/ 綱　134
classification /ˌklæsəfəˈkeʃən/ 分類　132
cleistogamy /klaɪsˈtɑgəmɪ/ 閉花受精　62
limacteric /klaɪˈmæktərɪk/ 果實成熟期　114
climatic factors /klaɪˈmætɪk ˈfæktəs/ 氣候因素　162
climax /ˈklaɪmæks/ 顛峰　152
climber /ˈklaɪmə/ 攀緣植物　136
cline /klaɪn/ 生態群　135
clone /klon/ 無性繁殖系　41
closed community /klozd kəˈmjunətɪ/ 密生群落　152
clubmoss /ˈklʌbmɔs/ 石松　127
CO_2 fixation /siɔ tu fɪksˈeʃən/ CO_2 固定作用　88
cocci /ˈkɑksaɪ/ 球菌　119
coccoid /ˈkɑkˌɔɪd/ 球形的　120
codon /ˈkodɑn/ 密碼子　53
coenobium /sinˈobɪəm/ 定型群體　119
coenzyme /koˈɛnzaɪm/ 輔酶　15
colchicine /ˈkɑltʃəsin/ 秋水仙素　47
coleoptile /ˌkolɪˈɑptɪl/ 胚芽鞘　100
collenchyma /kəˈlɛŋkɪmə/ 厚角組織　92
colonization /ˌkɑlənaɪˈzeʃən/ 群落形成　151
colony /ˈkɑlənɪ/ 群體　119
columella[1] /ˌkɑljuˈmɛlə/ 蒴軸　125
columella[2] /ˌkɑljuˈmɛlə/ 囊軸　169
commensalism /kəˈmɛnsəlɪzəm/ 片利共生　144
community /kəˈmjunətɪ/ 群落　149
companion cell /kəmˈpænjən sɛl/ 伴細胞　108
compatible /kəmˈpætəbl/ 親和的　63
competition /ˌkɑmpəˈtɪʃən/ 競爭　152
Compositae /kəmˈpɑzətɪ/ 菊科　131
composite /kəmˈpɑzɪt/ 菊科複合花序的　81
compound[1] (n) /ˈkɑmpaʊnd/ 化合物　10
compound[2] (adj) /ˈkɑmpaʊnd/ 複葉的　98
concentration /ˌkɑnsn̩ˈtreʃən/ 濃度　12
cone /kon/ 球果　68
conidium /koˈnɪdɪəm/ 分生孢子　168
conifer /ˈkɑnəfə/ 針葉松　128
conjugation /ˌkɑndʒəˈgeʃən/ 接合生殖　61
consumer /kənˈsumə/ 消費者　154

coriaceous /ˌkorɪˈeʃəs/ 草質的　99
cork /kɔrk/ 木栓　94
corm /kɔrm/ 球莖　60
corolla /kəˈrɑlə/ 花冠　70
corpus /ˈkɔrpəs/ 原體　109
correlation /ˌkɔrəˈleʃən/ 相關　173
cortex /ˈkɔrtɛks/ 皮層　89
corymb /ˈkɔrɪmb/ 繖房花序　80
cotyledon /ˌkɑtḷˈidn̩/ 子葉　86
crassulacean acid metabolism /krɑsuˈlesiən ˌæsɪd məˈtæbl̩ɪzəm/ 景天酸代謝作用　35
creeper /ˈkripɚ/ 葡匐植物　136
cristae /ˈkrɪsti/ 嵴　21
cross-fertilization /krɔs ˌfɛtl̩əˈzeʃən/ 異花受精　62
cross-pollination /ˌkrɔs palə'neʃən/ 異花受粉　74
crossing-over /ˈkrɔsɪŋ ˈovɚ/ 交換　47
crown /kraʊn/ 樹冠　92
crustose /ˈkrʌstos/ 殼狀的　147
cryophyte /ˈkraɪəfaɪt/ 冰雪植物　138
cryptogam /ˈkrɪptəˌgæm/ 隱花植物　128
cryptophyte /ˈkrɪptəfaɪt/ 隱芽植物　138
crystal /ˈkrɪstl̩/ 晶體　9
culm /kʌlm/ 稈　91
cuticle /ˈkjutɪkl̩/ 角質層　95
cutin /ˈkjutɪn/ 角質　96
cutting /ˈkʌtɪŋ/ 插條　69
cycad /ˈsaɪkæd/ 蘇鐵　129
cyclic phosphorylation /ˈsaɪklɪk ˌfɑsfərəˌleʃən/ 循環磷酸化作用　39
cyme /saɪm/ 聚繖花序　80
cytochromes /ˈsaɪtəˌkromz/ 細胞色素　37
cytokinesis /ˌsaɪtokɪˈnisɪs/ 胞質分裂　45
cytokinins /ˌsaɪtoˈkɪŋɪnz/ 細胞分裂素　113
cytology /saɪˈtɑlədʒɪ/ 細胞學　16
cytoplasm /ˈsaɪtəˌplæzəm/ 細胞質　18
cytoplasmic inheritance /ˌsaɪtəˌplæzəmɪk ɪnˈhɛrətəns/ 細胞質遺傳　43
cytosine /ˈsaɪtoˌsin/ 胞嘧啶　53

dark reaction /dɑrk rɪˈækʃən/ 暗反應　33
Darwin /ˈdɑrwɪn/ 達爾文　140

decay /dɪˈke/ 腐化　157
deciduous /dɪˈsɪdʒʊəs/ 落葉的；脫落的　136
decomposer /ˌdikəmˈpozɚ/ 分解者　157
deficiency /dɪˈfɪʃənsɪ/ 養分缺乏　111
dehisce /dɪˈhɪs/ 開裂　84
deme /dim/ 同類群　44
denitrifying bacteria /diˈnaɪtrəˌfaɪɪŋ bækˈtɪrɪə/ 反硝化細菌　154
Deuteromycetes /ˌdjutərəˈmaɪˈsitəs/ 半知菌綱　165
development /dɪˈvɛləpmənt/ 發育　109
diadelphous /ˌdaɪəˈdɛlfəs/ 兩體雄蕊的　73
diakinesis /ˌdaɪəkəˈnɪsɪs/ 肥厚期　49
diastase /ˈdaɪəˌstes/ 澱粉酶製劑　30
diatom /ˈdaɪətəm/ 矽藻；硅藻　121
dichogambous /daɪˈkagəmos/ 雌雄蕊異熟的　79
dichotomous /daɪˈkatəməs/ 二岐的　93
dicotyledon /ˌdaɪkɑtl̩ˈidn̩/ 雙子葉植物　131
dictyosome /ˈdɪktɪəsəm/ 高爾基體分裂片　20
differentiated /ˌdɪfəˈrɛn ʃɪˌetɪd/ 分化的　110
diffusion /dɪˈfjuʒən/ 擴散作用　102
digitate /ˈdɪdʒəˌtet/ 掌狀的　97
dihybrid inheritance /daɪˈhaɪbrɪd ɪnˈhɛrətəns/ 雙基因雜種遺傳　44
dikaryon /dɪˈkarɪən/ 雙核體　166
dimorphic /daɪˈmɔrfɪk/ 兩形的　73
dinoflagellate /ˌdaɪnəˈflædʒəˌlet/ 雙鞭甲藻　121
dioecious /daɪˈiʃəs/ 雌雄異株的　79
diploid /ˈdɪplɔɪd/ 二倍體的　50
diplont /ˈdɪplɑnt/ 二倍體的　64
diplotene /ˈdɪplətin/ 雙絲期　49
disaccharide /daɪˈsækəraɪd/ 雙醣；雙糖　29
disk /dɪsk/ 花盤　72
disk-floret /dɪsk ˈflorɪt/ 盤心花　81
dispersal /dɪˈspɚsl̩/ 散播　84
dissected /dɪˈsɛktɪd/ 分裂的　97
dissolve /dɪˈzɑlv/ 溶解　12
distribution /ˌdɪstrəˈbjuʃən/ 分佈　135
division /dəˈvɪʒən/ 門　134
DNA /di ɛn e/ 去氧核糖核酸　51
dolipore septum /ˌdɑləpor ˈsɛptəm/ 陷孔隔　167
dominant[1] /ˈdɑmənənt/ 顯性的　44

dominant² /ˈdɑmənənt/ 優勢的 150
dormant /ˈdɔrmənt/ 休眠的 117
double fertilization /ˈdʌbl̩ ˌfɚtl̩əˈzeʃən/ 雙受精作用 78
drip tip /drɪp tɪp/ 滴水葉尖 99
drupe /drup/ 核果 83
dry weight /draɪ wet/ 乾重 151

ecology /ɪˈkɑlədʒɪ/ 生態學 149
ecosystem /ˈikoˌsɪstəm/ 生態系統 149
ecotone /ˈikəton/ 群落交錯區 150
ecotype /ˈikətaɪp/ 生態型 135
ectotrophic /ˈɛktəˌtrɑfɪk/ 體外營養的 145
edaphic factors /ɪˈdæfɪk ˈfæktɚz/ 土壤因素 154
egg-cell /ɛg sɛl/ 卵細胞 61
elater /ˈɛlətɚ/ 彈孢絲 124
electron /ɪˈlɛktrɑn/ 電子 8
electron microscope /ɪˈlɛktrɑn ˈmaɪkrəˌskop/ 電子顯微鏡 171
electron transfer chain /ɪˈlɛtrɑn trænsˈfɚ tʃen/ 電子傳遞鏈 40
element /ˈɛləmənt/ 元素 8
embryo /ˈɛmbrɪˌo/ 胚 85
embryo sac /ˈɛmbrɪˈo sæk/ 胚囊 78
endemic /ɛnˈdɛmɪk/ 特有的 135
endocarp /ˈɛndoˌkɑrp/ 內果皮 83
endodermis /ˌɛndoˈdɚmɪs/ 內皮層 89
endogenous rhythm /ɛnˈdɑdʒənəs ˈrɪðəm/ 內源節律 116
endoplasmic reticulum /ˈɛndoˌplæzəmɪk rɪˈtɪkjələm/ 內質網 56
endosperm /ˈɛndəˌspɚm/ 內胚乳 86
endosperm mother cell /ˈɛndəˌspɚm ˈmʌðɚ sɛl/ 胚乳母細胞 79
endotrophic /ˌɛndoˈtrɑfɪk/ 內生的；體內營養的 145
entire /ɪnˈtaɪr/ 全緣 97
entomophily /ˌɛntəˈmɑfəlɪ/ 蟲媒 75
environment /ɪnˈvaɪrənmənt/ 環境 149
enzyme /ˈɛnzaɪm/ 酶 15
ephemeral /əˈfɛmərəl/ 短生的 117
epicarp /ˈɛpɪˌkɑrp/ 外果皮 83
epicotyl /ˈɛpɪˌkɑtl̩/ 上胚軸 86
epidermis /ˌɛpəˈdɚmɪs/ 表皮 90

epigeal /ˌɛpɪˈdʒiəl/ 出土的 87
epigynous /ɪˈpɪdʒənəs/ 上位的 72
epiphyll /ˈɛpəfɪl/ 葉附生植物 137
ethene /ˈɛθin/ 乙烯 114
ethylene /ˈɛθəˌlin/ see ethene 乙烯
etiolation /ˈitɪəˌleʃən/ 黃化現象 111
euglenoid /juˈglinɔɪd/ 裸藻；眼蟲藻 120
eukaryotic /ˌjukərɪˈɑtɪk/ 真核的 16
eutrophic /juˈtrɑfɪk/ 富營養的 161
eutrophication /juˌtrɑfɪˈkeʃən/ 富營養化 161
evaporation /ɪˌvæpəˈreʃən/ 蒸發作用 12
evapotranspiration /ɪˌvæpoˌtrænspəˈreʃən/ 蒸發；蒸騰作用 101
evergreen /ˈɛvɚˌgrin/ 常綠的 136
evolution /ˌɛvəˈluʃən/ 演化 139
exalbuminous /ˌɛksælˈbjumənəs/ 無胚乳的 86
excretion /ɪkˈskriʃən/ 排泄作用 112
exine /ˈɛgˌzɪn/ 外壁 74
exocarp /ˈɛksəˌkɑrp/ 外果皮 83
exodermis /ˌɛksoˈdɚmɪs/ 外皮層 90
exstipulate /ɛksˈstɪpjəˌlet/ 無托葉的 99
extant /ɪkˈstænt/ 現存的 141
extinct /ɪkˈstɪŋkt/ 滅絕的 141
extracellular /ˌɛkstrəˈsɛljulɚ/ 細胞外的 16
extrafloral /ˌɛkstrəˈflorəl/ 花外的 73
exudate /ˈɛsjuˌdet/ 滲出液 112
exude /ɪgˈzjud/ 滲出 112

F₁ generation /ɛf wʌn ˌdʒɛnəˈreʃən/ F_1 子一代 42
F₂ generation /ɛf tu ˌdʒɛnəˈreʃən/ F_2 子二代 42
facultative /ˈfækl̩ˌtetɪv/ 兼性的 148
FAD /ɛf e di/ 黃素腺嘌呤二核苷酸 24
family /ˈfæməlɪ/ 科 134
fatty acid /ˈfætɪ ˈæsɪd/ 脂肪酸 31
feedback /ˈfidˌbæk/ 反饋 14
female /ˈfimel/ 雌性的 61
fermentation /ˌfɚmenˈteʃən/ 醱酵作用 24
fern /ˈfɚn/ 真蕨類 126
ferredoxin /ˌfɛrəˈdɑksən/ 鐵氧化還原蛋白 37
fertile /ˈfɚtl̩/ 能育的 62
fertilization /ˌfɚtl̩əˈzeʃən/ 受精 62

fibre /ˈfaɪbɚ/ 纖維　91
filament /ˈfɪləmənt/ 花絲　73
filamentous /ˌfɪləˈmɛntəs/ 絲狀的　120
filmy fern /ˈfɪlmɪ fɝn/ 膜蕨　127
flagellum /fləˈdʒɛləm/ 鞭毛　121
flavoprotein /ˌflevoˈprotin/ 黃素蛋白　37
fleshy /ˈflɛʃɪ/ 多肉的　99
flora /ˈflorə/ 植物誌　135
floral diagram /ˈflorəl ˈdaɪəˌgræm/ 花圖式　71
floret /ˈflorɪt/ 小花　81
florigen /ˈflorɪdʒən/ 成花素　114
flower /ˈflaʊɚ/ 花　70
flowering plant /ˈflaʊərɪŋ plænt/ 有花植物　130
fluorescence /ˌfluəˈrɛsn̩s/ 螢光　38
foliage /ˈfolɪɪdʒ/ 葉叢　96
foliose /ˈfolɪˌos/ 葉狀的　147
follicle /ˈfɑlɪkl̩/ 蓇葖果　84
food chain /fud tʃen/ 食物鏈　153
food web /fud wɛb/ 食物網　153
foot /fʊt/ 基部　122
forest /ˈfɔrɪst/ 森林　158
fossil /ˈfɑsl̩/ 化石　142
free central /fri ˈsɛntrəl/ 分離中央的　177
frequency /ˈfrikwənsɪ/ 頻數　172
freshwater /ˈfrɛʃˌwɔtɚ/ 淡水的　161
frond /frɑnd/ 蕨葉；棕葉　126
fructose /ˈfrʌktos/ 果糖　29
fruit /frut/ 果實　83
fruiting body /ˈfrutɪŋ ˈbɑdɪ/ 子實體　165
fruticose /ˈfrutəˌkos/ 灌木狀的　147
function /ˈfʌŋkʃən/ 功能　170
fungi /ˈfʌndʒaɪ/ 真菌　163
Fungi Imperfecti /ˈfʌndʒaɪ ɪmpɚˈfɛktɪ/ 不完全菌　165
funicle /ˈfjunɪkl̩/ 珠柄　78
fusion /ˈfjuʒən/ 融合　61

gametangium /ˌgæməˈtændʒɪəm/ 配子囊　65
gamete /ˈgæmit/ 配子　71
gametophyte /gəˈmitoˌfaɪt/ 配子體　65
gamopetalous /ˌgæməˈpɛtələs/ 花瓣相連的　71
gamosepalous /ˌgæməˈsɛpələs/ 合萼的　71

gemmae /ˈdʒɛmi/ 孢芽　124
gender /ˈdʒɛndɚ/ 性別　61
gene /dʒin/ 基因　41
genecology /dʒinɪˈkɑlədʒɪ/ 基因生態學　41
gene pool /dʒin pul/ 基因庫　44
generation /ˌdʒɛnəˈreʃən/ 世代　63
genetic code /dʒəˈnɛtɪk kod/ 遺傳密碼　54
genetics /dʒəˈnɛtɪks/ 遺傳學　41
genome /ˈdʒiˌnom/ 基因組　41
genotype /ˈdʒɛnoˌtaɪp/ 基因型　41
genus /ˈdʒinəs/ 屬　134
geological epoch /ˌdʒiəˈlɑdʒɪkl̩ ˈɛpək/ 地質世　143
geological era /ˌdʒiəˈlɑdʒɪkl̩ ˈɪrə/ 地質代　143
geological period /ˌdʒiəˈlɑdʒɪkl̩ ˈpɪrɪəd/ 地質紀　143
geological time /ˌdʒiəˈlɑdʒɪkl̩ taɪm/ 地質年代　143
geophyte /ˈdʒiəˌfaɪt/ 地下芽植物　138
geotropism /dʒiˈɑtrəˌpɪzəm/ 向地性　115
germination /ˌdʒɝməˈneʃən/ 發芽；萌發　87
gibberellins /ˌdʒɪbəˈrɛlənz/ 赤霉素　113
gill /gɪl/ 菌褶　166
Ginkgoales /ˌgɪŋkoˈeliz/ 銀杏目　129
gland /glænd/ 腺體　112
glucose /ˈglukos/ 葡萄糖　28
glumes /glumz/ 穎片　82
glyceric acid-3-phosphate /glɪˈsɛrɪk ˈæsɪd θri ˈfɑsfet/ 3-二磷酸甘油酸　34
glycerol /ˈglɪsəˌrol/ 甘油　31
glycolysis /glaɪˈkɑləsɪs/ 糖酵解　22
glycoprotein /ˌglɪkəˈprɑtin/ 糖蛋白　58
glycoside /ˈglaɪkəˌsaɪd/ 葡萄糖苷　28
glycosidic bond /ˌglaɪkəˌsaɪdɪk bɑnd/ 糖苷鍵　29
Gnetales /ˈnɛtələs/ 買麻藤目　129
Golgi body /ˈgoldʒɪ ˈbɑdɪ/ 高爾基體　20
gradient /ˈgredɪənt/ 梯度　24
graft /græft/ 嫁接　69
grana /ˈgrenə/ 葉綠餅；基粒　33
grass /græs/ 禾草類　130
grassland /ˈgræsˌlænd/ 草本植被區　160
green algae /grin ˈældʒi/ 綠藻　120
growth /groθ/ 生長　109
GTP /dʒi ti pi/ 鳥嘌呤核苷三磷酸　24

INDEX 索引・187

guanine /ˈgwanin/ 鳥嘌呤　53
guard cells /gard ˈsɛlz/ 保衛細胞　96
gut flora /gʌt ˈflorə/ 腸道物區系　146
guttation /ˌgʌteʃən/ 吐水作用　112
gymnosperm /ˈdʒɪmnəˌspɚm/ 裸子植物　128
gynodioecious /dʒɪˈnadaɪˈifəs/ 雌花兩性花異株的　79
gynoecium /dʒɪˈnisɪəm/ 雌蕊群　75

habit /ˈhæbɪt/ 習性　136
habitat /ˈhæbəˌtæt/ 生境　149
haem /him/ 血紅素　37
halophyte /ˈhæləˌfaɪt/ 鹽土植物　138
haploid /ˈhæplɔɪd/ 單倍體　50
haplont /ˈhæplənt/ 單倍體的　64
haustorium /hɔˈtorɪəm/ 吸胞　110
heartwood /ˈhartˌwʊd/ 心材　93
helix /ˈhiliks/ 螺旋結構　51
hemicryptophyte /ˌhɛmɪˈkrɪptəˈfaɪt/ 半隱芽植物　138
hemiparasite /ˌhɛmɪˈpærəˌsaɪt/ 半寄生植物　144
hepatic /hɪˈpætɪk/ 地錢　123
herb /hɝb/ 草本植物　136
herbarium /hɝˈbɛrɪəm/ 植物標本室　133
herbivore /ˈhɝbɪˌvor/ 食草動物　153
heredity /həˈrɛdəti/ 遺傳　41
hermaphrodite /hɝˈmæfrəˌdaɪt/ 兩性花的　79
heterogamous /ˌhɛtəˈragəməs/ 異型配子的　61
heterophyllous /ˌhɛtərəˈfɪləs/ 具異型葉的　99
heterosis /ˌhɛtəˈrasɪs/ 雜種優勢　63
heterosporous /ˌhɛtəˈraspərəs/ 具異型孢子的　67
heterostylous /ˌhɛtərəˈstaɪləs/ 花柱異長的　76
heterothallic /ˌhɛtəroˈθælɪk/ 異宗配合的　168
heterotrophic /ˌhɛtərəˈtrafɪk/ 異養的；異營的　32
heterozygous /ˌhɛtəˈrazaɪgəs/ 雜合的　44
hexose /ˈhɛksos/ 己醣；己糖　28
hibernation /ˌhaɪbɚˈneʃən/ 冬眠　117
Hill reaction /hɪl rɪˈækʃən/ 希爾反應　36
hilum /ˈhaɪləm/ 種臍　85
histogram /ˈhɪstəˌgræm/ 柱形圖　172
homogamous /hoˈmagəməs/ 雌雄同熟的　79
Homologous /hoˈmaləgəs/ 同源的　46
homosporous /hoˈmaspərəs/ 具同型孢子的　67

homostylous /hoˈmostaɪləs/ 花柱同長的　76
homothallic /ˌhomoˈθælɪk/ 同宗配合的　168
homozygous /ˌhoməˈzaɪgəs/ 純合的　44
honey guides /ˈhʌnɪ gaɪdz/ 蜜指標　75
horizon /həˈraɪzn̩/ 層位　156
hormone /ˈhɔrmon/ 激素；生長素　112
horsetail /ˈhɔrsˌtel/ 木賊　127
host /host/ 宿主　144
humus /ˈhjuməs/ 腐殖土　156
hybrid /ˈhaɪbrɪd/ 雜種　63
hybrid vigour /ˈhaɪbrɪd ˈvɪgɚ/ 雜種優勢；雜種精壯　63
hydathode /ˈhaɪdəθod/ 水孔　112
hydrolysis /haɪˈdraləsɪs/ 水解作用　13
hydrophyte /ˈhaɪdrəˌfaɪt/ 水生植物　138
hydrotropism /haɪˈdratrəˌpɪzəm/ 向水性　115
hypanthium /hɪˈpænθɪəm/ 隱頭花序　72
hypertonic /ˌhaɪpɚˈtanɪk/ 高滲的　103
hypha /ˈhaɪfə/ 菌絲　163
hypocotyl /ˌhaɪpəˈkatl̩/ 下胚軸　86
hypodermis /ˌhaɪpəˈdɚmɪs/ 下皮　96
hypogeal /ˌhaɪpəˈdʒiəl/ 留土的　87
hypogynous /haɪˈpadʒənəs/ 下位的　72
hypotonic /ˌhaɪpəˈtanɪk/ 低滲的　103

IAA /aɪ e e/ 吲哚乙酸　112
imbibition /ˌɪmbɪˈbɪʃən/ 吸脹作用　87
impermeable /ɪmˈpɚmɪəbl̩/ 不滲透性的　102
inbreeding /ˈɪnˌbridɪŋ/ 近交　63
incompatible /ˌɪnkəmˈpætəbl̩/ 不親和的　63
incubous /ˈɪŋkjəbəs/ 蔽前式的　123
indehiscent /ˌɪndɪˈhɪsənt/ 不裂的　84
independent assortment /ˌɪndɪˈpɛndənt əˈsɔrtmənt/ 獨立分配律　43
individual /ˌɪndəˈvɪdʒuəl/ 個體　135
indole acetic acid /ˈɪndol əsitɪk ˈæsɪd/ 吲哚乙酸　112
indumentum /ˌɪnduˈmɛntəm/ 毛狀外被；表皮物　100
indusium /ɪnˈduzɪəm/ 囊群蓋　127
infection /ɪnˈfɛkʃən/ 感染　144
inferior ovary /ɪnˈfɪrɪɚ ˈovəri/ 下位子房　77
inflorescence /ˌɪnfloˈrɛsn̩s/ 花序　80
infraspecific /ˌɪnfrəˈspɪˈsɪfɪk/ 種以下的　135

inherit /ɪnˈhɛrɪt/ 遺傳　41
inhibition /ˌɪnhɪˈbɪʃən/ 抑制作用　14
inhibitor /ɪnˈhɪbətɚ/ 抑制劑　14
inorganic /ˌɪnɔrˈgænɪk/ 無機的　11
insectivorous /ˌɪnsɛkˈtɪvərəs/ 食蟲的　148
insoluble /ɪnˈsɑljəbl/ 不溶解的　12
integuments /ɪnˈtɛgjəmənts/ 珠被　78
interaction /ˌɪntɚˈækʃən/ 相互作用　144
intercalary /ɪnˈtɝkəˌlɛrɪ/ 居間的　109
intercellular space /ˌɪntɚˈsɛljəlɚ spes/ 細胞間隙　95
internode /ˈɪntɚˌnod/ 節間　90
interphase /ˈɪntɚˌfez/ 分裂間期　45
interspecific /ˌɪntɚspɪˈsɪfɪk/ 種間的　152
intracellular /ˌɪntrəˈsɛljəlɚ/ 細胞內的　16
intraspecific /ˌɪntrəspɪˈsɪfɪk/ 種內的　152
inulin /ˈɪnjəlɪn/ 菊芋多醣　30
involucre /ˈɪnvəˌlukɚ/ 總苞　81
ion /ˈaɪən/ 離子　9
isodiametric /ˌaɪsoˌdaɪəˈmɛtrɪk/ 等徑的　91
isogamous /aɪˈsɑgəməs/ 同配生殖的　61
isolation /ˌaɪsˈleʃən/ 隔離　44
isomers /ˈaɪsəmɚz/ 同分異構體　10
isotonic /ˌaɪsəˈtɑnɪk/ 等滲的　103

jungle /ˈdʒʌŋgl/ 熱帶植叢　158

karyogamy /ˌkærɪˈɑgəmɪ/ 核配合　166
kernel /ˈkɝnl/ 果仁　83
ketose /ˈkitos/ 酮醣；酮糖　29
kingdom /ˈkɪŋdəm/ 界　134
Krebs cycle /krɛbz ˈsaɪkl/ 克雷伯氏循環　24

lactic acid /ˈlæktɪk ˈæsɪd/ 乳酸　24
Lamarck /ˌlaˈmɑrk/ 拉馬克　140
lamellae /ləˈmɛli/ 片層　33
lamina /ˈlæmənə/ 葉面　96
lateral /ˈlætərəl/ 側面的　92
latex /ˈletɛks/ 乳汁　102
leaf /lif/ 葉片　95
leaf gap /lif gæp/ 葉隙　105
leaflet /ˈlɪflɪt/ 小葉　98

leaf trace /lif tres/ 葉跡　105
leafy liverwort /ˈlifɪ ˈlɪvɚˌwɝt/ 葉狀苔類　123
legume /ˈlɛgjum/ 莢果　84
Leguminosae /lɛgjuˈmɪnosi/ 豆科　131
lemma /ˈlɛmə/ 外稃　82
lenticel /ˈlɛntəˌsɛl/ 皮孔　91
leptocaul /ˌlɛptəˈkol/ 細瘦莖的　94
leptotene /ˈlɛptətin/ 細絲期　49
leucoplast /ˈljukəˌplæst/ 白色體　19
liana /lɪˈɑnə/ 籐本植物　137
lichen /ˈlaɪkɪn/ 地衣　147
life cycle /laɪf ˈsaɪkl/ 生活週期；生活史　64
light microscope /laɪt ˈmaɪkrəˌskop/ 光學顯微鏡　171
light reaction /laɪt rɪˈækʃən/ 光反應　36
lignin /ˈlɪgnɪn/ 木質素　93
ligule¹ /ˈlɪgjul/ 舌片　81
ligule² /ˈlɪgjul/ 葉舌　100
linkage /ˈlɪŋkɪdʒ/ 連鎖　42
Linnaeus /lɪˈniəs/ 林奈　133
lipid /ˈlaɪpɪd/ 脂類　31
litter /ˈlɪtɚ/ 枯枝落葉層　156
littoral /ˈlɪtərəl/ 潮汐區　161
liverwort /ˈlɪvɚˌwɝt/ 苔類植物　123
lobe /lob/ 裂片　97
locule /ˈlɑkjul/ 子房室　76
loculicidal /ˌlɑkjələˈsaɪdəl/ 室背開裂的　84
locus /ˈlokəs/ 基因座　44
long-day plant /lɔŋ de plænt/ 長日照植物　116
longitudinal section /ˌlɑndʒəˈtjudnḷ ˈsɛkʃən/ 縱切面　171
lysosome /ˈlaɪsəsʌm/ 溶酶體　20

macromolecule /ˌmækroˈmɑləˌkjul/ 大分子　9
male /mel/ 雄性的　61
margin /ˈmɑrdʒɪn/ 葉緣　97
matric potential /məˈtrɪk pəˈtɛnʃəl/ 襯質勢　103
matrix /ˈmetrɪks/ 基質　21
mean /min/ 平均值　172
mechanism /ˈmɛkəˌnɪzəm/ 機制　171
medium /ˈmidɪəm/ 培養基　171
medulla /mɪˈdʌlə/ 髓部　90
megaphyll /ˌmɛgəˈfɪl/ 巨型葉　106

megasporangium /ˌmɛgəspəˈrændʒɪəm/ 大孢子囊　68
megaspore /ˈmɛgəˌspor/ 大孢子　68
megasporophyll /ˌmɛgəˈsporəˌfɪl/ 大孢子葉　68
meiosis /maɪˈosɪs/ 減數分裂；成熟分裂　49
membranaceous /ˌmɛmbrəˈneʃəs/ 膜質的　99
membrane /ˈmɛmbren/ 膜　18
Mendel's laws /ˈmɛndlz loz/ 孟德爾定律　42
meristem /ˈmɛrɪˌstɛm/ 分生組織　109
mesocarp /ˈmɛsəˌkɑrp/ 中果皮　83
mesophyll /ˈmɛsəˌfɪl/ 葉肉　95
Mesozoic /ˌmɛsəˈzoɪk/ 中生代　143
metabolic pathway /ˌmɛtəˈbɑlɪk ˈpæθˌwe/ 代謝途徑　14
metabolic poison /ˌmɛtəˈbɑlɪk ˈpɔɪzn̩/ 代謝毒　26
metabolism /məˈtæblˌɪzəm/ 新陳代謝；代謝作用　14
metabolite /məˈtæbəˌlaɪt/ 代謝產物　14
metaphase /ˈmɛtəˌfez/ 中期　45
microclimate /ˈmaɪkroˌklaɪmɪt/ 小氣候　162
microfibril /ˌmaɪkroˈfaɪbrɪl/ 微纖絲　17
microorganism /ˌmaɪkroˈɔrgənˌɪzəm/ 微生物　118
microphyll /ˈmaɪkroˌfɪl/ 小型葉　106
micropyle /ˈmaɪkroˌpaɪl/ 珠孔　85
microsporangium /ˌmaɪkrospəˈrændʒɪəm/ 小孢子囊　67
microspore /ˈmaɪkrəˌspor/ 小孢子　67
microsporophyll /ˌmaɪkrəˈsporəfɪl/ 小孢子葉　67
microtubule /ˌmaɪkroˈtjubjul/ 微質管　21
middle lamella /ˈmɪdl̩ ləˈmɛlə/ 中膠層　17
midrib /ˈmɪdˌrɪb/ 中脈　97
mildew /ˈmɪlˌdju/ 霉菌　164
mitochondrion /ˌmaɪtəˈkɑndrɪəm/ 粒線體；線粒體　21
mitosis /mɪˈtosɪs/ 有絲分裂　45
mode /mod/ 眾數　172
modification /ˌmɑdəfəˈkeʃən/ 變態；誘發變異　170
molecule /ˈmɑləˌkjul/ 分子　9
monadelphous /ˌmɑnəˈdɛlfəs/ 單體雄蕊的　73
monocarpic /ˌmɑnəˈkɑrpɪk/ 結一次果的　83
monocotyledon /ˌmɑnəkɑtlˈidn̩/ 單子葉植物　130
monoecious /ˌmɑnəˈiʃəs/ 雌雄同株的　79
monohybrid inheritance /ˌmɑnəˌhaɪbrɪd ɪnˈhɛrətəns/ 單基因雜種遺傳　44
monomer /ˈmɑnəmə/ 單體　10
monopodial /ˌmɑnəˈpodɪəl/ 單軸的　93

monosaccharide /ˌmɑnəˈsækəraɪd/ 單醣；單糖　28
monotypic /ˈmɑnəˌtaɪpɪk/ 單型的；單種的　134
montane forest /ˈmɑnten ˈfɔrɪst/ 山地森林　158
mor /mɔr/ 酸性有機質　157
morph /mɔrf/ 型　135
morphogenesis /ˌmɔrfəˈdʒɛnəsɪs/ 形態發生　109
morphology /mɔrˈfɑlədʒɪ/ 形態學　88
moss /mɑs/ 蘚類植物　124
motile /ˈmotl̩/ 能游動的　121
mould /mold/ 黴菌　164
mull /mʌl/ 腐熟；腐殖質　157
multicellular /ˌmʌltɪˈsɛljələ/ 多細胞的　119
multi-enzyme complex /ˌmʌltɪˈɛnzaɪm ˈkɑmplɛks/ 多酶複合物　15
multinucleate /ˌmʌltɪˈnjuklɪɪt/ 多核的　163
mushroom /ˈmʌʃrum/ 菇類　165
mutagen /ˈmjutədʒən/ 誘變劑　54
mutant /ˈmjutənt/ 突變體　54
mutation /mjuˈteʃən/ 突變　54
mutualism /ˈmjutʃuəlɪzm̩/ 互利共生　144
mycelium /maɪˈsɪlɪəm/ 菌絲體　163
mycobiont /ˌmaɪkoˈbaɪˌɑnt/ 地衣共生菌　147
mycology /maɪkɑlədʒɪ/ 真菌學　163
mycorrhiza /ˌmaɪkəˈraɪzə/ 菌根　145
Myxomycetes /ˌmɪksomaɪˈsits/ 黏菌綱　169

NAD /ɛn ɛ di/ 菸鹼醯胺腺嘌呤二核苷酸　24
NADP /ɛn e di pi/ 菸鹼醯胺腺嘌呤二核苷酸磷酸　38
nastic movement /ˈnæstɪk ˈmuvmənt/ 感性運動　115
natural selection /ˈnætʃərəl səˈlɛkʃən/ 自然選擇　140
nectar /ˈnɛktə/ 花蜜　73
nectary /ˈnɛktərɪ/ 蜜腺　72
needle /ˈnidl̩/ 針狀葉　99
neo-Darwinism /ˌnioˈdɑrwɪnɪzm̩/ 新達爾文主義　140
neoteny /niˈɑtənɪ/ 幼態成熟　141
neuter /ˈnjutə/ 中性的　61
neutron /ˈnjutrɑn/ 中子　8
niche /nɪtʃ/ 生態區位　153
nitrate /ˈnaɪtret/ 硝酸根　13
nitrifying bacteria /ˈnaɪtrəfaɪŋ bækˈtɪrɪə/ 硝化細菌　154
nitrogen cycle /ˈnaɪtrədʒən ˈsaɪkl̩/ 氮循環　154

nitrogen fixation /ˈnaɪtrədʒən fɪksˈeʃən/ 固氮作用 146
node /nod/ 節 90
nodule /ˈnɑdʒul/ 根瘤 146
nomenclature /ˈnomənˌkletʃɚ/ 命名法；名稱 132
noncyclic phosphorylation /ˌnɔnsaɪklɪkˈfɑsfərəleʃən/ 非循環光合磷酸化 39
nonsense codon /ˈnɑnsɛnsˈkodɑn/ 無意義密碼子 53
nonvascular /nɑnˈvæskjəlɚ/ 無維管的 122
normal distribution /ˈnɔrml̩ˌdɪstrəˈbjuʃən/ 正態分佈 172
nucellus /nuˈsɛləs/ 珠心 78
nuclear membrane /ˈnjuklɪɚˈmɛmbren/ 核膜 19
nucleic acid /njuˈklɪɪkˈæsɪd/ 核酸 51
nucleolus /njuˈklɪələs/ 核仁 19
nucleoplasm /ˈnjuklɪɚˌplæzm/ 核質 19
nucleotide /ˈnjuklɪɚˌtaɪd/ 核苷酸 52
nucleus /ˈnjuklɪəs/ 細胞核 19
nut /nʌt/ 堅果 84
nutrient /ˈnjutrɪənt/ 營養素；養分 111
nutrition /njuˈtrɪʃən/ 營養 111

obligate /ˈɑbləˌget/ 專性的 148
offspring /ˈɔfˌsprɪŋ/ 後代；後裔 44
oligosaccharide /ˌɑlɪgoˈsækəˌraɪd/ 寡糖 30
oligotrophic /ˌɑləgoˈtrɑfɪk/ 寡營養的 161
ontogeny /ɑnˈtɑdʒənɪ/ 個體發育 109
oogamous /oˈɑgəməs/ 卵配生殖的 61
oogonium /ˌoəˈgonɪəm/ 藏卵器 169
Oomycetes /oəmaɪˈsitəs/ 卵菌 168
oosphere /ˈoəsˌfɚ/ 卵球 169
oospore /ˈoəˌspɔr/ 卵孢子 169
open community /ˈopən kəˈmjunətɪ/ 開闊群落 152
operculum /oˈpɚkjuləm/ 萌蓋 125
opposite /ˈɑpəzɪt/ 對生的 98
orchid /ˈɔrkɪd/ 菌 130
order /ˈɔrdɚ/ 目 134
organ /ˈɔrgən/ 器官 88
organelle /ˌɔrgənˈɛl/ 細胞器 16
organic /ɔrˈgænɪk/ 有機的 11
organism /ˈɔrgənˌɪzəm/ 有機體；生物體 118
ornithophily /ˌɔrnɪˈθəfɪlɪ/ 鳥媒 75
orthophosphate /ˌɔrθofɑsˈfet/ 正磷酸根 13

orthotropic /ˌɔrθəˈtrɑpɪk/ 直生的 93
orthotropous /ɔrˈθɑtrəpəs/ 直生的 78
osmosis /ɑzˈmosɪs/ 滲透作用 102
osmotic potential /ɑzˈmɑtɪk pəˈtɛnʃəl/ 滲透勢 103
osmotic pressure /ɑzˈmɑtɪkˈprɛʃɚ/ 滲透壓 102
outbreeding /ˈaʊtˌbridɪŋ/ 遠交 63
ovary /ˈovərɪ/ 子房 76
ovule /ˈovjul/ 胚珠 78
ovum /ˈovəm/ 卵 61
oxidation /ˌɑksəˈdeʃən/ 氧化作用 11
oxidative phosphorylation /ˈɑksəˌdetɪvˌfɑsfərəˈleɪʃən/ 氧化磷酸化 26

pachycaul /ˌpɑkɪkol/ 粗短莖的 94
pachytene /ˈpækɪtin/ 粗絲期 49
palaeobotany /ˌpelɪoˈbɑtnɪ/ 古植物學 142
Palaeozoic /ˌpelɪəˈzo·ɪk/ 古生代 143
palea /ˈpelɪə/ 內稃 82
palisade parenchyma /ˌpæləˈsed pəˈrɛŋkɪmə/ 柵狀薄壁組織 95
palm /pɑm/ 棕櫚 130
palmate /ˈpælmet/ 掌狀的 98
palynology /ˌpæləˈnɑlədʒɪ/ 孢粉學 142
panicle /ˈpænɪkl̩/ 圓錐花序 80
pappus /ˈpæpəs/ 冠毛 84
parallel /ˈpærəˌlɛl/ 平行的 110
paramylum /ˈpærəmɪləm/ 裸藻澱粉 120
paraphyses /pəˈræfəsɪz/ 側絲 125
parasite /ˈpærəˌsaɪt/ 寄生物 144
parenchyma /pəˈrɛŋkɪmə/ 薄壁組織 90
parenchymatous /pəˈrɛŋkɪmətəs/ 薄壁組織的 120
parietal /pəˈraɪətl̩/ 側膜的 77
parthenocarpic /ˌpɑrθənoˈkɑrpɪk/ 單性結實的 83
pathogen /ˈpæθədʒən/ 病源體 144
peat /pit/ 泥炭土 157
pectin /ˈpɛktɪn/ 果膠 17
pedicel /ˈpɛdəsl̩/ 花梗 80
peduncle /pɪˈdʌŋkl̩/ 總花梗 80
pentose /ˈpɛntos/ 戊醣；戊糖 28
peptide /ˈpɛptaɪd/ 肽 56
perennation /pəˈrɛnəʃən/ 多年生性 117

perennial /pəˈrɛnɪəl/ 多年生的　117
perfect /ˈpɝfɪkt/ 完全花的　79
perianth /ˈpɛrɪˌænθ/ 花被　70
pericarp /ˈpɛrɪˌkɑrp/ 果皮　83
periclinal /ˈpɛrəˌkaɪnl/ 平周的　110
pericycle /ˈpɛrəˌsaɪkl/ 中柱鞘　89
periderm /ˈpɛrɪdɝm/ 周皮　94
perigynous /pəˈrɪdʒənəs/ 周位的　72
peristome /ˈpɛrɪˌstom/ 蒴齒層　125
perithecium /ˌpɛrəˈθiʃɪəm/ 子囊殼　147
permeable /ˈpɝmɪəbl/ 滲透性的　103
peroxisome /pəˈrɑksəˌsom/ 過氧化酶體　21
petal /ˈpɛtl/ 花瓣　70
petiole /ˈpɛtɪˌol/ 葉柄　96
PGA /piˈdʒiˈe/ 磷酸甘油酸　33
phage /fedʒ/ 噬體　118
phanerogam /ˈfænərəˌgæm/ 顯花植物　128
phanerophyte /ˈfænərəˌfaɪt/ 高位芽植物　138
phellem /ˈfɛləm/ 木栓層　94
phelloderm /ˈfɛləˌdɝm/ 栓內層　94
phellogen /ˈfɛlədʒən/ 栓皮形成層；木栓生長帶　94
phenology /fɪˈnɑlədʒɪ/ 物候學　162
phenotype /ˈfinəˌtaɪp/ 表現型　41
phloem /ˈfloɛm/ 韌皮部　108
phosphate /ˈfɑsfet/ 磷酸根　13
phosphoglyceric acid /ˌfɑstəˌglɪsɛrɪk ˈæsɪd/ 磷酸甘油酸　33
phospholipid /ˌfɑsfəˈlɪpɪd/ 磷脂　31
phosphorescence /ˌfɑsfəˈrɛsn̩s/ 磷光　38
phosphorylation /ˈfɑsfərəˌleʃən/ 磷酸化　26
photolysis of water /foˈtɑləsɪs əv ˈwɑtɝ/ 水的光解　36
photoperiod /ˌfotəˈpɪrɪəd/ 光周期　116
photoperiodism /ˌfotəˈpɪrɪəˌdɪzm̩/ 光周期性　116
photophosphorylation /foˈtɑfɑsfərəˌleʃən/ 光合磷酸化作用　39
photorespiration /ˌfotoˌrɛspəˈreʃən/ 光呼吸　26
photosynthesis /ˌfotəˈsɪnθəsɪs/ 光合作用　32
phototropism /foˈtɑtrəˌpɪzm̩/ 向光性　115
phycobiont /ˌfaɪkoˈbaɪˌɑnt/ 藻共生體　147
Phycomycetes /ˌfaɪkoməˈsitɪz/ 藻菌綱　164
phyllode /ˈfɪlod/ 葉狀柄　99

phyllotaxy /ˈfɪləˌtæksɪ/ 葉序　98
phylogeny /faɪˈlɑdʒənɪ/ 系統發育　141
physiology /ˌfɪzɪˈɑlədʒɪ/ 生理學　111
phytoalexin /ˌfaɪtoəˈlɛksɪn/ 植物抗毒　148
phytochemistry /ˌfaɪtoˈkɛmɪstrɪ/ 植物化學　8
phytochrome /ˈfaɪtəˌkrom/ 光敏色素　116
phytopathology /ˌfaɪtopəˈθɑlədʒɪ/ 植物病理學　144
phytoplankton /ˌfaɪtoˈplæŋktən/ 浮游植物　121
phytosociology /ˌfaɪtoˌsoʃɪˈɑlədʒɪ/ 植物社會學　150
pigment /ˈpɪgmənt/ 色素　36
piliferous layer /paɪˈlɪfərəs ˈleɝ/ 根毛層　89
pinna /ˈpɪnə/ 羽片　98
pinnate /ˈpɪnet/ 羽狀的　98
pinnule /ˈpɪnjul/ 小羽片　98
pioneer /ˌpaɪəˈnɪr/ 先鋒植物　151
pistil /ˈpɪstl/ 雌蕊　75
pistillate /ˈpɪstlˌet/ 雌蕊的　75
pit /pɪt/ 紋孔　107
pith /pɪθ/ 髓部　92
placenta /pləˈsɛntə/ 胎座　77
placentation /ˌplæsənˈteʃən/ 胎座式　77
plagiogeotropism /ˌpledʒɪˈɑtrəˌpɪzm̩/ 斜向地性　115
plagiotropic /ˌpledʒɪəˈtrɑpɪk/ 斜生的　93
plant /plænt/ 植物　118
plasmagene /ˈplæzməˌdʒin/ 細胞質基因　43
plasmalemma /ˌplæzməˈlɛmə/ 原生質膜　18
plasma membrane /ˈplæzmə ˈmɛmbren/ 質膜　18
plasmodesmata /ˌplæzməˈdɛzmətə/ 胞間連絲　21
plasmodium /plæzˈmodɪəm/ 原質團　169
plasmogamy /plæzˈmɑgəmɪ/ 胞質配合　166
plasmolysis /plæzˈmɑləsɪs/ 質壁分離　104
plastid /ˈplæstɪd/ 質體醌　18
plastochrone /ˌplæstoˈkron/ 間隔期　110
plastocyanin /ˈplæstəˈsaɪəˌnɪn/ 質體藍素　36
plastoglobuli /ˌplæstəˈglɑbjulɪ/ 質體球體　18
plastoquinone /ˌplæstoˈkwɪnən/ 質體醌　36
pleiotropic /plaɪˈɑtrəˌpɪk/ 多效性的　41
Pleistocene /ˈplaɪstəˌsin/ 更新世　143
pleurocarpous /ˌplurəˈkɑrpəs/ 側生蒴的　125
plumule /ˈplumjul/ 胚芽　86
pod /pɑd/ 豆莢　84

polar /ˈpolɚ/ 極地的　162
pollen /ˈpalən/ 花粉　74
pollen diagram /ˈpalən ˈdaɪəˌgræm/ 花粉圖式　142
pollen sac /ˈpalən sæk/ 花粉囊　74
pollen tube /ˈpalən tjub/ 花粉管　74
pollination /ˌpaləˈneʃən/ 授粉作用　74
pollinium /paˈlɪnɪəm/ 花粉塊　75
polyadelphous /ˌpalɪəˈdɛlfəs/ 多體雄蕊的　73
polygamous /pəˈlɪgəməs/ 雜性的　79
polymer /ˈpalɪmɚ/ 聚合體　10
polymorphism /ˌpalɪˈmɔrfɪzm̩/ 多態性　135
polypeptide /ˌpalɪˈpɛpˌtaɪd/ 多肽　56
polypetalous /ˌpalɪˈpɛtl̩əs/ 離瓣的　71
polyploid /ˈpalɪˌplɔɪd/ 多倍體的　50
polysaccharide /ˌpalɪˈsækəˌraɪd/ 多醣；多糖　30
polysepalous /ˌpalɪˈsɛpələs/ 離萼的　71
polysome /ˌpalɪˈsom/ 多核糖體　56
pome /pom/ 梨果　83
population /ˌpapjəˈleʃən/ 種群　135
pore /por/ 孔　19
porphyrin /ˈpɔrfərɪn/ 葉卟啉　38
potential energy /pəˈtɛnʃəl ˈɛnɚdʒɪ/ 勢能　11
potometer /pəˈtɔmɪtɚ/ 蒸騰計　101
prairie /ˈprɛrɪ/ 高草原　160
pressure potential /ˈprɛʃɚ pəˈtɛnʃəl/ 壓力勢　103
primary production /ˈpraɪˌmɛrɪ prəˈdʌkʃən/ 初級生產量　150
primary productivity /ˈpraɪˌmɛrɪ ˌprodʌkˈtɪvətɪ/ 初級生產力　150
primary thickening /ˈpraɪˌmɛrɪ ˈθɪkənɪŋ/ 初生增厚　94
primary vegetation /ˈpraɪˌmɛrɪ ˌvɛdʒəˈteʃən/ 原生植被　150
primitive /ˈprɪmətɪv/ 原始的　141
primordium /praɪˈmordɪəm/ 原基　110
producer /prəˈdjusɚ/ 生產者　150
progeny /ˈpradʒənɪ/ 後代　59
prokaryotic /ˌprokærɪˈɑtɪk/ 原核的　16
propagation /ˌprapəˈgeʃən/ 繁殖　69
propagule /ˈprapəgjul/ 繁殖體　59
prophase /ˈproˌfez/ 前期　45
prop root /prap rut/ 支持根　89

protandrous /proˈtændrəs/ 雄蕊先熟的　79
protein /ˈprotiɪn/ 蛋白質　56
protein structure /ˈprotiɪn ˈstrʌktʃɚ/ 蛋白質結構　58
protein synthesis /ˈprotiɪn ˈsɪnθəsɪs/ 蛋白質合成　57
prothallus /prəˈθæləs/ 原葉體　122
protogynous /proˈtadʒənəs/ 雌蕊先熟的　79
proton /ˈprotan/ 質子　8
protonema /ˌprotəˈnimə/ 原絲體　125
protoplasm /ˈprotəˌplæzəm/ 原生質　16
protoplast /ˈprotəˌplæst/ 原生質體　18
pseudocarp /ˈsjudəkarp/ 假果　83
pteridophytes /ˈtɛrədoˌfaɪts/ 蕨類植物　126
pubescent /pjuˈbɛsn̩t/ 被短柔毛的　100
pulp /pʌlp/ 果肉　83
pure line /pjʊr laɪn/ 純系　44
purine /ˈpjʊrin/ 嘌呤　52
pyramid of numbers /ˈpɪrəmɪd ɑv ˈnʌmbɚz/ 數量金字塔　153
pyrene /ˈpaɪrin/ 小堅果　83
pyrenoid /paɪˈrinɔɪd/ 澱粉核　119
pyrimidine /ˈpɪrəməˌdin/ 嘧啶　53
pyrrole /ˈpɪrˌol/ 吡咯　38
pyruvic acid /paɪˈruvɪk ˈæsɪd/ 丙酮酸　24

quadrat /ˈkwadrət/ 樣方　162
Quaternary /kwəˈtɚnərɪ/ 第四紀　143
quiescent centre /kwaɪˈɛsn̩t ˈsɛntɚ/ 靜止中心　89
raceme /reˈsim/ 總狀花序　80
rachis /ˈrekɪs/ 葉軸　98
radical /ˈrædɪkl̩/ 根的　88
radicle /ˈrædɪkl̩/ 胚根　86
rain forest /ren ˈfɔrɪst/ 雨林　158
raphe /ˈrefɪ/ 種脊　85
ray /re/ 射線　90
ray-floret /re ˈflorɪt/ 盤邊小花　81
reaction /rɪˈækʃən/ 反應　11
receptacle /rɪˈsɛptəkl̩/ 花托　72
recessive /rɪˈsɛsɪv/ 隱性的　44
recombination /riˌkambəˈneʃən/ 重組　47
red algae /rɛd ˈældʒi/ 紅藻　120
redox /ˈridɑks/ 氧化還原作用　11

reduction /rɪ'dʌkʃən/ 還原作用 11
reduction division /rɪ'dʌkʃən də'vɪʒən/ 減數分裂 49
reductive pentose pathway /rɪ'dʌktɪv 'pɛntos 'pæθ,we/ 還原性戊醣途徑 33
regeneration[1] /rɪ,dʒɛnə'reʃən/ 再生 111
regeneration[2] /rɪ,dʒɛnə'reʃən/ 更新 150
replication /,rɛplə'keʃən/ 複製 54
reproduction /,riprə'dʌkʃən/ 生殖；繁殖 59
reproductive isolation /,riprə'dʌktɪv ,aɪsl'eʃən/ 生殖隔離 142
respiration /,rɛspə'reʃən/ 呼吸作用 22
reticulate /rɪ'tɪkjəlɪt/ 網狀的 97
rhachis /'rekɪs/ 葉軸 98
rhizine /'raɪzɪn/ 假根 147
rhizoid /'raɪzɔɪd/ 假根 122
rhizome /'raɪzom/ 根狀莖 60
rhizosphere /'raɪzə'sfɪə/ 根圍 157
riboflavin /,raɪbə'flevɪn/ 核黃素；維生素 B_2 37
ribosome /'raɪbə,som/ 核糖體 56
ribulose-diphosphate /'rɪbələs daɪ'fɑsfet/ 核酮糖二磷酸 33
ribulose-diphosphate carboxylase /'rɪbələs-daɪ'fɑsfet kɑr'bɑksəles/ 核酮糖二磷酸羧化酶 33
ripe /raɪp/ 成熟的 83
RNA /ɑr ɛn e/ 核糖核酸 51
root /rut/ 根 88
root-cap /rut kæp/ 根冠 89
root hair /rut hɛr/ 根毛 89
root pressure /rut 'prɛʃɚ/ 根壓 102
rootstock /'rut,stɑk/ 根砧木 69
rosette /ro'zɛt/ 蓮座葉叢 99
runner /'rʌnɚ/ 纖匐枝；走莖 60
rust /rʌst/ 銹病菌 167

salt marsh /sɔlt mɑrʃ/ 鹽沼 161
samara /'sæmərə/ 翅 84
sample /'sæmpl/ 樣本 173
sap /sæp/ 汁液 102
sapling /'sæplɪŋ/ 幼樹 136
saprophyte /'sæpro,faɪt/ 腐生植物 137
sapwood /'sæp,wʊd/ 邊材 94

saturate /'sætʃə,ret/ 飽和的 31
savanna /sə'vænə/ 稀樹乾草原 160
saxicolous /sæk'sɪkələs/ 岩生的 137
scalariform /skə'lærə,fɔrm/ 梯紋的 107
scale /skel/ 鱗片 100
scape /skep/ 花葶 80
scatter diagram /'skætɚ 'daɪə,græm/ 散佈圖 173
schizocarp /'skɪzə,kɑrp/ 離果 84
sclereid /'skɪrɪd/ 石細胞 91
sclerenchyma /sklɪ'rɛŋkɪmə/ 厚壁組織 91
sclerophyllous /,sklɪrə'fɪləs/ 硬葉的 91
scrub /skrʌb/ 密灌叢 160
seaweed /'si,wid/ 海藻 122
secondary thickening /'sɛkən,dɛrɪ 'θɪkənɪŋ/ 次生增厚 94
secondary vegetation /'sɛkən,dɛrɪ ,vɛdʒə'teʃən/ 次生植被 150
secretion /sɪ'kriʃən/ 分泌作用 112
sedge /sɛdʒ/ 薹草 130
sedoheptulose /,sɛdə'hɛptules/ 景天庚酮糖 35
seed /sid/ 種子 85
seed leaf /sid lif/ 種子葉；子葉 86
seedling /'sidlɪŋ/ 幼苗 87
seed plant /sid plænt/ 種子植物 128
segregation /,sɛgrɪ'geʃən/ 分離律 42
self-compatible /'sɛlf,kəm'pætəbl/ 自交親和的 63
self-fertilization /'sɛlf,fɝtlə'zeʃən/ 自體受精 62
self-incompatible /'sɛlf,ɪnkəm'pætəbl/ 自交不親和的 63
self-pollination /'sɛlf,pɑlə'neʃən/ 自花受粉 75
semipermeable /,sɛmə'pɝmɪəbl/ 半透性的 103
senescence /sə'nɛsn̩s/ 衰老 117
sepal /'sipl/ 萼片 70
septum /'sɛptəm/ 隔膜 163
sequence /'sikwəns/ 順序 170
sere /sɪr/ 演替系列 152
sessile /'sɛsl/ 無柄的 100
seta /'sitə/ 蒴柄 123
sex cell /sɛks sɛl/ 性細胞 61
sexual /'sɛkʃʊəl/ 有性的 59
sheath /ʃiθ/ 葉鞘 100
shoot /ʃut/ 枝條 90
short-day plant /'ʃɔrt,de plænt/ 短日照植物 116

shrub /ʃrʌb/ 灌木　136
sieve element /sɪv ˈɛləmənt/ 篩管分子　108
sieve plate /sɪv plet/ 篩板　108
sieve tube /sɪv tjub/ 篩管　108
siliceous skeleton /sɪˈlɪʃəs ˈskɛlətn̩/ 矽質骨骼　121
silicula /sɪˈlɪkjulə/ 短角果　84
siliqua /ˈsɪlɪkwə/ 長角果　84
simple /ˈsɪmpl̩/ 單葉的　97
siphoneous /ˌsaɪfəˈnefəs/ 管藻狀　120
skewed distribution /skjud ˌdɪstrəˈbjuʃən/ 偏斜分佈　173
slime moulds /slaɪm moldz/ 黏菌　169
soil profile /sɔɪ ˈprofaɪl/ 土壤剖面　156
solitary /ˈsɑləˌtɛrɪ/ 單生的　82
soluble /ˈsɑljəbl̩/ 可溶解的　12
solute /ˈsɑljut/ 溶質　12
solution /səˈluʃən/ 溶液　12
solvent /ˈsɑlvənt/ 溶劑　12
somatic /soˈmætɪk/ 體細胞的　45
sorus /ˈsɔrəs/ 子囊群　126
spadix /ˈspedɪks/ 佛焰花序　82
spathe /speð/ 佛焰苞　82
specialized /ˈspɛʃəlˌaɪzd/ 特化的　170
speciation /ˌspiʃɪˈeʃən/ 物種形成　142
species /ˈspiʃiz/ 物種　134
spermatophyte /ˈspɝmətəˌfaɪt/ 種子植物　128
spermatozoid /spɝˈmætəzɔɪd/ 游走精子　65
spike /spaɪk/ 穗狀花序　80
spikelet /ˈspaɪklɛt/ 小穗　82
spindle /ˈspɪndl̩/ 紡錘體　46
spine /spaɪn/ 刺　100
spiral /ˈspaɪrəl/ 旋生的　98
spirochaete /ˈspaɪrəˌkit/ 螺旋體　119
spongy mesophyll /ˈspʌndʒɪ ˈmɛsəˌfɪl/ 海綿葉肉　95
sporangiophore /spəˈrændʒɪəˌfor/ 孢囊柄　66
sporangiospore /spəˈrændʒɪˌspor/ 孢囊孢子　168
sporangium /spəˈrændʒɪəm/ 孢子囊　66
spore /spor/ 孢子　66
spore mother cell /spor ˈmʌðɚ sɛl/ 孢子母細胞　66
sporogenous /spoˈrædʒənəs/ 產孢子的　66
sporogonium /ˌsporəˈɡonɪəm/ 孢子體　122
sporophyll /ˈsporəfɪl/ 孢子葉　67

sporophyte /ˈsporəˌfaɪt/ 孢子體　65
sporopollenin /ˌsporəˈpolənɪn/ 孢粉質　74
sporulation /ˌsporjuˈleʃən/ 孢子形成　66
stain /sten/ 染色劑　171
stamen /ˈstemən/ 雄蕊　73
staminate /ˈstæmənɪt/ 雄蕊的　73
staminode /ˈstæmənod/ 退化雄蕊　73
starch /startʃ/ 澱粉　30
statolith /ˈstætˌlɪθ/ 平衡石　115
stele /ˈstili/ 中柱　105
stem /stɛm/ 莖　90
steppes /stɛps/ 乾草原　160
sterile /ˈstɛrəl/ 不育的　62
stigma /ˈstɪɡmə/ 柱頭　76
stilt root /stɪlt rut/ 支柱根　89
stimulus /ˈstɪmjələs/ 刺激　170
stipe /staɪp/ 菌柄　166
stipule /ˈstɪpjul/ 托葉　99
stolon /ˈstolən/ 匍匐莖　60
stoma /ˈstomə/ 氣孔　96
stone /ston/ 核　83
stone cell /ston sɛl/ 短石細胞　91
strain /stren/ 品系　135
strobilus /ˈstrɑbələs/ 孢子葉球　68
stroma /ˈstromə/ 基質　32
structure /ˈstrʌktʃɚ/ 結構　170
style /staɪl/ 花柱　76
subdivision /ˌsʌbdəˈvɪʒən/ 亞門　134
suberin /ˈsjubərɪn/ 木栓質　94
subsoil /ˈsʌbˌsɔɪl/ 底土　156
subspecies /ˌsʌbˈspiʃiz/ 亞種　135
substrate[1] /ˈsʌbstret/ 受質；基質　15
substrate[2] /ˈsʌbstret/ 基質　154
subtropical /sʌbˈtrɑpɪkl̩/ 亞熱帶的　162
succession /səkˈsɛʃən/ 演替　151
succubous /ˈsʌkjəbə/ 蔽後式的　123
succulent /ˈsʌkjələnt/ 肉質的　99
sucker /ˈsʌkɚ/ 根出條　60
sucrose /ˈsukros/ 蔗糖　29
sugar /ˈʃuɡɚ/ 糖　28
superior ovary /səˈpɪrɪɚ ˈovərɪ/ 上位子房　77

Shapes of simple leaves 單葉的形狀

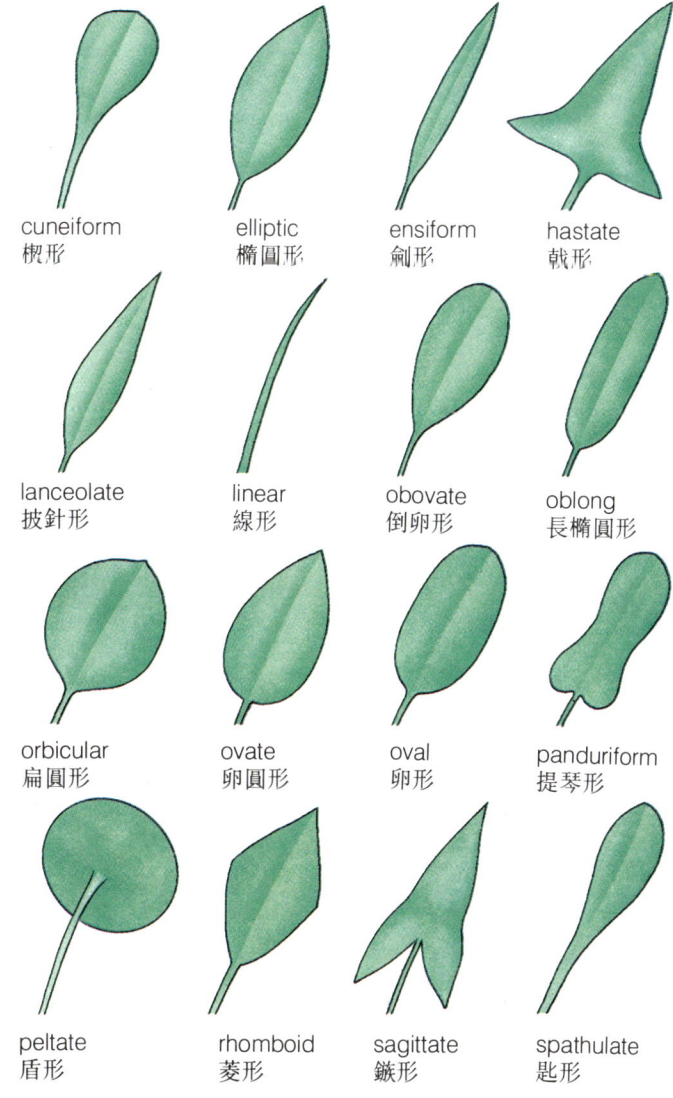

出　版　者：沈維賢
著　　　者：ANDREW SUGDEN
審　　　訂：關培生
翻　　　譯：張宏達、任善相
編　　　輯：陳繼勤

unicellular /ˌjunɪˈsɛljələ/ 單細胞的 119
unit /ˈjunɪt/ 單元；單位 170
unsaturated /ʌnˈsætʃəˌretɪd/ 不飽和的 31
uptake /ˈʌpˌtek/ 吸收 101
uracil /ˈjurəsəl/ 尿嘧啶 53
uredospore /juˈridospɔr/ 夏孢子 167

vacuolar sap /ˈvækjuələ sæp/ 液泡液 20
vacuole /ˈvækjuˌol/ 液泡 20
variable /ˈvɛrɪəbl/ 變量 173
variation /ˌvɛrɪˈeʃən/ 變異 135
variegated /ˈvɛrɪˌgetɪd/ 彩斑的 99
variety /vəˈraɪətɪ/ 變種 135
vascular /ˈvæskjələ/ 維管的 122
vascular bundle /ˈvæskjələ ˈbʌndl/ 維管束 105
vascular cylinder /ˈvæskjələ ˈsɪlɪndə/ 維管柱 105
vascular system /ˈvæskjələ ˈsɪstəm/ 維管系統 105
vector¹ /ˈvɛktə/ 傳粉媒介 75
vector² /ˈvɛktə/ 媒介 144
vegetation /ˌvɛdʒəˈteʃən/ 植被 150
vegetative /ˈvɛdʒəˌtetɪv/ 營養生長的 60
vegetative growth /ˈvɛdʒəˌtetɪv groθ/ 營養生長 109
vegetative reproduction /ˈvɛdʒəˌtetɪv ˌriprəˈdʌkʃən/ 營養體生殖 60
vein /ven/ 葉脈 97
velamen /vɪˈlemɛn/ 根被 89
venation /viˈneʃən/ 脈序 97
vernalization /ˌvɝnlɪˈzeʃən/ 春化作用 117
vesicle /ˈvɛsɪkl/ 泡囊 20
vessel /ˈvɛsl/ 導管 107
vessel element /ˈvɛsl ˈɛləmənt/ 導管分子 107
viable /ˈvaɪəbl/ 有生活力的 62
vine /vaɪn/ 籐本植物 136
virus /ˈvaɪrəs/ 病毒 118
vitamin /ˈvaɪtəmɪn/ 維生素 15

water potential /ˈwɔtə pəˈtɛnʃəl/ 水勢 103
wavelength /ˈwevlɛŋkθ/ 波長 38
wax /wæks/ 蠟 96
whorl /hwɝl/ 輪生 98

wild type /waɪld taɪp/ 野生型 41
wilt /wɪlt/ 萎蔫 104
wood /wud/ 木質部 93
woodland /ˈwudˌlænd/ 林地 159

xanthophyll /ˈzænθəfɪl/ 葉黃素 37
xeromorphic /ˌzɪrəˈmɔrfɪk/ 旱生結構的 137
xerophyte /ˈzɪrəˌfaɪt/ 旱生植物 137
xylem /ˈzaɪlɛm/ 木質部 106

yeast /jist/ 酵母 104

zoospore /ˈzoəˌspor/ 游動孢子 168
zygomorphic /ˌzaɪgəˈmɔrfɪk/ 兩側對稱的 71
Zygomycetes /ˌzaɪgoˈmaɪsitiz/ 接合菌綱 168
zygospore /ˈzaɪgəˌspor/ 接合孢子 168
zygote /ˈzaɪgot/ 合子 61
zygotene /ˈzaɪgotin/ 偶絲期 49

survival of the fittest /sɚˈvaɪvl̩ ɑv ðə fitɛst/ 適者生存　140
suspensor¹ /səˈspɛnsɚ/ 珠柄　86
suspensor² /səˈspɛnsɚ/ 囊柄　168
sward /swɔrd/ 草地　160
symbiont /ˈsɪmbaɪˌɑnt/ 共生生物　144
symbiosis /ˌsɪmbaɪˈosɪs/ 共生　144
symmetrical /sɪˈmɛtrɪkl̩/ 對稱的　71
sympatric /sɪmˈpætrɪk/ 同域的　142
sympetalous /sɪmˈpɛtələs/ 合瓣的　71
symplast /ˈsɪmplæst/ 共質體　102
sympodial /sɪmˈpodɪəl/ 合軸的　93
synapsis /sɪˈnæpsɪs/ 聯會　47
syncarpous /sɪnˈkɑrpəs/ 合心皮的　76
synecology /ˌsɪnəˈkɑlədʒɪ/ 群體生態學　149
synergids /sɪˈnɝdʒɪdz/ 助細胞　78
synergistic /ˌsɪnɚˈdʒɪstɪk/ 增效的　114
synthesis /ˈsɪnθəsɪs/ 合成作用　13
systematics /ˌsɪstəˈmætɪks/ 系統分類學　132

tannins /ˈtænɪnz/ 丹寧　148
tap root /tæp rut/ 直根　88
taxon /ˈtæksɑn/ 分類單元　133
taxonomy /tæksˈɑnəmɪ/ 分類學　133
teleutospore /təˈlutəˌspɔr/ 冬孢子　167
telome theory /ˈtɛləm ˈθɪərɪ/ 頂枝學說　140
telophase /ˈtɛləˌfez/ 末期　45
temperate /ˈtɛmprɪt/ 溫帶的　162
tendril /ˈtɛndrɪl/ 捲鬚　136
tepal /ˈtipəl/ 花被片　70
Tertiary /ˈtɝʃɪˌɛrɪ/ 第三紀　143
testa /ˈtɛstə/ 外種皮　85
tetrad¹ /ˈtɛtræd/ 四分體　47
tetrad² /ˈtɛtræd/ 四分體　66
tetraploid /ˈtɛtrəˌplɔɪd/ 四倍體　50
thalloid liverwort /ˈθæloɪd ˈlɪvɚˌwɝt/ 似葉狀體苔類　123
thallus /ˈθæləs/ 葉狀體　122
thigmotropism /θɪɡˈmɑtrəpɪzm/ 向觸性；向實體性　115
thorn /θɔrn/ 棘刺　100
thylakoid /ˈθaɪləˌkɔɪd/ 類囊體　33
thymine /ˈθaɪmin/ 胸腺嘧啶　53
tiller /ˈtɪlɚ/ 分蘖　60

tissue /ˈtɪʃu/ 組織　88
tissue culture /ˈtɪʃu ˈkʌltʃɚ/ 組織培養　69
toadstool /ˈtɑdˌstul/ 毒蕈　165
tomentose /təˈmɛntos/ 被綿毛的；被絨毛的　100
tonoplast /ˈtonəplæst/ 液泡膜　20
topsoil /ˈtɑpˌsɔɪl/ 表土　156
torus /ˈtorəs/ 花托　72
toxin /ˈtɑksɪn/ 毒素　148
trace element /tres ˈɛləmənt/ 微量元素　111
tracheid /ˈtrekɪɪd/ 管胞　107
trait /tret/ 特質　41
transcription /trænˈskrɪpʃən/ 轉錄　56
transect /trænˈsɛkt/ 狹樣區　162
translation /trænsˈleʃən/ 轉譯　56
translocation /ˌtrænsloˈkeʃən/ 轉移作用　101
transpiration /ˌtrænspəˈreʃən/ 蒸騰作用　101
transpiration stream /ˌtrænspəˈreʃən strim/ 蒸騰流動　101
transverse section /trænsˈvɝs sɛkʃən/ 橫切面　171
tree /tri/ 喬木　136
tree fern /tri fɝn/ 樹蕨　127
tree line /tri laɪn/ 林木線　159
tribe /traɪb/ 族　134
tricarboxylic acid cycle /traɪˌkɑrbɑkˈsɪlɪk ˈæsɪd ˈsaɪkl̩/ 三羧酸循環　24
trichome /ˈtrɪˌkom/ 毛狀體　100
triose /ˈtraɪos/ 丙醣；丙糖　28
triplet code /ˈtrɪplɪt kod/ 三聯體密碼　54
triploid /ˈtrɪplɔɪd/ 三倍體的　50
trophic level /ˈtrɑfɪk ˈlɛvl̩/ 食性層次；營養級　153
tropical /ˈtrɑpɪkl̩/ 熱帶的　162
tropism /ˈtropɪzəm/ 向性　115
trunk /trʌŋk/ 主幹　92
tuber /ˈtjubɚ/ 塊莖　60
tunica /ˈtjunɪkə/ 原套　109
turgid /ˈtɝdʒɪd/ 緊脹的　104
turgor /ˈtɝɡɚ/ 膨壓　104
type /taɪp/ 模式標本　133

umbel /ˈʌmbl̩/ 繖形花序　80
understorey /ˈʌndɚˌstorɪ/ 下層林木　158
undifferentiated /ˌʌndɪfəˈrɛnʃɪˌetɪd/ 未分化的　110